DEVELOPMENTS IN FOOD SCIENCE 6

TESTING METHODS IN FOOD MICROBIOLOGY

Edited by

ISTVÁN KISS

Central Food Research Institute
Budapest, Hungary

ELSEVIER
Amsterdam–Oxford–New York–Tokyo 1984

This book is the revised version of
MIKROBIOLÓGIAI VIZSGÁLATI MÓDSZEREK AZ ÉLELMISZERIPARBAN
1. Mennyiségi vizsgálatok
and
2. Minőségi vizsgálatok (a mikroorganizmusok kimutatása)
Mezőgazdasági Kiadó, Budapest 1977 and 1978

Translated by E. FARKAS

English text revised by K. VAS

Translation Editor: Mark COWAN

Contributors: J. FARKAS, K. INCZE, I. KISS, L. ORMAY, G. PULAY, J. TAKÁCS, J. VÖRÖS,
            E. ZUKÁL

Joint edition published by

Elsevier Science Publishers, Amsterdam, The Netherlands and Akadémiai Kiadó, the Publishing House
of the Hungarian Academy of Sciences, Budapest, Hungary

The distribution of this book is being handled by the following publishers

for the U.S.A. and Canada

Elsevier Science Publishing Co., Inc.
52 Vanderbilt Avenue
New York, New York 10017, U.S.A.

for the East European countries, Korean People's Republic, Cuba, People's Republic of Vietnam and
Mongolia

Kultura Hungarian Foreign Trading Co. P.O. Box 149 H-1389 Budapest 62, Hungary

for all remaining areas

Elsevier Science Publishers
Molenwerf, 1
P.O. Box 211, 1000 AE Amsterdam, The Netherlands

Library of Congress Cataloging in Publication Data

Testing methods in food microbiology.
(Developments in food science; 6)
Rev. English version of: Mikrobiológiai vizsgálati módszerek az élelmiszeriparban.
Bibliography: p.
Includes index.
1. Food—Microbiology—Technique.   I. Kiss, István, 1923—   II. Vas, K. (Károly)
III. Farkas, József, Dr.   IV. Title.   V. Series.
QR115.M4913   1984   664'.07   83-11725

ISBN 0-444-99648-6 (Vol 6)
ISBN 0-444-41688-9 (Series)

© Akadémiai Kiadó, Budapest 1984

Printed in Hungary

Developments in Food Science 6

TESTING METHODS IN FOOD MICROBIOLOGY

DEVELOPMENTS IN FOOD SCIENCE

CONTENTS

16

FOREWORD

The microbiological characteristics, like the physical, chemical and organoleptic characteristics, are now regarded as an integral part of the quality of foodstuffs, and play an ever-increasing role in quality grading. Moreover, in most branches of the food industry the inactivation of microbes, and in other cases the promotion of microbial activity, are fundamental aspects of food processing.

These considerations underline the technological importance of microbiological testing, not only of raw materials, but also of foods in the course of processing, and finished products. Testing must include the qualitative characterization and identification of microbes, as well as a quantitative determination of their frequency of occurrence and concentration.

The authors have attempted to assist those engaged in food-microbiological testing by presenting a synthesis of methods currently in use. These have been selected as being easy to carry out during routine microbiological testing in laboratories in factories, and as providing as much useful information as possible for the food industry as a whole. The procedures are well known to the authors, and therefore, may be regarded as both reliable and convenient.

The work of selecting and thoroughly checking the methods has been a lengthy labour of love, and it is hoped that the resulting manual will be found extremely useful by all food microbiologists, both in industrial and in research laboratories. It is certainly a book that has been needed for some time, and I feel that the authors' efforts will be crowned with success.

Károly Vas

1. THE MICROBIOLOGICAL LABORATORY

1.1   Laboratory design and equipment

1.1.1   Placing and illumination

Good illumination and easily cleaned surfaces are important requirements in the design of a microbiological laboratory.

Microscopy and other laboratory activities are disturbed by direct sunlight. To ensure approximately even illumination during the hours of daylight, windows should face north. Among artificial light sources, those providing indirect illumination are preferable.

Cleanliness and aseptic conditions cannot be maintained in a contaminated environment such as that of a dusty city street. For this reason, and to ensure better natural illumination, the placing of laboratories on upper stories is desirable.

1.1.2   Ventilation

The windows of an up-to-date microbiological laboratory remain sealed; filtered air of extremely low germ count is drawn in to produce a slight overpressure inside the building, with air motion kept as weak as possible.

1.1.3   The arrangement of the laboratory

Figure 1 shows a plan of a fully independent microbiological unit.

The relative size and number of rooms are determined by the nature of the work. Rooms, equipment and furniture should be designed with the aim of achieving a continuous flow of instruments and culture media with minimal transport. Glazed swing-doors (except those communicating sterile rooms and incubator rooms) opening from a central corridor, and uninterrupted, even flooring help to ensure a smooth flow of traffic.

For the disinfection of instruments it is advisable to build a horizontal double-door autoclave of large capacity in the wall separating the contaminated rooms

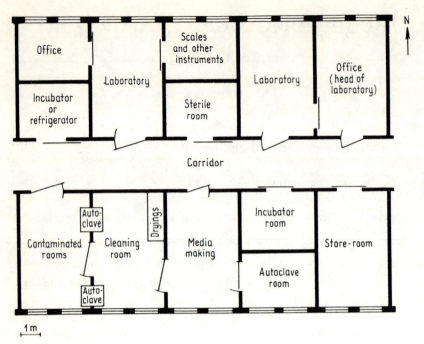

Fig. 1. Diagram of an independent microbiological unit

from the clean area. By this means contaminated instruments placed in the autoclave and sterilized are accessible from the clean part of the laboratory.

In the floors of rooms to be used for the preparation of culture media and the cleaning of instruments, drainage points are indispensable; these allow the whole surface of the floor to be flooded.

The cleaning and drying of the many vessels used in a microbiological unit requires much room and equipment (washing-machines, pipette rinsers, etc.). In calculating the capacity of a laboratory washing unit one should reckon on the necessity of rising large flasks among other glassware. Running water is absolutely necessary. Also, the rinsing of glassware with deionized water should not be left out of consideration.

The walls of those rooms in which culture media are prepared and sterilized and in which instruments are cleaned and sterilized should be tiled to a height of at least 2 m. The same rooms must be equipped with a dust extractor.

Microbiological work requires adequate incubator capacity; large units should be provided with walk-in incubator(s), smaller ones with thermostatic cabinets. In either case, at least three temperature levels must be available for the cultivation of microbes requiring different optimum temperatures.

Since microbiological samples and preparations are perishable, they must be kept at a low temperature. This makes for large demands on refrigerator (or cold room) capacity. For special purposes deep-freezing is also indispensable.

1.1.4   The basic equipment

Doors and windows should be free of dust-collecting grooves and surfaces and all furniture should have smooth surfaces with rounded-off edges and corners. Cupboards and drawers should be of laminated construction. Backed chairs of adjustable height are preferable.

The floor should withstand daily scrubbing, wiping, and/or oiling. Equipment served by gas, electricity and water should be easily accessible to facilitate repair.

The surfaces of microscopy benches should be at a height of 80 to 90 cm above the floor; for other purposes, a height of between 90 and 100 cm is the most practicable. The surfaces should be resistant to chemicals including disinfectants, as well as being heat-resistant.

Inoculation of media is carried out in small rooms provided with ventilation as outlined above. The wall surfaces of these rooms should withstand condensation and frequent cleaning with disinfectant. Electric cables as well as gas and water pipes should be built into the wall. For the sterilization of walls and working-surfaces, appropriately-placed ultraviolet (UV) radiators are the most suitable. If UV radiation is used, the walls must be covered with a UV-resistant substance such as chlorocaoutchouc, instead of the usual oil-paint which is less resistant. The working-surfaces should be U-shaped on plan; in this way instruments can be placed within arm's reach (De Man et al., 1974).

If the risk of infection is particularly high, sterile boxes with rounded internal edges and corners may be used on the bench. The box is constructed of metal except for a glass window through which the operator can view the working area. In each of the side walls of the box, there is a hole with a long rubber glove for the operator's hands tightly sealed around the lip of the hole. Electricity, gas and water services are led through the back wall of the box and it is recommended that an atomizer be built in for aerosal disinfection. Fluorescent lighting and a UV tube or tubes, appropriately arranged inside the box, complete the equipment of the box.

Aseptic work is facilitated and the routine sterilization of the conventional sterile box can be avoided by using the so-called laminar-flow cabinet (Favero and Berquist, 1968; McDade et al., 1969). In this cabinet which is open at the front, filtered air is continuously flowing in a laminar stream, i.e. free of turbulence. The filtering equipment of the cabinet consists of so-called HEPA (high-efficiency particulate air) filter plates, a system that retains more than 99·9% of all particles larger than 0·3 µm in diameter. According to the direction of the air-flow, laminar-flow cabinets of vertical flow and horizontal flow are distinguished. An example of the latter type of cabinet is illustrated in *Figure 2*. For high-risk procedures laminar-flow cabinets have been produced in which the air flowing over the working-surface does not come into contact with the operator and the air flowing out of the cabinet is conducted through another HEPA filter system. This system protects the material

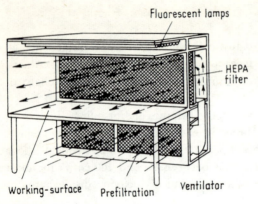

Fluorescent lamps

HEPA filter

Working-surface   Prefiltration   Ventilator

Fig. 2. Laminar-flow cabinet with horizontal air-flow

under test from being contaminated from outside as well as protecting the operator from infection by pathogens present in the material.

Laminar-flow systems have come into wider use outside the laboratory, in medical and industrial establishments requiring dust-free air and aseptic conditions. This system eliminates all the previously unsolved problems arising in connection with microbiological culturing, namely those of the air supply, undesirable rise of temperature, and shortcomings with respect to cleaning and sterilizing. It is a further advantage of this system that many culture procedures need no flaming if performed in a laminar-flow cabinet.

A beneficial by-product of the laminar-flow system is that the air of the laboratory in which the box is functioning is purified by its use.

1.1.5 Apparatus and instruments commonly used in microbiological laboratories

In this Chapter only those instruments and items of apparatus are mentioned which are indispensable to microbiological laboratory procedures. Those necessary for special procedures are described in more comprehensive books (e.g., Janke and Dickscheit, 1967; Norris and Ribbons, 1969).

1.1.5.1 Glassware for storage of media and cultivation of micro-organisms

Test-tubes. For microbiological purposes rimless tubes 160 or 180 mm in length and 16 or 18 mm in diameter are preferred. The mechanical stability of these so-called microbiological tubes surpasses that of the rimmed tubes used in chemical laboratories.

24

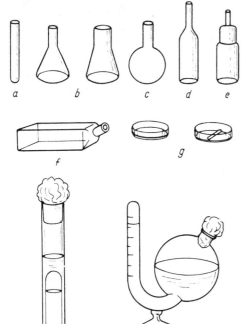

Fig. 3. The most important items of glassware for the storage of culture media and cultivation of micro-organisms

a — rimless thick-walled test-tube; b — Erlenmeyer flasks; c — standing flask; d — media flask; e — Freundenreich flask; f — Roux flask; g — Petri dishes; h — Durham tube; i— Einborn saccharometer

Erlenmeyer flasks. These flasks, which are conical in shape, are practical and easy to clean. The culture medium incubated in an Erlenmeyer flask is in contact with the air over a large surface area. This favours growth in aerobic cultivation, but accelerates exsiccation of the medium. Erlenmeyer flasks are available with various widths of the neck.

Standing flasks. In these flasks evaporation is less rapid than in Erlenmeyer flasks, but standing flasks are less easy to clean. Large flasks of 1–5 litres are mainly used for preparing culture media.

Media flasks. Being cylindrical in shape, media flasks make more economical use of space and thus are preferable for the sterilization and storage of media.

Freundenreich flasks. These are small flasks cylindrical in shape, ending in a narrow glass neck and a removable cap. The cap is fitted with a glass tube 2 cm long and 0·2 cm in diameter which is plugged with cotton wool.

Roux bottles. Roux bottles are quadrangular or rounded in profile, and are suitable for incubation in a horizontal position, i.e., for culturing micro-organisms over a large surface area of solid medium.

Petri dishes. A pair of circular dishes, each 1 cm in height forms a culture vessel. The medium is poured into the smaller dish, which is about 9 cm in diameter, the larger one being used as a loosely-fitting lid.

25

Petri dishes divided into segments by walls are also available.

Durham tubes. These are small glass tubes sealed at one end and open at the other. To indicate gas formation in a culture, a Durham tube is placed with the open end facing downwards in a test-tube containing liquid medium. During moist-heat sterilization of the medium, the Durham tube will be completely filled with the medium. Growth of a gas-forming bacterium in the medium will be indicated by a gas bubble in the Durham tube.

Einhorn's tube. This is a two-armed fermentation saccharometer of which the vertical, graduated arm is sealed at its top end. The other, bulb-like arm is open. The medium is poured in through the bulb to fill the scaled arm. A known proportion of the gas produced by the growing bacterium will accumulate in the graduated tube.

The glassware described above is shown in *Figure 3.*

1.1.5.2   Inoculating devices

Pipettes. The most commonly used pipettes are graduated ones, 1 ml full scale. Some pipettes are designed to be blown out, others (differential pipettes) have a zero marking. To prevent overdrawing, a cotton plug is placed in the upper end of the pipette before sterilization—a procedure which has been mechanized in some laboratories (De Man et al., 1974). In laboratories in which pathogenic micro-

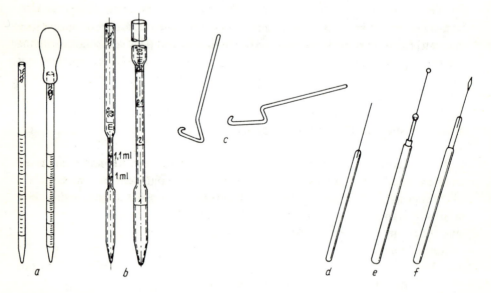

Fig. 4. Devices commonly used in the microbiological laboratory

a — simple microbiological pipettes; *b* — Demeter's pipettes; *c* — bent glass rods; *d* — inoculating wire fused into a glass rod; *e* — inoculating loop fastened in a Kolle holder; *f* — deltoid spatula

organisms are handled, pipetting by mouth should be avoided and a compressible rubber cap may be used instead. Demeter's pipette is suitable for distributing 1·0 and 0·1 ml volumes from 2·2 or 1·1 ml samples.

Glass rods. Bent glass rods are used for spreading suspensions of micro-organisms on solid media.

Straight wires (needles) and *wire loops* for inoculation. These are made of platinum or chromium nickel and are fixed in a holder. Depending on the use of the loop, its diameter may vary between 2 mm and 8 mm. Loops of 4 mm, 3 mm and 2 mm are capable of holding about 7·5 µl, 5 µl, and 2 µl liquid, respectively, although this also depends on the physical properties of the liquid. Loops which are used for quantitative purposes are calibrated. Straight wires and loops are usually fused into a glass rod, fixed with silver solder to a bronze rod, or mounted in a so-called Kolle holder, which is a metal rod provided with a screwed retaining nut and an ebonite handle.

Devices used for the inoculation of media are shown in *Figure 4*.

1.1.5.3   Disposable presterilized equipment

The use of disposable plastic apparatus including pipettes, Petri dishes, vials, culture tubes, as well as universal containers of approx. 28 ml volume and screw-capped tubes of approx. 8 ml volume is time- and work-saving. Disposable items are presterilized with ethylene oxide or by irradiation. Screw-capped tubes can also be used as centrifuge tubes. Petri dishes that are divided into sectors, culture dishes that can be used in automated systems (Repli-Schalen), and the space-saving square Petri dishes (Seidler, 1976) are all preferably made of presterilized plastic.

1.2   Portable microbiological units and mobile microbiological laboratories

It is often necessary to carry out microbiological work in the field and this type of work is greatly facilitated by the use of portable equipment or entire mobile microbiological laboratories. For further information on this subject, see Schultz et al. (1955), Hobson et al. (1968) and Fulghum (1971).

2.   MICROBIOLOGICAL PROCEDURES

2.1   Sterilization, cleaning and disinfection

Sterility means the absence of any kind of living micro-organism and all methods of sterilization must achieve this condition.

2.1.1   Sterilization by dry heat

Many methods of heating are in common use for sterilizing heat-stable apparatus in the microbiological laboratory.

2.1.1.1   Heating to redness

Wire needles and loops used for inoculation can be sterilized by heating the wire to redness. The straight wire and the loop should be held obliquely in the upper (oxidizing) part of the Bunsen flame, both before and after use, until they begin to glow. Every part of the tool that may come in contact with the inner part of the culture vessel should be heated in a flame. The wires and loops used in subculturing should be introduced into the flame gradually to avoid splittering of droplets which may contain viable micro-organisms.

2.1.1.2   Flaming

When glass vessels, test-tubes or flasks are opened, their mouths and metal caps should be flamed to prevent microbial contamination of cultures. Scalpels, spatulas and glass rods should be immersed in alcohol before flaming.

2.1.1.3   Sterilization by dry heat

Empty metal and glass vessels, as well as waterfree oil of high boiling-point, etc., can be sterilized in the dry state in a hot-air oven, but apparatus with rubber components must not be treated in this way. Bacterial spores are safely killed by

28

exposure to temperatures of 160 to 165 °C for 2 h, or 170 to 180 °C for 1 h, the period of heating commencing from the time the oven temperature reaches 160 °C.

Standard electric drying cabinets equipped with an electric thermoregulator can be used as hot-air sterilizers, provided it is kept in mind that in these cabinets the temperature depends not only on the control setting, but also on the location and the amount of apparatus inside (Darmady et al., 1961).

In circulating flow hot-air sterilizers, ambient air is passed through a heater into the double-walled jacket, wherefrom a proportion is taken back to the heater, the remainder being expelled through holes in the top of the sterilizer, together with vapour and gas that is occasionally produced during sterilization.

In dry-heat sterilizers, items of glassware are wrapped in filter paper, arranged individually, or loaded into metal cases. It should be borne in mind that paper and cotton become charred, and paper covers tend to disintegrate at temperatures higher than 160 °C. Furthermore, substances which inhibit microbial growth, and/or fatty acids may be released from paper at temperatures over 160 °C. These substances condense on the less hot surfaces during cooling and build up a contaminant layer unless the sterilizer is cleaned properly (Meynell and Meynell, 1965).

Only the all-glass type of syringe may be sterilized in dry air; the adhesive of those syringes made of metal and glass does not withstand hot air.

Powders, vaseline, paraffin, etc. are usually sterilized in dry air in the form of a layer not more than 5 mm thick.

2.1.2   Sterilization by moist heat

The most common method of sterilizing materials that can withstand humidity and high temperature, is heating in hot steam. Culture media are usually sterilized by this method.

Micro-organisms are substantially more sensitive to moist heat than to dry heat of the same temperature. In accord with the characteristics of the thermal destruction of micro-organisms, there is a close negative correlation between the applied temperature and the time required for sterilization at that temperature. The shortest time for effective sterilization is that required for destruction of the most heat-resistant micro-organisms; at an approximately neutral pH, bacterial spores are the most resistant.

It is an essential requirement of sterilization by moist air that all parts of the equipment, etc. must be in contact with saturated steam of the required temperature throughout the necessary sterilization period.

2.1.2.1 Sterilization in flowing steam at atmospheric pressure

Short treatment with flowing steam at atmospheric pressure suffices for the sterilization of media whose microflora is sensitive to heat, and that may be spoiled by high temperature treatment. The latter media, however, may only be successfully sterilized by this means if thermoresistant cells cannot multiply, e.g., because the medium has a low pH.

Prolonged moist heat treatment at atmospheric pressure is also sufficient for the sterilization of fluids of neutral pH, but prolonged heating may cause undesirable changes in the medium. Such changes can be avoided by intermittent sterilization (tyndallization). The material to be sterilized is exposed to short periods of steaming at intervals of 12–14 h. During the intervals the media are kept at a temperature favourable to spore germination. Most of the spores that have survived the first steaming will germinate during the subsequent interval, thus becoming sensitive to the next heat treatment. Three consecutive steamings, each of 15–20 minutes duration and a temperature during the intervals of 25–30 °C, are usually satisfactory for achieving microbiological stability in media intended for the cultivation of mesophilic bacteria. With volumes of 2 or more litres, each period of steaming must be extended to 30–60 minutes. Of course, only those media that support the germination of spores can be sterilized by tyndallization.

2.1.2.2 Sterilization by moist heat above atmospheric pressure

Sterilization in pressure-sealed chambers by saturated steam at a temperature over 100 °C requires much shorter heat treatment. Usually, 115 °C for 30 minutes or 121 °C for 20 minutes is more than satisfactory.

2.1.2.3 Apparatus for steam treatment in the laboratory

2.1.2.3.A Steam sterilizers ("steamers")

These, including Koch's steamer and Arnold's steamer, function at 100 °C.

Steamers are of metal plate construction. Water is held in the bottom of the steamer and the load, preferably arranged in wire baskets, is supported on a grid above the water level. Water may be added to the steamer as needed or conducted through a siphon that maintains a constant water level. Excess steam from the boiling water is discharged through a hole in the lid. Steamers are usually equipped with a thermometer exposed to the steam.

In some varieties of steamer, the steam is not allowed to escape, but condenses instead on a metal cooler from which the water is led back into the water reservoir. In this way, the unfavourable effects of steam in the laboratory are avoided.

In double-walled steamers, the heat is more efficiently utilized and the load becomes less moist. The steam is formed in the jacket and is conducted into the chamber through holes at the top of the water reservoir, so that the steam reaches the load from above.

The use of a central steam-supply system is time-saving.

Media must not be allowed to cool in the steamer, since the cooled cotton plugs become moistened by the steam and may allow condensation to pass into the medium.

2.1.2.3B   Autoclaves

Autoclaves are used for sterilization under pressure. The autoclave is a pressurized metal container in which steam can be generated, or steam may be introduced from an external source.

The autoclave is equipped with a manometer, a thermometer, an admission valve, an exhaust valve and a safety valve. Its lid is pressure-sealed. In autoclaves functioning by internal steam generation, electric or gas heating is built in. Horizontal and vertical autoclaves are equally widely used. A diagram of a horizontal autoclave with external steam supply is shown in *Figure 5*.

Autoclaves equipped with an automatic, programmable control system and recording indicators provide more reproducible results and are easier to operate. Such autoclaves, however, are more expensive.

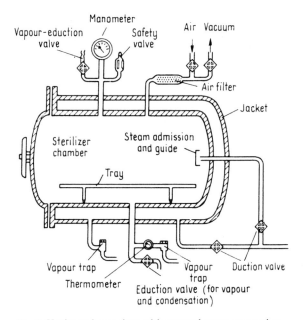

Fig. 5. Horizontal autoclave with external steam generation

Any autoclave can be used with its exhaust valve open, operating at 100 °C as a steamer.

Practical rules governing the use of autoclaves.

When using an autoclave with internal steam production, it is necessary to ensure that there is sufficient water in the reservoir, otherwise saturation of the atmosphere with steam may not be attained, and consequently, the bactericidal effect will be weaker than that of saturated steam at the same temperature (see *Table 1*).

Table 1

Autoclave temperatures and corresponding steam pressures

Temperature °C	Steam pressure		Overpressure att
	mmHg	atm	
100·0	760	1·0	0·0
102·7	836	1·1	0·1
105·2	912	1·2	0·2
107·5	988	1·3	0·3
109·7	1064	1·4	0·4
111·7	1140	1·5	0·5
113·7	1216	1·6	0·6
115·5	1292	1·7	0·7
117·3	1368	1·8	0·8
119·0	1447	1·9	0·9
120·6	1520	2·0	1·0
128·0	1900	2·5	1·5
133·9	2280	3·0	2·0
139·0	2660	3·5	2·5
144·0	3040	4·0	3·0
180·3	7600	10·0	9·0

Opposite screws of the autoclave lid should be tightened at the same time to ensure that tightening is effective and that the sealing ring is not damaged.

Before sterilization commences, steam must be allowed to escape from the exhaust valve for at least 5 minutes to drive out the air from the autoclave. If this is not done, the steam in the autoclave chamber will not become saturated.

After sterilization the steam valve must not be opened until the internal pressure falls to atmospheric pressure. A sudden release of pressure can cause boiling of the medium and blowing-out of cotton plugs.

Media containing gelatin must not be sterilized at over-atmospheric pressure.

In the case of glycerol, oil, or liquid paraffin, it is advisable to add some water (approx. 0·1%) before sterilizing in the autoclave.

Whether a steamer or an autoclave is used, the time at which the full sterilization temperature is reached should be regarded as the start of the sterilization period. This period very much depends on the size of the autoclave and the nature of the load. A culture medium in large flasks requires a much longer time than the same quantity of medium distributed in test-tubes. It is therefore best to sterilize units of approximately the same size in any one loading and to load the autoclave to the same extent on different occasions.

The times required for the warming up to 121 °C of media placed in conical flasks of different volumes, from the time the autoclave chamber becomes saturated with steam at 121 °C, are as follows:

Volume of liquid (ml)	Time required (minutes)
30	1
200	3
500	8
1000	12
2000	20

It is an advantage to keep autoclaves of different sizes in the laboratory because the use of a small autoclave is time-saving and economical when a small load is to be sterilized.

Electrically-heated, automated bench-top agar sterilizers are very convenient and distribute the sterilized medium into Petri dishes or other glassware.

2.1.2.4   Indicators of the effect of heat treatment

Heat sterilization can be monitored by physical, chemical and biological methods. Among the biological indicators, spores of *Bacillus stearothermophilus* are the most reliable for general use (Costin and Grigo, 1976). A suspension of a known number of heat-resistant spores, dried on filter paper or on aluminium foil, or in measured quantities in ampoules, is included as part of the loading of the autoclave. After sterilization, the viability of the indicator spores is tested (Costin and Grigo, 1974).

A commercial biological indicator system is available (Attest biological sterilization indicator) that consists of dried spores and culture medium in an ampoule. After sterilization the ampoule is broken so that the spores are suspended in the medium.

Biological indicator methods require a considerable time of incubation after sterilization. It has to be taken into account that heat-injured spores revive slowly. Thus attempts have been made to introduce more rapid physical and chemical monitoring systems. Among the physical methods, the placing of a thermopile amongst the load with continuous recording of the load temperature during sterilization is a widely-employed technique. Chemical methods are based on a colour change of an indicator material at a known temperature. The so-called

Browne tubes are held to be the most reliable chemical indicator (Kelsey, 1958). These are minute sealed glass tubes in which a red fluid turns green if the heat treatment has been satisfactory. Various types are available, indicating 15 minutes in moist air at 121 °C or 115 °C, or 1 h in dry air at 160 °C. Heat treatment indicators are also available in the form of adhesive tape, on which the word "sterile" appears after the necessary heat treatment has been applied. Chemical indicators should be kept in the refrigerator before use, since the chemical reaction proceeds at a slow rate, even at room temperature (Brown and Ridout, 1960).

2.1.3   Irradiation

Both UV and ionizing radiations have a microbicidal effect. In microbiological laboratories, however, only UV radiation has been widely applied as a means of sterilization. Since the penetration capacity of UV light is very low, this kind of radiation is effective only for the sterilization of clean rooms and laboratory surfaces. Radiation of wavelength 253 to 265 nm is the most effective. Care must be taken to protect the skin and eyes.

2.1.4   Sterilization using gases

Ethylene oxide and formaldehyde are the most widely used gases for sterilization.

2.1.4.1   Sterilization with ethylene oxide

Ethylene oxide, an alkylating agent, is often used for the sterilization of heat-sensitive materials such as plastics. Ethylene oxide is very reactive, forming an explosive mixture with the air. The risk of explosion can be eliminated by using the gas in a mixture with carbon dioxide, nitrogen or freon. The sterilizing effect of ethylene oxide requires considerable time to develop, the required exposure time depending very much on the temperature and humidity of the air. The temperature coefficient (Q_{10} value) of the death rate of micro-organisms is 2·7 in the temperature range 10–40 °C. Owing to the toxicity of the gas, its use requires special equipment.

In ethylene-oxide sterilizers of different types the relative humidity varies between 50% and 85%.

One of the procedures used in gas sterilization involves low-concentration (slow) treatment at an ethylene-oxide concentration of 200 gm^{-3} (10% v/v) for 18 h at a temperature not lower than 20 °C. Another method involves short high-temperature treatment (3–4 h exposure at between 55 and 60 °C); here, the ethylene-oxide concentration ranges from 800 to 1000 gm^{-3}. Using pressurized apparatus at a gauge pressure of 5–6 atmospheres, and with a temperature of 50 to

34

60 °C, the time of sterilization may be reduced to 1 or 2 h. For laboratory purposes, sterilizers of 25 to 1000 litres capacity are available. Sterilization by ethylene oxide has been described in detail by Vitéz (1971).

It should be borne in mind that ethylene oxide does not effect micro-organisms that are present in the hermetically closed cavities of instruments, or locked within crystalline substances such as salts and sugars. Thus ethylene-oxide sterilization is inferior to sterilization by heat as regards microbiological safety. The efficiency of the procedure can be tested by the biological indicator (spore test) described in 2.1.2.4 or with so-called Royce bags (Royce and Bowler, 1959). Royce bags are small polyethylene bags containing a solution of $MgCl_2$, HCl and an acid-base indicator. An adequate exposure to ethylene oxide neutralizes the acid in the bag and changes the colour of the indicator.

2.1.4.2   Sterilization with formalin vapour

Though the diffusion rate of formaldehyde is less than that of ethylene oxide, and the vapour readily condenses, formaldehyde vapour is convenient for the disinfection of clean rooms and incubators. Condensation and polymerization are negligible provided that the formaldehyde concentration in the air does not exceed 3 mg dm^{-3} and the relative humidity is maintained above 60%.

The microbiocidal effect of formaldehyde vapour is greatest between 50% and 90% relative humidity. To generate the vapour, formaldehyde solution may be evaporated, or sprayed into the air.

2.1.5   Sterilization by filtration

Microbial cells can be removed from fluids and gases by filtration. In industrial microbiological processes heat-sensitive fluids as well as the laboratory air are sterilized by filtration.

2.1.5.1   Filtration of air and other gases

The cotton plugs uses in flasks and tubes serve as filters, preventing microbiological media from becoming contaminated with airborne micro-organisms. Gases entering microbiological cultures are sterilized by passage through glass tubes or metal tubes containing cotton wool, glass wool or asbestos.

Large volumes of air, such as are required in clean rooms and laminar-flow cabinets, are sterilized by filters as described in 1.1.4. Dryness is a prerequisite of the effectiveness of any air filter.

2.1.5.2 Filtration of liquids

The following types of filter are those used most commonly for the sterilization of liquids.

(1) *Membrane filters* consisting of cellulose derivatives. These vary in pore diameter and size. They withstand acids and alkalies, but are attacked by organic solvents. Both presterilized and unsterilized membrane filters are available commercially; they can be sterilized in steam either alone or in the filtration apparatus assembly. Membrane filters are elastic and flexible, and can be stored for unlimited periods. They also find application in various methods of germ counting (see 5.2.2.2), and the range of their industrial uses is wide.

(2) *Seitz filters*. These are essentially asbestos plates, which are also used on a large scale in the wine, beer and pharmaceutical industries.

(3) *Sintered glass filters* are made of finely ground glass of well-defined particle size fused sufficiently to form a very fine filter. Glass filters fitted into funnels made of heat-resistant glass can be carefully sterilized both in dry air and in steam.

The filtration of liquids can be speeded up by applying a vacuum pump to the receiving vessel, or applying pressure to the non-sterile side of the filter. The latter method is preferable because micro-organisms are sometimes drawn in between the stopper and neck of the collector glass when a vacuum is applied.

2.1.6 Cleaning and disinfection

2.1.6.1 Cleaning

It is important that surface-active substances are removed by thorough rinsing after cleaning and disinfection. These substances, especially the detergents, are strongly adsorbed on to glass, and thus may interfere with bacteriological and serological work.

Before a new detergent is used in the laboratory, it is advisable to find out whether microbial growth is disturbed in those vessels that have been cleaned with the detergent.

Soaking in detergent and washing with a brush is the most thorough method of cleaning glassware and no part of the surface of the vessel should escape the action of the brush in this procedure. It is, therefore, necessary to remove the internal liners from screwed caps before sterilization.

Cleaned glassware and instruments should be rinsed with tap water followed by several rinses with distilled water, and after draining, dried in a drying cabinet.

2.1.6.2   Disinfectants

In microbiological laboratories, disinfectants are used to reduce the numbers of micro-organisms on surfaces, or in the cleaning of instruments.

Among those disinfectants based on the microbiocidal effect of heavy metal ions, corrosive sublimate ($HgCl_2$) is mainly of historical interest to the microbiologist. Since it has been superseded by equally effective disinfectants that are less toxic to man or can be substituted by physical methods, the use of $HgCl_2$ is preferably avoided.

Among oxidizing disinfectants, calcium hypochlorite or sodium hypochlorite, as well as chloramine I (p-toluol-sulphone-chloramide sodium) and Neomagnol (benzene-sulphone-chloramide sodium) deserve mention. A 2% solution of hypochlorite contains 0·5% active chlorine. Chloramine I and Neomagnol can be stored for a longer time and smell less strongly than hypochlorites; 0·2–1·0% solutions of chloramine I and Neomagnol are used for disinfecting surfaces.

Iodophores are aqueous solutions of iodine and non-ionic surface-active compounds such as Iosan (CIBA, Basle; Phylaxia, Budapest), and have been recently induced as disinfectants with improved properties. The iodine which is loosely bound to the surface-active substance, retains its disinfective activity although in this form it is less volatile and more soluble than free iodine. Exhaustion of an iodophore solution is indicated by the fading of its colour. The iodophores are used in acid solution, the effective concentration of Iosan being ∼0·3%.

Formaldehyde is used as a disinfectant in 1–5% solutions.

Ethanol is widely used as a disinfectant for metal instruments, working-surfaces, hands, rubber tubes and rubber stoppers. It is most effective as an aqueous solution of about 70%. However, bacterial spores can resist this concentration and, therefore, the alcohol that remains after dipping on the surface of non-inflammable materials should be ignited; 96% and 50% solutions of ethanol have almost no disinfectant value.

Many quaternary ammonium compounds, e.g. Zephirol (Bayer), and the Hungarian product Sterogenol (EGYT, Budapest) (Mono-nitrogenol; cetylpyrimidium bromide) are widely used as disinfectants in laboratories and in industry. Used in the form of the warm (40–50 °C) 0·2–0·3% solutions, they do not corrode metal surfaces.

However, the effect of quaternary ammonium compounds is reduced in the presence of alkalies, soap, or large amounts of organic substances such as fats and protein. Furthermore, quaternary ammonium salts are less effective than the above-mentioned disinfectants against moulds and have practically no effect on some viruses.

2.2 Preparation and storage of culture media

There is no universal medium for the cultivation of all micro-organisms, because different micro-organisms require different nutrients and environmental conditions. The requirements of some organisms may even be completely incompatible with the requirements of others. This is why the introduction of standardized methods in microbiological practice is of importance in order to improve the reproducibility of investigations.

Culture media can be divided into two main groups according to the breadth of the range of micro-organisms that can grow on the medium. Accordingly there are general media, i.e. those supporting the growth of a wide range of micro-organisms, and selective media, i.e. those suitable for the culture of only a few species.

The composition of the medium provides another basis for classifying media. Thus, natural and artificial media are distinguished. For the preparation of the former, natural organic material of animal and/or plant origin is used; the chemical composition of these is usually complex and not well understood.

According to their physical condition, media solidified by agar (or other gel-forming substance) and liquid media are distinguished. For the composition of the most important media, see Chapter 9.

2.2.1 Requirements relating to the composition of media

The water used in natural media may be tap water, whereas artificial media require deionized or distilled water. Distilled water prepared in an apparatus with metal components contains traces of heavy metal ions, and these, particularly copper ions, are toxic to micro-organisms and may interfere with exact biochemical tests. Therefore, distilled water—double-distilled if possible—from a glass apparatus is preferable.

The ingredients of a medium should be dissolved consecutively, i.e. an ingredient should not be added until those previously added have completely dissolved. Ingredients of which only traces are to be used are added in the form of solutions.

For solidifying media agar-agar (agar) is the most widely used geling agent, since it is inert as far as the great majority of micro-organisms is concerned. It swells in cold water, but a homogeneous gel requires heating to 100 °C and subsequent cooling to 40–50 °C. Agar is commercially available in lamellar, fibrous and powdered forms. Solidification of media requires a concentration of 1–3% agar, depending on the quality of the agar preparation. Since commercial products may contain considerable amounts of calcium and magnesium salts which may inhibit microbial growth, it is recommended that the agar product is thoroughly washed before use; fibrous agar is usually rinsed in flowing water for three days, then soaked in distilled water for 8 days with daily changes of fresh distilled water. The

38

pH of the medium is not influenced by the presence of an agar of good quality, and, therefore, it is advisable to dissolve the agar after all the other components of the medium have been dissolved and the pH has been adjusted to the desired value.

In preparing highly acid media (e.g. of pH 3·5), sterilization should precede acidification, otherwise the agar may be hydrolysed during the heat treatment and lose its gel-forming capacity.

Gelatin, a proteinaceous substance, is another gel-forming material used in microbiology; 10–14% solutions are used for the solidification of media. The characteristic liquefaction of a 10% gelatin gel by proteolytic organisms is of great importance in microbiological diagnostics.

Gelatin loses its gel-forming capacity when hydrolysed, and, therefore, intense heating of gelatin solutions must be avoided. It is reasonable to sterilize gelatin-containing media by steaming after the medium has been poured out into glassware presterilized by dry heat.

In order to achieve reproducible results, the pH of any medium should be controlled precisely, and if necessary, re-adjusted after sterilization since the pH may have decreased during autoclaving. The commercially produced dehydrated agar media are generally composed in such a way that the correct pH is attained after dissolution and sterilization are carried out as prescribed by the manufacturer.

Microbiological culture media should not be turbid and should contain no sediment. Suspended matter can be removed by centrifugation and/or filtration. For filtration, cotton wool or one or two sheets of folded filter paper are usually sufficient. Solidified media can be filtered in the fluid state in a heated funnel. Melted solidified media can be purified by sedimentation at a suitable temperature followed by decantation. Alternatively, the medium is allowed to solidify in a cylindrical vessel whereupon the medium is allowed to slip out of the cylinder as a block so that the part of the block containing the sediment can be cut off. The clear upper section is then melted and resolidified before use.

Any turbidity that is not removed by filtration can be eliminated by clarifiers, for example egg white whipped to a stiff foam is thoroughly stirred into the medium which is pre-cooled to a temperature of about 50 °C (the white of one egg is added to 2 litres of medium). The medium is then heated, and the contaminant causing the turbidity is removed in the coagulating albumen.

Fatty acids which inhibit microbial reproduction may be derived from the ingredients of some media, or from the cotton plug (e.g. during sterilization of the glassware by dry heating; see 2.1.1.3). These substances can be removed by adsorbents, namely, 0·5% activated charcoal, 0·15% soluble starch, or an anion-exchange synthetic resin (Meynell and Meynell, 1965).

2.2.2   Commercial media

Much time and work can be saved by using commercially produced media, especially if small amounts of many different media are in use.

Commercial media are available in powdered, granulated, or liquid form, and as tablets. In many countries, the composition of commercial preparations is kept under tight control, and thus they are of constant quality which is essential for reproducible results. Special care should be taken in the storage and processing of commercial media, for example, accidental wetting or dissolving before sterilization should be avoided. Dehydrated media and hygroscopic ingredients must be stored under air-tight conditions, and should not be exposed either to variations of temperature or to intense radiation.

In the near future, the further standardization of microbiological methods will be accompanied by a widening use of commercial media.

2.2.3   Distribution of media

Before direct use or prior to storage, media are distributed in vessels of small or large capacity. It is not good practice to store ready-made media in large units because contamination of a large unit would render a large volume of medium unusable.

The distribution of media is greatly facilitated by using dispensers, one type of which is electrically operated and automatic (Engelbrecht, 1974; Leistner and Hechelman, 1974). The dispensing parts of these instruments can be sterilized, making them suitable for aseptic work.

Contamination of the mouths of tubes and flasks with medium should be avoided because it leads to adhesion of the cotton plug to the glass. Dispensers should be thoroughly cleaned immediately after they are used.

For the preparation of agar slopes, about 5 ml, (for long slopes, about 10 ml) of melted medium are measured into each test-tube. The still liquid medium is allowed to solidify on a horizontal surface on which glass rods or other supports are placed so as to raise the plugged ends of the tubes. The agar slopes thus obtained cover one side of the tube up to about mid-height. The oblique surface is large enough for spreading bacteria.

If many plates are poured with the same medium, the medium can be sterilized in vessels of larger capacity than test-tubes (e.g. media glasses or conical flasks).

Test-tubes and flasks are usually closed with cotton plugs or metal caps. Cotton plugs are preferably made with defatted cotton, which glows less readily after ignition than untreated cotton. Loose paper wadding, which is sometimes used instead of cotton plugs, does not fit properly into the mouths of tubes and flasks, even when it is made of finely-ruffled wood cellulose.

If media are to be stored in cotton-plugged vessels, it is recommended that the plug is covered with aluminium foil or celophane secured with a rubber band around the neck of the vessel.

The handling of vessels with cotton plugs is more cumbersome than the manipulation of those with metal caps. In a vessel with a metal cap, the passage of air is confined to an S-shaped path. Thus, the risk of dust penetration, and the entry of microbes carried with the dust, is substantially reduced. Metal caps, in contrast to cotton plugs, can be used repeatedly, but can only be used in conjunction with rimless tubes or flasks. Caps should not fit too tightly, otherwise condensation may become trapped between the inner surface of the cap and the outer surface of the neck, and allow the entry of micro-organisms. Furthermore, tightly fitting caps are sometimes difficult to remove.

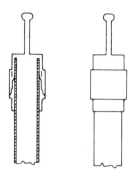

Fig. 6. Kapsenberg's cap

Kapsenberg's handled caps (Fig. 6) are more complex than the oxidized aluminium type of cap. The former also consist of an aluminium alloy, but by virtue of a plate-spring inside, have a better fit on the necks of vessels. Caps must be stain-resistant and must not contain any substances that might pass into the medium during sterilization. Also, plastic caps that withstanding sterilization in the autoclave are available commercially.

Media should not be exposed to any more heating than is necessary, and should not be stored at high temperatures. Microbial reproduction is adversely influenced by any kind of heating of the medium before use.

2.2.4   Storage

It should always be borne in mind that culture media are not simple solutions of inert chemicals, but are sensitive, labile, organic-chemical systems.

The composition of the medium, the method of preparation, the pH, the conditions of sterilization, and the date of preparation should be indicated on the label of the medium container. In large laboratory units it is advisable to keep a diary or a card file on the preparation of media.

It is good practice to store media for several days at the temperature at which they are eventually to be used. At the end of this incubation, any microbial contamination can be identified.

The rooms in which media are stored should be dustfree, cold (about 4 °C), dark, and not too humid. In rooms with high humidity, cotton plugs may become mouldy and the mould organisms may enter the medium. It is advisable to keep capped vessels in a vertical position.

Screw-capped glass bottles of various sizes (e.g., the narrow-necked McCartney bottles of 30 ml capacity and the wide-necked Universal bottles) are widely used. They have aluminium caps or plastic caps, both of which withstand autoclaving. Screw-capped bottles are particularly suitable for storing media because they can be tightly closed, and make efficient use of available space. During the sterilization of media and in the cultivation of aerobes, screw-capped bottles must not be closed tightly.

Media stored for periods longer than 6 months should be checked with respect to quality before use.

2.3   Sampling, sample-storage and -processing

2.3.1   Aspects of sampling

The main task of food microbiology is to search for and identify micro-organisms in foodstuffs, and to determine the numbers of organisms present.

Sampling of food means that a small part of a batch of food is taken for qualitative and quantitative investigation. The sample should exhibit all the quantitative and qualitative properties of the microflora of the food, in other words it must be representative; furthermore, the microbiological status of the sample must not change between sampling and testing. These requirements are very difficult to meet, and, therefore, effective sampling may present problems more complex than those of the testing itself. Foods are usually heterogeneous substances in which micro-organisms are unevenly distributed. Representative sampling becomes difficult if a large quantity of food is to be given a qualitative microbiological examination. The mathematical-statistical bases of sampling from large lots are dealt with in detail in Chapter 3.

Here, only the basic rules of sampling are presented together with storage considerations and the principles of the processing of samples. Detailed procedures for sampling in food quality control can be found in the appropriate branches of the literature (see: Duncan, 1965; Thatcher and Clark, 1968; Hobbs et al., 1973; Olson, 1973; Ingram et al. 1974; Mossel, 1975).

The exact time of sampling for microbiological testing should be carefully considered. If the sampling time cannot be pre-planned, the age and history of the material should be noted, since most foods are perishable, i.e. they are microbiologically unstable.

2.3.2 Sampling and sampling instruments

Samples should be taken aseptically and transferred with sterile tools into sterile vessels. Wide-mouthed bottles which are glass-stoppered, screw-capped or rubber-stoppered can be used for this purpose. The tool to be used (a spoon, knife, pipette, etc.) must be clean and tightly wrapped. The sample should be labelled carefully and precisely, indicating the place and the exact time of sampling, as well as the purpose of the examination. In the case of samples taken for official purposes (e.g. for public authorities) even the testing schedule is prescribed.

2.3.3 Transport and storage of samples

Efforts should be made to process material as quickly as possible after sampling. Samples should be transported at low temperature in a secure package. However, they should not be frozen either during shipment or storage, except in the case of samples of frozen food which must be kept frozen until tested.

If it is impossible to process samples immediately, they may be stored at 0–4 °C overnight (Mossel et al., 1972). Storage for still longer periods obviously will allow changes to occur with respect to the microbiological state of a sample with the possible exception of tinned or dried material. Samples to be tested for anaerobes should not be exposed to the air, and may be stored in an environment of low redox potential.

2.3.4 Processing of samples

Samples should be processed under aseptic conditions. The surface of air-tight sample containers should be sterilized by immersion of the area where the package is to be opened in alcohol, and flaming. Deep-frozen products should be allowed to thaw at 2–5 °C overnight and then processed immediately.

2.3.4.1 Homogenization

Samples should be homogenized before examination. Homogenization is also an important step before the extraction of microbial toxins, since toxins may be unevenly distributed in the sample, and homogenization also facilitates extraction.

Liquid food samples can be homogenized simply by shaking. Samples of solid, combined solid and liquid, or semi-liquid consistency are homogenized by grinding, rubbing, triturating in a mortar, shaking with glass beads, or blending in an electrical homogenizer.

The requirements of homogenizers to be used in microbiological practice are that they should

(a) properly homogenize the sample without heating it excessively;

(b) disperse aggregates without damaging individual microbial cells;

(c) not endanger the operator (e.g. by emitting an aerosol);

(d) be easy to use, withstanding continuous use including cleaning and sterilization.

The homogenization process often requires the addition of sterile fluid to the sample (see below) and the quantity of the diluent must therefore be taken into account in any calculation.

Because of the uneven distribution of the microflora in foods, it is desirable to homogenize at least 10 g of each sample, usually in 90 ml diluent, or preferably, 50 g in 450 ml diluent (a 10-fold dilution).

The use of blenders of 6,000–12,000 r.p.m. improves the reproducibility of microbiological test results and raises the germ count values that are obtained compared with trituration in a brayer mortar (Barraud et al., 1967). A specialist committee of the International Organization of Standardization (I.S.O.) has recommended a total of 15,000 to 20,000 revolutions within a period not longer than 2·5 minutes. During homogenization, the temperature of the sample must be kept low so that the viability of even the most heat-sensitive microbial cells is not impaired.

Highly viscous homogenates are difficult to handle in a pipette, except at dilutions which may give false negative results. Nevertheless, homogenization is indispensable if the presence of micro-organisms is suspected deep within sample fragments. If it seems certain that only the surface of the sample is contaminated, and the micro-organisms can be washed successfully into the diluent, homogenization may be unnecessary. In such cases, an ultrasonic vibrator or a Whirlimixer-type mixer can be employed (Sharpe and Kilsby, 1970).

The ultrasonic homogenization of food samples must not be carried out at very high intensities because over-intense sonication may denature proteins and disintegrate microbial cells. It is advisable to add a non-microbiocidal surface-active agent (e.g. 0·02% Pluronic L 61, sodium sulpholaurate, or Tween 80) to the diluent before sonication, to increase the wetting effect of the suspension fluid. In this way the tendency for aggregation is reduced and an appropriate degree of dispersion can be achieved.

Recently, a homogenizer developed by Sharpe and Jackson (1972) called a "Stomacher" has come into wide use. The sample is measured into sterile disposable plastic bags $160 \times 200 \times 300$ mm in size, together with aliquots of sterile diluent. In the homogenizer the bag is pressed by two alternately moving metal plates against the inner face of the front plate of the apparatus. Pressing and shearing forces thus arise which produce a mincing and homogenizing effect. The plastic bag suspended inside the apparatus is clasped in place by the upper rubber lace of the front plate when the latter is closed. The bags are sterilized by "cold" procedures, that is, with ethylene oxide or by irradiation. For homogenization, it is advisable to cut the sample into nut-sized pieces with sterile scissors. In comparative experiments, the viable cell counts obtained after 0·5–5 minutes homogenization in an "Ato-Mix" blender or an "Ultra-Turrax" homogenizer were usually in good agreement with those obtained with the "Stomacher" (Sharpe and Jackson, 1972; Baumgart, 1973). The "Stomacher" is less noisy than the other homogenizers and does not heat the sample to any significant extent; its main advantage is that owing to the use of disposable bags the work and time involved in sterilization are omitted, and an aerosol is not formed, the latter being an important advantage when pathogenic and/or toxigenic micro-organisms are being handled (Lötzsch et al., 1973; Leistner and Hechelmann, 1974). Furthermore, the plastic bag makes a very suitable sample container. However, samples containing fragments capable of piercing the plastic bag (e.g. bone fragments) cannot be processed in this way.

According to Sharpe (1973), suspensions obtained after processing in the "Stomacher" apparatus contain less debris (the presence of which interferes with colony counting), than those obtained by any other available method; furthermore, more than one bag can be processed in the "Stomacher" at the same time.

2.3.4.2 Preparation of serial dilutions

Decimal (1 : 10) dilution series are the most commonly used *(Fig. 7)*, although twofold, hundred-fold and other series are occasionally used.

As a diluent, sterile saline [45] containing 0·85% NaCl may be used, and more recently a solution [109] containing 0·1% peptone has gained preference. If placed in water, the vegetative forms of some bacteria may in large measure perish (Straka and Stokes, 1957; King and Hurst, 1963).

For the culture of anaerobic bacteria, it is advisable to use an anaerobic medium of low redox potential as a diluent. For counting the colonies of strictly halophilic bacteria, a 15% NaCl solution may be used.

Successive dilutions must be homogeneous, since any errors due to inhomogeneity may be multiplied in the dilution series.

For making serial dilutions, pipettes of 0·1 ml graduation, calibrated for blowing out, are most suitable. The tip of the pipette is dipped to a depth of 1·5 to 2·0 cm in the suspension that is to be diluted, and the suspension is drawn in to the uppermost mark and released several times. After the last filling of pipette, the tip is raised from the liquid and it is touched on the glass wall of the vessel so that fluid adhering to the

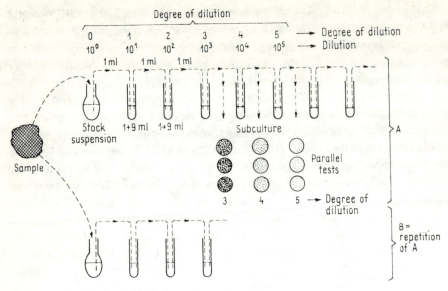

Fig. 7. Preparation of tenfold dilution series

outer surface of the pipette can drain away. Then the pipette is transferred to a vessel containing diluent, the tip is touched on the wall of the vessel above the liquid, and the contents of the pipette are allowed to flow out, or if the pipette is calibrated, the remaining suspension is blown out (but not by mouth!). A fresh sterile pipette must be used at each step of the dilution. At each step, the suspension which is to be further diluted is thoroughly stirred with a test-tube stirrer; vortex stirrers are produced for this purpose by several firms (see, for example, Ramming et al., 1971; Leistner and Hechelmann, 1974). Alternatively, the test-tube is simply rotated between two palms. The remaining steps of the serial dilution are carried out in the same manner.

Each suspension on the 10-fold dilution series is usually designated by the logarithm of the overall dilution factor. 1·0 ml of the original dilution added to 9 ml of the diluent, and the procedure is repeated until a dilution is reached which is expected to contain no viable micro-organism (the limiting-dilution technique), or until a dilution which contains a countable number (30–300) of colony-forming micro-organisms per unit volume is attained (the colony-counting technique). 0·1 ml of each dilution is spread on the surface of a solid medium, whereas in the case of plate-pouring technique the usual volume of the inoculum is 1·0 ml. The volume of the inoculum has to be taken into account in the final calculations.

The total viable count and/or the viable count for a particular group of micro-organisms is determined by inoculating appropriate media with each dilution in the series.

46

The number of dilutions can be reduced by half by diluting 1 part of suspension in 99 parts of diluent (1 : 100 dilution). The so-called Demeter pipette is suitable for carrying out two different dilutions simultaneously from the same basal suspension by transferring 1·0 and 0·1 ml, respectively, into 10 ml diluent. This procedure is widely used in the dairy industry.

In the preparation of serial dilutions, one should take special care to maintain asepsis. Pipettes are individually wrapped in paper and sterilized in metal or glass cylinders. Before a pipette is used, the paper is broken near the upper end, and holding this end of the pipette in one hand, the paper is pulled down with the other and the surface of the pipette is flamed. When a flask or test-tube is opened, the neck is flamed twice, once after the plug or cap has been removed and again before it is replaced.

The dilutions must be inoculated on to a medium within half an hour of preparation.

2.3.4.2.A Dilution errors

The accuracy of serial dilutions was investigated by Jennison and Wadsworth (1940). The errors arising when 1-ml pipettes were used with 9-ml volumes of diluent, are presented in *Table 2;* the data show that errors grow enormously if serial dilution is carried out with inaccurate measuring devices or a lack of care.

Table 2

Error of the serial dilution procedure as a function of the accuracy of pipetting and the diluent volume

Diluting Volume ml	Pipette S.D. of the volume ml	Diluent volume ml	S.D. of the diluent volume ml	Error arising from the dilution expressed in per cent S.D. at dilution levels			
				10^2	10^4	10^6	10^8
1	±0·01	9	±0·1	±2·8	± 5·7	± 8·5	±11·3
1	±0·01	99	±1	±1·4	± 2·8	± 4·2	± 5·7
1	±0·03	9	±0·3	±8·4	±17·1	±25·5	±33·9
1	±0·03	99	±3	±4·2	± 8·4	±12·6	±17·1

2.4   The use of the microscope

The microscope is an indispensable part of the microbiological laboratory equipment. In food microbiology, "microscope" usually refers to the light microscope. The electron microscope, owing to its higher resolving power, provides a means of investigating the ultrastructure of micro-organisms. In this chapter neither the theoretical aspects nor the detailed practical problems of microscopy are touched on. For details of these subjects, see Meynell and Meynell (1965), Barabás and Vadász (1966), and Quesnel (1971).

2.4.1 The construction of the light microscope

The optical system of the microscope consists of two lens systems—the objective, which produces a magnified image of the object, and the ocular (or eyepiece), which further magnifies this image *(Fig. 8)*. (This type of microscope is also referred to as the compound microscope, in contrast to the simple microscope which has a single lens or lens system.) The image seen through the ocular is a magnified inverted image of the object. The magnifying power of a microscope is the product of the magnifying power of the objective and that of the eyepiece. The object to be examined is usually placed on the stage of the microscope between two glass plates, namely a slide and a coverslip. If such a preparation is placed beneath a high-power objective, the amount of light reaching the eye from the object is very small, and therefore the object must be strongly illuminated. Light of the required intensity is provided by an illuminator system and condenser. The light emitted by the source (the microscope lamp) is reflected by means of a mirror through an iris diaphragm into the substage condenser, which concentrates the light at the position of the

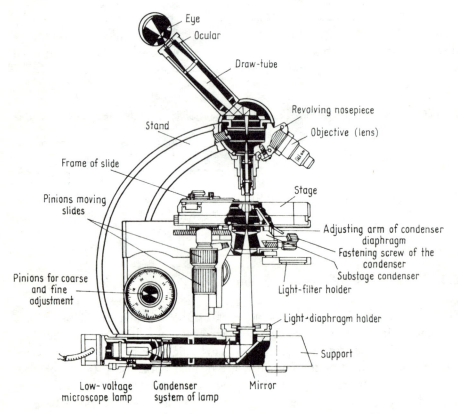

Fig. 8. Diagram of the Zeiss (Oberkochen) microscope "Standard RA"

48

object. After transmission through the object the beam of light passes along the microscope tube into the ocular, and finally arrives at the eye of the observer.

A microscope possessing a single ocular lens system is called a monocular microscope, and a binocular microscope is one in which the object can be observed simultaneously with both eyes.

Modern microscopes are equipped with exchangeable ocular sets and a revolving nosepiece on which at least three objectives are mounted. Objectives can be turned into the path of the light transmitted through the object by a simple movement.

The magnifying powers of the most widely used objectives are $\times 10$, $\times 40$ and $\times 95$ or $\times 100$; those of the oculars are $\times 5$, $\times 10$, $\times 15$ and $\times 20$.

Objectives of $\times 10$ magnifying power can only be used for the examination of relatively large micro-organisms such as protozoa and algae. Bacteria, being of the order of 1 µm in diameter, need to be observed with objectives of much higher magnifying power.

2.4.2   Resolving power and numerical aperture

The resolving power of a microscope refers to the clarity of the image produced (i.e. with high resolution, two points situated very close together are seen more clearly as separate points). Theoretically, the shortest distance (d) between two points which can still be discerned as separate points, depends on the wavelength of the illuminating light (λ), the refractive index (η) of the medium between the glass slide and the first lens of the objective, and the greatest angle of divergences from the optical axis of the microscope (μ) of rays passing through the lens:

$$d = \frac{\lambda}{\eta \cdot \sin\mu}.$$

Accordingly, the resolving power of a microscope becomes greater as the wavelength of the radiation source is decreased. (This is the reason that UV light and electron rays are used in high magnification work.)

Resolution can also be improved by increasing the refractive index (r.i.) of the medium from which light enters the first lens of the objective, and by widening the angle μ. The product $\eta \sin \mu$ is called the numerical aperture, and is directly related to the resolving power. Theoretically, the highest possible value of μ may be 90°, when $\sin \mu$ reaches a maximum value of 1.

The relationships between the numerical aperture of the objective, the resolving power, and the so-called useful magnification of a microscope are shown in *Table 3*. The useful magnification obtainable with an objective of given numerical aperture is the highest magnification that can be used without losing sharpness of the image. The data in *Table 3* help in the selection of the most suitable ocular for work with a given objective.

4 Kiss

Table 3

The range of useful magnification at given values of numerical aperture and resolution

Numerical aperture	Resolution* d (μm)	Range of useful magnification
0·10	2·75	× 50 — 100
0·20	1·37	× 100 — 200
0·30	0·92	× 150 — 300
0·40	0·69	× 200 — 400
0·50	0·55	× 250 — 500
0·60	0·46	× 300 — 600
0·70	0·39	× 350 — 700
0·80	0·34	× 400 — 800
0·90	0·31	× 450 — 950
1·00	0·27	× 500 — 1000
1·10	0·25	× 550 — 1150
1·20	0·23	× 600 — 1250
1·30	0·21	× 650 — 1350

* Measured in green light of 550 nm.

2.4.3   Increasing the resolving power by use of the immersion objective

The highest practical value of μ is 72° (sin 72° = 0·95), so that the numerical aperture cannot be higher than 0·95 if air fills the space between the coverslip and the first lens. However, the numerical aperture and thus also the resolving power can be increased by increasing the value of η. Thus a fluid of high r.i. may be placed on the coverslip so that the lens of the objective becomes immersed in the fluid. Cedar oil and anisol (methyl-phenylether) are the most widely used immersion oils, the r.i. of these fluids being about 1·5, considerably higher than that of air (1·00) or water (1·33).

One advantage of anisol is that it can be removed from the lens without a solvent, simply by wiping. Cedar oil, on the other hand, may dry on the lens and has then to be removed with a solvent.

The immersion oil should be removed from the surface of the objective immediately after use. For this purpose soft cloth dipped in xylol, or a special lens-cleaning paper may be used. When out of use, the microscope should be protected from dust, either in its case or under a suitable cover.

2.4.4 Illuminating systems

2.4.4.1 Köhler illumination

Köhler's system is the most frequently used illumination system.

If the source of light is not built in, adjustment is carried out as follows. The microscope lamp is placed with its iris diaphragm 15 cm from the microscope. The opaque glass in front of the lamp is taken out and the iris diaphragm of the lamp is gradually closed until the light beam falls on a small area in the centre of the mirror. Then the diaphragm below the condenser is partially closed and the light is directed to the centre of the condenser's diaphragm.

The lamp is moved in its case to find the position at which an image of the incandescent filament is sharply formed on the condenser diaphragm. Then both diaphragms are opened, the preparation to be examined is placed on the microscope stage, and its image is brought into focus. The diaphragm is slowly narrowed while the illuminated spot in the field of vision is simultaneously observed. The narrowing spot of light must be in the centre of the field of vision when the diaphragms are closed, and any necessary adjustment can be made by means of the mirror. With both diaphragms closed, the small blurred image of the lamp's iris diaphragm appears and the condenser is moved upwards or downwards until sharp outline of the iris diaphragm and the image of the incandescent filament are obtained. In the next step the iris diaphragm of the lamp is opened until the edge of the diaphragm coincides with the edge of the field of vision. If the diaphragm is opened further, undesirable reflections may appear.

If one of the eyepieces is removed from the microscope tube, the image of the condenser diaphragm can be seen. The width of this diaphragm must not be less than one quarter of the objective's aperture. If the aperture of the diaphragm is too wide, the image will be blurred; if it is too narrow, light-diffraction phenomena will occur.

The illuminated area is regulated by the iris diaphragm of the lamp, and the intensity of the illumination and the depth of the focus are adjusted with the condenser diaphragm.

2.4.4.2 Bright field microscopy

In food microbiology, bright field microscopy is the most widely used microscope technique. In this system, the object is viewed against a bright field, and is visible by virtue of differences in light refraction and light absorption in different parts of the object. However, with this type of illumination the details of a transparent object cannot be recognized because the contrast between object and background is poor. To increase the contrast, various procedures have been developed, the earliest of these being colouring (i.e. staining) of the object. The dyes used for this purpose stain the various structures within the object differentia. The early staining procedures were not, however, suitable for living cells or tissues, and, therefore, methods were sought for increasing the contrast by means of new illumination systems and further improvements in the construction of the objective.

2.4.4.3 Oblique illumination and dark field microscopy

Oblique illumination means that the source of light is an off-centre point or a ring, which results in shadow phenomena arising on the surface of the object, the detail of which, therefore, becomes more pronounced. In dark field microscopy, on the other hand, the light is emitted from the condenser at such a large angle that only dispersed light from the object reaches the objective. Consequently the edges of the object are very bright while the background remains dark. The great disadvantage of this system is that little light is transmitted through the microscope, and the object may be damaged by the heat of the illumination. Oblique illumination and dark field microscopy can be combined.

2.4.4.4 Interference microscopy and phase-contrast microscopy

Interference and phase-contrast microscopy are both superior to the oblique and dark field techniques. Both make use of the fact that the light that has passed through an object is out of phase with the light that has passed through the surroundings of the object *(Fig. 9)*. Interference between the two out-of-phase

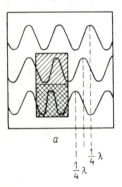 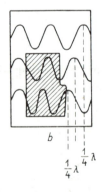

Fig. 9. Prerequisites of the phase-contrast phenomenon

a — the refractive index of the cell or cell components exceeds that of the surrounding medium; *b* — uneven thickness of the object (the effect of light absorption on light intensity is not taken into account)

52

beams of light in the plane of the image brings about differences in light intensity, so that some points of the object look darker and others lighter than the background. The interference technique requires a special microscope whereas most modern microscopes can be fitted with phase-contrast components.

Phase-contrast equipment consists of special objectives, the so-called phase condenser, and an auxiliary telescope (also referred to as an "auxiliary microscope" "auxiliary focusing magnifier" or "centring telescope") which can be put on the place of the ocular. There is an annular phase-shifting element (phase-plate) in each phase-contrast objective and corresponding annual diaphragms in the phase-contrast condenser system. The phase-shifting element of the objective and the annular diaphragm of the condenser should exactly superimpose one another for a phase-contrast observation.

The condenser is adjusted as follows.

— The revolving disc containing the phase-condensers is attached below the microscope stage. (On the pinion of this disc, a number indicates which condenser is in the path of the light. "O" indicates bright-field illumination. The revolving disc is turned to the latter position.)

— The position of the condenser is adjusted so that its front lens is 3 mm below the upper plane of the miscroscope stage.

— The oculars and the phase objectives to be used are fitted to the microscope.

— The object to be examined is mounted on the microscope.

— An image of the object is focussed as for Köhler illumination and the diaphragm of the condenser is opened as wide as possible.

— The auxiliary telescope is inserted in place of (one of) the ocular(s) and its telescopic section is adjusted until the image of the objective ring is focussed altering the position neither of the objective nor the condenser.

— The revolving condenser-disc is turned until the phase condenser is moved into the light path which corresponds to the phase-contrast objective to be used. The right position is indicated when the number corresponding to the magnification of the objective appears in the aperture of the revolving condenser-disc.

A yellow-green filter is inserted in the illuminating system to enhance the contrast.

— The image of the ring diaphragm of the phase condenser is made to superimpose that of the phase plate of the objective with the aid of adjusting wrenches. Inexact adjustment in this respect makes for poor contrast.

— The auxiliary telescope is removed and replaced by the ocular. The phase-contrast image of the object can now be viewed in the microscope.

In microscopes with the pancratic system (e.g. the NF-type microscope of VEB Zeiss, Jena), the phase condenser is not used instead, a ring diaphragm is attached to the pancreatic condenser.

Phase-contrast illumination produces rich contrast in the images of unstained preparations and thus makes it possible to examine the inner structure of the living cell, provided that the preparation does not exceed 15 μm in thickness.

Because of the bright halo that appears around cells, phase-contrast illumination cannot be used for measuring cell dimensions (see 2.6.5).

In the phase-contrast system, every glass surface in the light path must be kept particularly clean, and slides and coverslips must be free of dust, finger prints, etc.

2.4.4.5   Fluorescence microscopy

Fluorescence microscopy, a technique of considerable diagnostic value, is based on a special property of some dyes. When illuminated with UV light, these dyes emit visible light, i.e. they fluoresce (Paton and Jones, 1971). If mixed bacteria are treated with a fluorescent dye, those to which the dye becomes bound will fluoresce in the dark field of the fluorescence microscope. The technique can be utilized, for example, for demonstrating the presence of *Mycobacterium tuberculosis*. Mycobacteria take up the fluorescent dye Auramine-O, whereas most other bacteria do not.

Fluorescence microscopy is widely used in serological tests. Antibodies conjugated with a fluorescing dye such as fluorescein (labelled antibodies) link up with the corresponding antigen, thus causing the bacteria or other particles containing the antigen to emit light. Fluorescent antibody reactions are highly specific and sensitive.

2.5   Isolation, enrichment, cultivation and maintenance of strains

2.5.1   Methods of isolation

Isolation refers to the separation from a mixed microflora of a strain or species of microbe and the establishment of this organism in pure culture (in the absence of other micro-organisms).

Microbes can be isolated on plates of solid media and in single-cell cultures.

2.5.1.1   Plate cultures

Isolation on solid media can be achieved by surface streaking the plates or by making pour plates.

2.5.1.1.A   Isolation by surface streaking

Plating essentially involves the spreading out of a suspension of cells on the surface of a sterile agar plate usually solidified in a Petri dish. For this purpose, melted agar medium is cooled to a temperature of about 45 °C and poured under aseptic conditions into a sterile Petri dish which is unwrapped immediately before use. The lid of the Petri dish is raised and held in a tilted position while the medium is poured in cautiously *(Fig. 10)*, avoiding any spillage of the medium down the outside of the dish; medium adhering to the wall of the dish would provide means for contaminating micro-organisms to enter the culture.

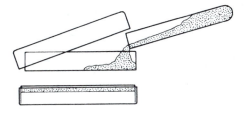

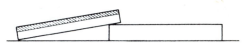

Fig. 10. Pouring of plate for surface-streaking (Proszt and Varga, 1965)

Fig. 11. Drying off the surface of an agar plate before inoculation

Machines have been constructed for the automatic pouring of agar into plastic Petri dishes, and the results obtained by the use of these have proved reliable (Albertz et al., 1972; Leistner and Hechelmann, 1974).

Before inoculation by streaking, the water of condensation must be removed from the solid agar surface of the plate by evaporation, otherwise confluence of colonies may occur. Drying off the surface of the agar plates is usually carried out in an incubator at 37 °C. The two parts of the Petri dish can be dried separately with their outer surfaces facing upward, or the dishes can be kept closed and placed upside down in the incubator until the droplets of the condensation on the lid disappear *(Fig. 11)*. The long period of incubation of the evaporation process makes it possible to recognize and discard any contaminated plates.

There are two commonly used methods of spreading the inoculum:

(i) one drop of the cell suspension is placed on the agar plate and spread over the entire surface of the plate with a sterile, bent glass rod *(Fig. 12);*

(ii) a loopful of the suspension is smeared over the plate by drawing a wavy line or parallel straight lines on the surface of the medium *(Fig. 13)*.

The object of either method is to obtain separate colonies each developing from a single cell.

Automated inoculating and streaking techniques have been reported for agar plates (Kaneko et al., 1977).

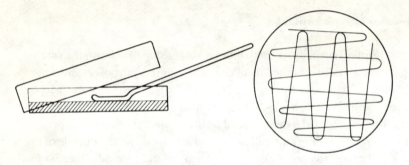

Fig. 12. Inoculation with a bent glass rod (Proszt and Varga, 1965)

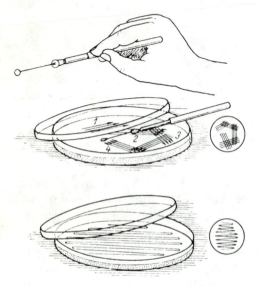

Fig. 13. Modes of speading with an inoculating loop

2.5.1.1.B   Pour plates

The isolaticn of micro-organisms can be carried out by the poured plate method, as follows. An aliquot (usually 1 ml) of the microbial suspension is pipetted into a sterile Petri dish. Before the cells have settled, liquid agar medium pre-cooled to 45 °C is poured on top of the suspension. The Petri dish is then closed and the suspension is evenly mixed with the medium by moving the dish around in a circle, alternately clockwise and anticlockwise, on a horizontal bench. The medium is allowed to solidify, the Petri dish then being upside down so that the surface of the medium dries. In this position evaporation is relatively slow, and the medium itself does not lose water. Another advantage of the inverted Petri dish is that condensate

56

on the lid does not drip back on to the surface of the medium where it would allow colonies to spread out laterally.

Whichever method of isolation is used, an even distribution of the cells and their suspension fluid is of great importance.

2.5.1.1.C   Preparation of pure cultures

An apparently separate colony developing on a plate cannot be regarded as having originated from a single cell if the total inoculum consisted of more than 10–100 viable cells (depending on the surface area of the plate). To obtain a suitably sized inoculum, the original inoculum must be diluted serially (see 2.3.4.2), and one (or preferably more than one) of the dilutions spread by either of the above methods.

To develop a pure culture, the inverted Petri dish is raised from the lid so that the medium is accessible from below. One separate colony is touched with an inoculating needle *(Fig. 14)*, and the cells thus removed are spread on a solid medium, or inoculated into a liquid medium. The purified culture thus obtained can be further purified by repeating the above procedure.

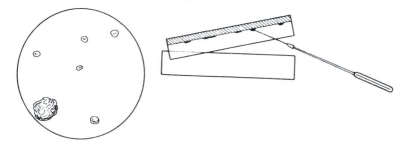

Fig. 14. Subculturing from a Petri-plate culture (Proszt and Varga, 1965)

2.5.1.2   Isolation techniques based on single-cell culture

These methods, including Hansen's and Lindner's techniques, are used mainly for the isolation of yeasts and moulds.

2.5.1.2.A   Hansen's technique

A sterile glass ring with the edges smeared with vaseline is placed centrally on a sterile glass slide, and a drop of sterile water is pipetted in the centre of the ring. The microbial suspension diluted in a melted agar medium is dropped on to a square-

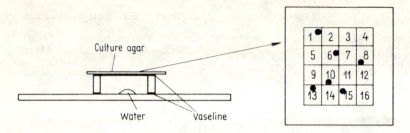

Fig. 15. Hansen's culture chamber (Proszt and Varga, 1965)

meshed coverslip and smeared over it with a loop to form a thin film. The coverslip is inverted and placed on the vaseline-coated rim of the glass ring. In this way an air-tight, so-called wet chamber is set up with nearly 100% relative humidity developing inside. The chamber is examined under the microscope, and each square that contains a cell is mapped *(Fig. 15)*. The cell density of the suspension is important, since the whole of the squared field should contain about 10 cells. After incubation for an appropriate period, the preparation is examined again to find out which of the solitary cells have multiplied. Pure cultures originating from a single cell can then be obtained by subculturing from the micro-colonies.

2.5.1.2.B Lindner's technique

In this procedure, the bottom part of the wet chamber is a sterile hollowed slide. Sterile water is dropped into the depression and the rim of the hollow is smeared with vaseline. Droplets of cell suspension are placed on each square of a meshed coverslip. The area covered by one drop must not extend beyond the field of vision when under the microscope at the required magnification. The device used to spot the suspension on the squares is a pen fused with a glass tube. Before use, the pen is sterilized by dipping in alcohol and flaming. The coverslip is inverted and placed on the hollow slide to form the top of a wet chamber *(Fig. 16)*. Those squares in which the drop contains a single microbial cell are then noted, and after an appropriate

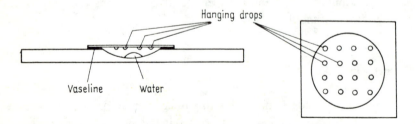

Fig. 16. Lindner's culture chamber (Proszt and Varga, 1965)

58

period of incubation, the same drops are re-examined to find out whether cell multiplication has taken place. Drops in which multiplication from a single cell has occurred are taken up in a sterile filter paper strip 2 × 10 mm, which is then placed in a test-tube containing sterile liquid medium. A pure culture is obtained after incubation.

2.5.2   Enrichment procedures and the use of elective and selective culture media

The plating technique described in 2.5.1.1.A is impracticable if it is desired to isolate an organism of which only a few cells are present amongst a mass of other micro-organisms. It is, however, possible to increase the numbers of the former at the expense of the contaminants, by altering the environmental conditions in favour of the micro-organisms which are to be isolated, and to the disadvantage of the contaminant flora. This effect can be achieved, for example, by making changes to the pH and the incubation temperature, by heating, or by adding chemicals which inhibit the contaminant flora, and nutrients which stimulate growth of the microorganism to be isolated. Isolation by plating may then be carried out in the usual way.

A culture medium providing all the nutrient requirements of the micro-organism (or group of micro-organisms) to be isolated, but not those of the contaminant flora is called an elective medium.

A selective medium, on the other hand, is inhibitory to both, but the inhibitory effect on the contaminant flora is greater. Selective inhibitors include dyes, antibiotics and other antimicrobial substances (Beech and Carr, 1960). Also, media of modified pH used for the cultivation of acid- and alkali-tolerant species may be regarded as selective media.

Media on which micro-organisms form characteristic colonies different from those of other microbes are called differential media. Examples of the use of selective and differential media are given in Chapters 3 and 5. (See also Reuter, 1970, and handbooks such as the Oxoid Manual, Difco Manual, etc.)

2.5.3   Methods of culturing

Some micro-organisms (aerobes) require air, or more precisely oxygen, for growth (aerobic cultivation), whereas others (anaerobes) require anaerobic conditions of cultivation.

2.5.3.1   Aerobic cultivation

In aerobic cultivation, the culture is in contact with sterile air (see 2.1.5.1). Cultures with a large oxygen demand may require aeration, which is achieved by shaking the culture in the presence of sterile air or bubbling sterile air through the culture.

2.5.3.2   Anaerobic cultivation

Some of the obligate anaerobes do not grow at all in media with a redox potential lower than 0·33 V. Since the redox potential of water in equilibrium with the air at atmospheric pressure is 0·80 V (Hungate, 1969), culture media made up with water exposed to air are not suitable for anaerobic cultivation until the redox potential is reduced by excluding air from the medium, and/or reductant(s) are added to the medium.

2.5.3.2.A   Lowering the redox potential by excluding air

Air can be excluded or the oxygen can be removed by various means (Dowel and Hawkins, 1968).

Cultures can be placed in an air-tight container (e.g. a dessicator) from which the air is dumped out, or in an anaerobic container specially made for this purpose *(Fig. 17)*. The vacuum is indicated on a manometer.

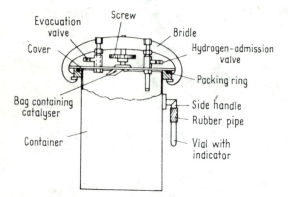

Fig. 17. Anaerobic culture container (a BTL product)

The maintenance of a vacuum for long incubation periods may meet with difficulties. As an alternative, it is possible to introduce a neutral gas such as nitrogen which displaces the air, or to use a palladium catalyst. The latter is placed into the container before evacuation, and then hydrogen is introduced through the reduction valve until atmospheric pressure is restored. Owing to the presence of the catalist, the residual oxygen reacts with the hydrogen in a non-violent manner (Heller, 1954; Willis, 1969). It should be noted that the catalyst is inactivated by moisture, and, therefore, it must be dried after use and stored in a warm dry place.

The need for a hydrogen gas cylinder can be obviated by using the GasPak anaerobic system developed in the U.S.A.; in this system, hydrogen is formed at a slow rate in a closed container. A substance capable of forming hydrogen and carbon dioxide is placed in a small bag (Brewer and Allgeier, 1966) in which gas formation commences when the bag becomes moist. The hydrogen is produced from sodium boro-hydride with $CoCl_2$ as a catalyst. For routine work, this method is more expensive than the use of hydrogen from a cylinder. The carbon-dioxide and hydrogen mixture, however, has further advantages in that 2–10% CO_2 in the gas mixture has been found to stimulate anaerobic growth, the gas mixture is relatively cheap, and the high density of the CO_2 gas is also a favourable factor.

The GasPak method has been used with success also in a large-volume incubator (Gardner and Martin, 1971).

The oxygen of the air can also be removed with alkaline pyrogallol, 1g of pyrogallol being capable of binding 260 ml oxygen at normal temperature and pressure. The pyrogallol method can also be used in individual anaerobic test-tube cultures (the anaerobic systems of Burri and of Preisz) *(Fig. 18)*. The end of the

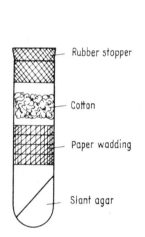

Fig. 18. Burri's method of anaerobic cultivation

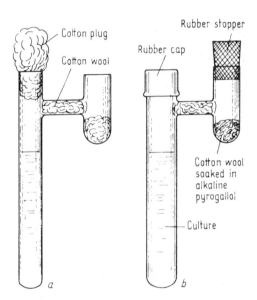

Fig. 19. Pankhurst's tube for anaerobic cultivation

a — during sterilization; *b* — during cultivation

paper wadding plug projecting from the test-tube is cut off, the remaining part then being flamed and pushed into the tube with a pincette towards the surface of the medium. Cotton wool is pushed into the test tube so that a small gap remains between the paper wadding and the cotton wool. 1 ml of a 20% pyrogallol solution and 1 ml of a 20% KOH solution are poured on to the cotton plug and the test-tube is immediately rubber-stoppered. The oxygen in the tube becomes bound by the alkaline pyrogallol, which in the process takes on a brown colour.

61

Subculturing from such an anaerobic culture is sometimes difficult on account of the pyrogallol plug, and there is a possibility of pyrogallol contaminating the culture. These problems can be avoided by using Pankhurst's anaerobic tube (Pankhurst, 1967), which consists of two vertical test-tubes, a shorter and a longer one, connected together by a horizontal glass tube. Alkaline pyrogallol is placed in the shorter tube *(Fig. 19)*. In this system, removal of the oxygen takes a longer time than de-oxygenation in Burri's system.

One disadvantage of the use of pyrogallol is that traces of carbon monoxide, which is inhibitory to some bacteria, are formed during the oxidation of the pyrogallol. Another disadvantage is that carbon dioxide, a gas that stimulates anaerobic growth, is absorbed by the alkaline solution.

Further methods of removing oxygen have been described in the literature (Harrigan and McCance, 1976; Vedamuthu and Reinbold, 1967; Willis, 1969, etc.); being more cumbersome, these are rarely used. The absence of oxygen in anaerobic containers can be checked with the oxygen indicator of Fildes and McIntosh, consisting of a mixture of equal volumes of 0·015% methylene blue, 0·024% sodium hydroxide and 6% glucose. The air is expelled from the mixture by boiling, as a result of which the methylene blue turns into leuco-methylene blue (a colourless derivative). If a small amount of the colourless indicator is placed in the anaerobic vessel before evacuation, the indicator remains colourless as long as oxygen is absent from the vessel (Cruickshank, 1965).

Lucas' anaerobic indicator is available in sealed ampoules (Willis, 1969).

2.5.3.2.B   Lowering the redox potential with reductants

As well as the removal of oxygen, it is advantageous to add a substance of low redox potential which, at the concentration used, is non-toxic to the micro-organism under study. Substances of high sulfhydryl(-SH) content are suitable for this purpose. Natural media such as liver broth [88] and Robertson's meat medium [116], and artificial media containing sodium thioglycolate (0·01–0·05%) or cysteine (≤0·05%) may be used.

It has been recommended that the dissolved air present in anaerobic media be removed by boiling and cooling (with minimal disturbances), immediately before inoculation. To keep air out, sterile liquid paraffin, "vaspar" (a 1: 1 mixture of vaseline and paraffin), or melted paraffin may be added on top of the medium. Semi-solid anaerobic media (containing 0·02–0·03% agar) and anaerobic media solidified with agar are also used, the latter being inoculated in the melted state. Solidification with agar restricts the movement of oxygen in the medium to the slow process of diffusion. Less fastiduous anaerobes can multiply in the presence of traces of oxygen (see above). In these cases, too, it is best to minimize the amount of space directly over the culture medium. For plate culture, for example, a Petri dish with a so-called Brewer's lid can be used. This touches the edge of the culture medium and so encloses a gas space approximately 1 mm in thickness *(Fig. 20)*. In this way the reductant in the medium has a relatively long-lasting effect.

The deep-agar test-tube for the cultivation of anaerobes is described in Chapter 5.

The roll-tube technique (see 5.2.2.3) is a simple, widely approved method of cultivating obligate anaerobes. Before inoculation the space over the agar film must be flushed with carbon dioxide (Hungate, 1969).

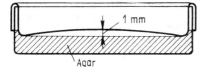

Fig. 20. Petri dish with Brewer's lid

The reducing environment can be checked by adding 0·0002% methylene blue to the medium. The medium is then colourless in the absence of oxygen and blue if oxygen has leaked in.

2.5.4 Storage of microbial cultures

Various methods have been developed for maintaining microbial cultures (strains). The rate of ageing and deterioration of cultures depend both on properties of the medium and on external conditions of the environment. Some microbes must be subcultured at intervals of a few days to be kept in a viable condition, but most of the bacteria that are of importance in the food industry can be maintained by subculturing every 6th to 8th week. Spore-forming bacteria can be maintained for much longer periods without subculturing. Sporulating cultures of moulds and yeasts retain their viability for a few years, even at room temperature and in water (McGinnis et al., 1974).

A temperature of $+4-+5$ °C is the most suitable for the storage of microbial cultures. Their metabolic rate, and consequently the rate of accumulation of toxic substances, as well as the rate of drying out of the medium all slow down at lower temperatures.

Cultures can be prevented from drying out by keeping them under liquid paraffin. A nutrient agar that supports the growth of the micro-organism to be stored is prepared in such a way that the exposed surface area is small; for example, the melted agar is solidified in a test-tube held in a slightly slanting position, and the medium is inoculated in the centre of the exposed surface. When the colony has grown to 2–4 mm in diameter, it is covered with a layer of sterile paraffin 1–2 cm thick. The agar medium should be solidifed not long before inoculation, otherwise the small degree of drying out that can occur may cause the medium to become detached from the wall of the vessel so that cells, especially yeast cells, slip underneath the medium. Some cells can multiply in this position and those forming gas at the same time may push the nutrient agar up above the paraffin layer. A large variety of micro-organisms can be stored at room temperature for a few years in correctly prepared paraffin-covered stock cultures.

Moulds can be stored in mini-cultures covered with mineral oil in small ampoules (Elliott, 1975).

Microbial cultures rapidly frozen in glycerol can be kept viable in the deep-freeze for several years. To prepare material for deep freezing, 2 ml of a 15% aqueous solution of neutral glycerol are pipetted on to agar slant cultures not more than 24 h old, and the cultures are suspended by stirring with a loop; an aliquot is transferred to a sterile 0·5 ml ampoule and frozen in a mixture of alcohol and dry ice at —70 °C. The cultures can be stored in the same mixture or in a deep-freeze at the same temperature (Sirockin and Cullimore, 1969).

Before subculturing from the stored culture, the ampoule is placed in a water bath at 45 °C for several minutes. If necessary, the purity of the culture may be checked by plating out (see 2.5.1).

Freeze-drying (lyophilization) is a commonly used method of keeping live strains. The microbial cell suspension is frozen in a small open ampoule and the water content of the suspension is sublimated in vacuum without allowing the ice to thaw. The ampoule is then sealed in vacuum. With regard to the long-term viability of micro-organisms preserved in this way, the absence of oxygen and the presence of protective substances in the medium are important factors. It is recommended that the lyophilized cultures be stored at a temperature of about +4 °C in darkness.

Lyophilized cultures obtained from a strain collection are handled in the following way. The ampoule is scratched with a file at the level of the cotton plug and cautiously broken in two. 0·2 ml of a liquid medium is pipetted into the dry culture, and after several minutes during which the culture becomes rehydrated, the contents of the ampoule are poured into a test-tube containing medium, and incubated.

2.6　　Preparation for microscopy

2.6.1　　Unfixed, unstained preparations

On drop of saline (0·9% NaCl) is pipetted or run from a glass rod on to a clean glass slide degreased by flaming. A loopful is taken from the culture or colony to be examined and the micro-organisms are mixed with the drop of water on the slide. The suspension is then covered with a coverslip. Air bubbles trapped under the coverslip are an advantage, since they allow the preparation to focus more rapidly and in the case of aerobes, allow bacterial viability to continue. The fluid that is occasionally expelled from the edges of the coverslip is blotted up with a strip of filter paper. (The used filter paper should be assumed to be contaminated with microbes.) Vaseline is smeared along the edges of the coverslip to protect the preparation from dessication; some vaseline is taken up in a loop and melted by holding above a flame for a short time. Then the loop with the melted vaseline is drawn along the edges of the coverslip to form an air-tight seal.

For long-term sealing, paraffin, liquid adhesive, or tinted varnish can be used.

The hanging drop is another type of live-culture preparation. The cell suspension is dropped on to a coverslip which is then inverted over the well of a hollowed slide. The hanging drop must be protected from drying.

Preparations of living material such as those described above are also suitable for observing the active movement, or motility of bacteria. However, some bacteria move too quickly to be observed at all in a living preparation. In these cases the movement can be slowed down by adding methylcellulose (Pijper, 1947).

In a living preparation the original shape and size of the cells can be observed, whereas in fixed and stained preparations the cells are always deformed.

2.6.2   Vital staining

The cells in living preparations are poorly visible under ordinary bright field illumination, and the use of phase-contrast gives a greatly superior image. If phase-contrast is not available, living microbes are made more easily visible by vital staining techniques. Bacteria are able to withstand large dilutions of some dyes (e.g. a 1 : 2000 dilution of crystal violet), and having taken up the dye, may become more visible, sometimes also still exhibiting motility.

2.6.3   Fixed, stained preparations

In such preparations cells are fixed to a slide by means of a physical or chemical techniques. The fixed cells thus remain in place during the staining procedure, although they are killed by fixation.

The simplest and commonest method of fixation is by heating; the preparation is usually drawn over a Bunsen flame several times. As a result, the protein in the cells coagulates and fixes the cells to the slide. Adhesion can be improved by mixing some horse serum to the cell suspension before heat fixation.

Fixation by chemical agents ($HgCl_2$, formalin, phosphotungstic acid, trichloro-acetic acid) is based on the drastic protein-coagulating effect of these agents.

Staining may be positive or negative. In positive staining, the cytoplasm of the cells takes up colour from a dilute dye solution. In negatively stained preparations, on the other hand, the cytoplasm does not take up the dye, which concentrates around the cells instead. Consequently, the cell appears unstained against a dark background.

Basic fuchsin, methylene blue, methyl violet and safranin are the most commonly used stains.

From the powdered dye, usually a 1% stock solution is prepared and the solution is allowed to stand for a few days before filtering with filter paper. The stock solution is diluted with distilled water before use.

Simple staining of a microbiological preparation is carried out as follows.

Grasp the slide in forceps and pass it through a Bunsen flame 4 to 5 times to degrease the surface. Place on the centre of the slide a loopful of the sample to be examined and spread it to form a thin film. (Use a flamed and cooled platinum loop).

Allow the smear to dry in the air.

Pass the slide, film-side up, through a Bunsen flame 8 to 10 times.

Immerse the fixed preparation in a staining solution for 1–2 minutes.

Take the slide out of the staining solution and add water drop to the upper edge of the obliquely held slide so that the water runs down the slide washing away the dye. Continue washing until the water becomes clear.

Rinse the slide in water, blot up excess water with a strip of filter paper, and dry the preparation in the air.

Place a droplet of water on the preparation, cover it with a coverslip and examine under the microscope. Alternatively, the preparation can be examined without a coverslip. Uncovered, dry preparations can also be examined with the immersion objective, the immersion oil being added directly to the dried smear.

2.6.4   Differential stains

Differential staining procedures make it easier to distinguish different intracellular structures, or different types of cell.

Among the differential stains, Gram's stain and Ziehl–Neelsen's stain are the most widely used. The so-called double stains produce a contrast effect between the colours of each component.

2.6.4.1   Gram's stain [55]

This has been the most important differential stain used in the taxonomic designation of bacteria. Some bacterial cells (Gram-positive bacteria), or more specifically the Mg ribonucleate in them, bind strongly to the iodopararosaniline stain (crystal violet and Lugol's solution), while others (Gram-negative bacteria)

66

take up the stain without strong binding, so that it can be extracted with 96% ethanol. The so-called Gram-variable bacteria are Gram-positive when young, but soon become Gram-negative. Most yeast cells and the granules in the hyphae of moulds are Gram-positive. Eukaryotic cells are Gram-negative.

Cells not staining permanently with iodopararosaniline are usually after-stained with safranin, which imparts to the cells a pale red colour.

The staining procedure is as follows. Prepare a fixed smear from a 24h microbial culture as described in 2.6.3.

Cover the smear with 1 or 2 drops of crystal violet solution. Allow the stain to remain on the preparation for 1 minute.

Pour off the stain and wash the preparation cautiously with tap water. Blot up the water on the slide and add 1 or 2 drops of Lugol's solution to the preparation, allowing it to stand for 1 minute.

Wash the preparation with an acetone–alcohol mixture until the fluid running off the slide is colourless.

At this stage, Gram-positive cells appear violet and Gram-negative cells appear pink.

An automated Gram staining apparatus has now been developed (Fung, 1974, Wilkins and Mills, 1975).

2.6.4.2   Ziehl–Neelsen's differential stain [65]

This method, too, has diagnostic importance. It is used for the detection of acid-fast bacteria including *Mycobacterium tuberculosis*.

Acid-fast bacteria react little to the common staining methods, but fuchsin binds strongly to them if the preparation is heated in the presence of phenol. The stain once it has penetrated, cannot be extracted from the cells, not even by washing with hydrochloric acid–alcohol.

The staining is carried out as follows. Cover the fixed preparation with a strip of filter paper as large as the slide.

Pour carbol-fuchsin solution on to the preparation so that the whole slide is covered. Heat the slide gently from underneath until steam is seen to rise from the solution. Repeat the heating twice at intervals of 5 minutes.

Rinse the preparation thoroughly and decolorize with hydrochloric acid–alcohol.

Add methylene blue as a counterstain.

Acid-fast organisms appear red against a blue background. Bacteria that are not acid-fast and other particles appear blue.

2.6.4.3   Endospore stain

Species of the genera *Bacillus* and *Clostridium* form endospores that are resistant to ordinary staining. In stained preparations, the spores are visible as colourless bodies in the stained sporangium. If, however, the preparation is subjected to drastic heating so that dye penetrates the spore wall, this dye is difficult to extract.

The method of staining is as follows (Bartholomew and Mittwer, 1950).

Prepare a smear on a slide and fix the smear thoroughly by passing the slide through a Bunsen flame.

Cover the smear with a saturated (low temperature saturation) solution of malachite green for 10 minutes.

Rinse the smear carefully for 10 seconds with cold water.

Add a layer of 0·25% safranin solution to the stained preparation and leave to stain for 15 seconds.

Wash the preparation with water and blot dry.

Examine under the immersion objective.

The spores stain green, the vegetative cells red.

2.6.4.4   Capsule-staining procedures

Many bacteria are surrounded by a capsule (see 2.7.2.1) of high water content. The refractive index of the capsule is almost equal to that of water, and, therefore, the capsule cannot be observed in simple aqueous preparations. In order to make the capsule visible, one of the capsule-staining methods can be used.

Capsule staining with Congo red [69] (Gebhardt, 1970) is carried out by the following steps.

Place a loopful of solution 1 on a clean slide.

Using a flamed and cooled loop, take a loopful of liquid culture of the bacterium to be examined and mix it with the stain by making a circular movement of the loop.

Allow the mixture to dry in the air.

Put a drop of solution 2 on the slide and make sure that the slide is covered with the solution for 1–3 minutes.

Pour off the solution and wash the slide carefully in water.

Allow the slide to dry in the air after blotting up the water with filter paper, and examine the preparation under the immersion lens.

Negative-staining can be effected by embedding in Indian ink, as follows.

Dilute good quality Indian ink to 10-fold. Allow the dilution to stand for a day while the coarse granules settle out.

Place a loopful of the finely granular supernatant on a degreased slide and mix the cells that are to be examined with the diluted Indian ink, using an inoculating wire.

Cover the preparation with a coverslip and examine it under the microscope. The mucilaginous sheaths will appear as a wide halo against a grayish background.

2.6.4.5 Flagella staining

The staining of bacterial flagella is a delicate technique, requiring some skill. Because of this, the procedure is often dispensed with in favour of observation of the motility of the bacterium in a live preparation (see 2.6.1).

The slides used in the staining of bacterial flagella must be degreased very thoroughly. They should be cleaned by washing in dichromate–sulphuric acid, followed by rinsing in 96% alcohol. The slides are allowed to dry in the vertical position, and each slide is heated in a gas flame until the flame becomes orange-red in colour. The slides are then cooled. A culture of the bacterium 8–22h old is suspended carefully in 2–3 ml sterile distilled water, and the suspension is poured out on to the edge of the slide and spread by rotation of the slide. The slide is allowed to dry in a slightly tilted position.

Solutions required for flagella stain:

Solution A; a mixture of 18 ml of 10% aqueous tannic acid solution with 6 ml of 6% ferric chloride solution.

Solution B; a mixture of 3·5 ml of solution A, 0·5 ml of 5% alcoholic fuchsin solution, 0·5 ml of concentrated hydrochloric acid, and 2·0 ml of saturated formaldehyde solution.

The cells dried on to the slide are treated with solution A for 3·5 minutes, after which the solution is poured off without washing. The preparation is treated with solution B for 7 minutes and the slides are then carefully washed in distilled water and stained with carbol-fuchsin according to Ziehl–Neelsen (see 2.6.4.2). The preparation is finally washed and examined under the immersion objective.

2.6.5 Determining size

2.6.5.1 Determining the size of microbial cells

For this purpose a microscope equipped with ocular and objective micrometers is used.

The objective micrometer is a special slide with a small glass disc at its centre. On the disc a circle is marked with a 1-mm-long scale divided into 100 equal parts. The distance between successive marks is thus 10 μm.

The ocular micrometer is a transparent disc with a finely-marked scale *(Fig. 21)* which can be placed on the diaphragm of the ocular so that the scale is in the plane

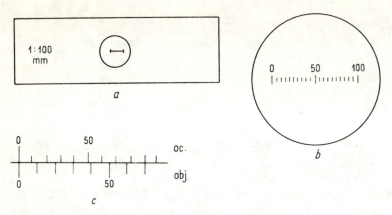

Fig. 21. Calibration of ocular micrometer (Proszt and Varga, 1965)
a — objective micrometer; *b* — ocular micrometer; *c* — calibration

of the magnified image. Since the graduation on this micrometer is arbitrary, it must be calibrated against the objective micrometer.

In Filar's micrometer ocular, a screwed micrometer can be moved in the plane of the image by means of cross-hairs and small movements can be read with a high degree of accuracy from the scale of the micrometer screw.

Calibration. The length of the graduation on the ocular micrometer is determined at a given magnification by viewing it against the superimposed image of the objective micrometer. The objective micrometer is placed on the microscope stage and the ocular micrometer is inserted in the ocular. The low-power objective is used to locate the circle on the disc of the objective micrometer together with the scale at the centre. The scale is moved exactly into the centre of the field of view.

The low-power objective is then replaced by an objective of the desired magnification, and the image of the scale of the objective micrometer is adjusted with respect to its position in the field of view; (a magnification of approximately 1000-fold is used).

A mark on the objective micrometer is exactly adjusted to coincide with a mark on the ocular micrometer, and then coincidence of any other marks on the two scales is noted. The distance between the two pairs of coincident marks is then the number of intervening graduations of the objective micrometer scale multiplied by 10 μm; this distance is divided by the corresponding number of graduation units of the ocular micrometer.

For example, if 10 graduation units (100 μm) of the objective micrometer scale correspond to 20 graduations of the ocular micrometer, then one graduation of the latter represent a distance of 5 μm. If a measurement taken on the object is expressed, for example, as 8·5 graduations of the ocular micrometer, its absolute value is $8·5 \times 5 = 42·5$ μm.

Fig. 22. The measurement of cells under the micro-scope (Proszt and Varga, 1965)

Measurements of cells are carried out with calibrated micrometers. The sealed preparation that contains the cells to be measured is attached to the microscope stage and the image is sharply focussed at the same magnification as that used for the micrometer calibration. Then the image of the cell to be measured is positioned tangentially against a mark on the ocular micrometer by moving the stage and rotating the ocular. In this position, the number of graduations between two points on the cell are counted *(Fig. 22)*. In this way any characteristic dimensions of the cell (length, diameter, etc.) can be calculated.

2.6.5.2   Determination of the diameter of the field of view

The diameter of the field of view requires to be known when counting is carried out by the so-called slide-chamber method (see 5.1.1.1.D). The diameter can be determined as follows. The scale of the objective micrometer is positioned diametrically of the field of view and the scale is focussed sharply. A mark on the scale is made to coincide with one edge of the field, number of graduation marks across the field of view is determined *(Fig. 23)*.

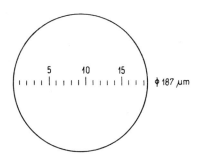

Fig. 23. Measurement of the diameter of field

71

2.7 Morphological characterization

2.7.1 The importance of morphological studies, and the problems of morphological evaluation

Microscopic morphological characters include the shapes of cells and the characteristic arrangements of cells in groups. These gross morphological characteristics are complemented by those of the fine-structure, i.e. the anatomy of microbes, which is investigated by electron microscopy. Electron microscopy is a technique of great scientific value, but is too expensive for routine diagnostic use.

In addition to the above characteristics, the macroscopic properties of pure cultures can also be considered as among the morphological characteristics of microbes, including, for example, the colour, shape, odour, and rate of growth of colonies.

The morphological characterization of micro-organisms, especially moulds, plays an important part in their identification. Bacteria and yeasts usually cannot be identified in terms of their morphology, different species and strains of these often being indistinguishable from one another on a purely morphological basis. Their identification is based mainly on biochemical and physiological characteristics.

It should be borne in mind that although the morphological and biochemical properties of a micro-organism are usually reliably constant under stable conditions, they may change considerably if the circumstances of their cultivation are altered. Such phenotypic changes do not imply any effect on the genotype of the micro-organism.

It is perhaps unfortunate that until recently the general anatomical characterization of micro-organisms has only been qualitative, although as in the case of other (non-morphological) properties, morphological characters such as the size and frequency of occurrence of cellular features may vary, even within the same strain. The differentiation of taxons, the individual cells of which are regarded as being indistinguishable in purely qualitative morphological terms, would be greatly facilitated by full biometric characterization, i.e. by presenting as part of the descriptions of taxons, mean values, average frequencies, as well as standard deviations with respect to morphological characters. It should be taken into account that morphological characters may vary depending on environmental factors (see Janke and Janke, 1959).

2.7.2   The morphology of bacteria

2.7.2.1   Microscopic characters

According to cell shape, a bacterial species can be classified as one of three main types, namely:

    spherical to ellipsoidal,

    cylindrical or rod-shaped,

    spiral or helicoid.

Spherical and ellipsoidal bacteria (cocci; sing.: coccus) occur singly, in pairs (diplococci), in chains (streptococci), in irregular grapelike clusters (staphylococci), as groups of four cocci in a quadratic arrangement (tetrade form), or as package-like cubes of eight cocci (sarcina).

The development of these different formations is a result of the characteristic modes of division of different cocci *(Fig. 24)*.

The arrangement of rod-like bacteria (bacilli) is a less reliable morphological character. It may vary within the same culture from one phase of cultivation to

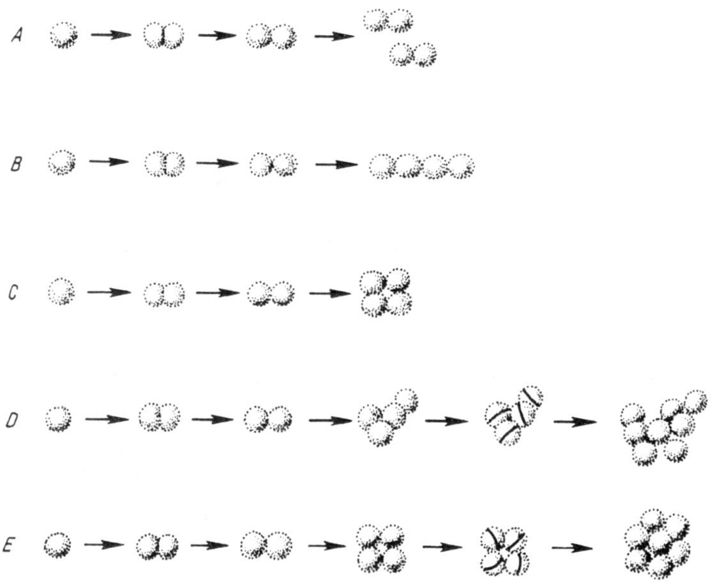

Fig. 24. Characteristic groups formed by spherical bacterial cells, and their relationships with particular modes of cell division (Pelczar and Reid, 1965)

A — *Diplococcus:* cells dividing in a single plane, most remaining in pairs; B — *Streptococcus:* cells dividing in a single plane and remaining connected to form a chain; C — *Tetracoccus:* cells dividing in two planes and forming groups each consisting of four cells; D — *Staphylococcus:* cells dividing in three planes, forming irregular groups resembling clusters of grapes; E — *Sarcina:* cells dividing regularly in three planes, producing a cuboid arrangement of cells

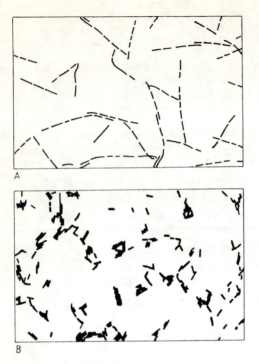

Fig. 25. The dependence of the shape of *Lactobacillus leichmannii* cells on the composition of the culture medium

A — cultivation in a complex medium containing thymidine; *B* — cultivation in a complex medium containing vitamin B_{12} (Deibel et al., 1956)

another, or it may vary according to the medium used *(Fig. 25)*. Single cells are often seen, but many cells occur in pairs (diplobacilli) or form chains (streptobacilli). *Figure 26* shows several types of rod-like bacteria, illustrating the variety of lengths and widths encountered. *Corynebacterium diphteriae* occurs in groups of cells resembling a parallel arrangement of match sticks (palisade). *Mycobacterium tuberculosis* often occurs in groups of three, resembling the branches of a tree.

Spiral organisms (spirillum; pl. spirilla) mostly occur as single cells. The length of the cells, the number and radius of the spiral turns, and the rigidity of the cells are all species-dependent factors. A curved, comma-like bacterium shorter than a single spiral turn is referred to a vibrio. Long slender cells with several spiral turns are called spirochaetes.

The members of the *Actinomycetales* family exhibit both filamentous growth and branching. Members of the genus *Streptomyces* produce long branches whereas the branching mycelia of *Nocardia* (i.e. *Proactinomyces*) are shorter and soon disintegrate into rod-like or coccoid forms.

The presence of flagella (sing.: flagellum) is a characteristic feature of some bacteria. Flagella are thin extensions responsible for locomotion of motile bacteria.

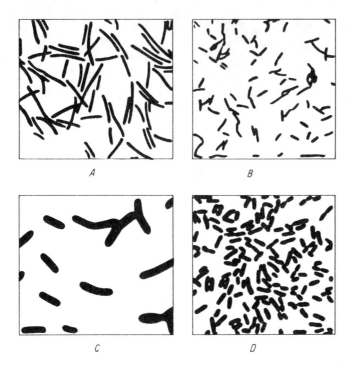

Fig. 26. Types of rod-shaped bacteria

A — Clostridium sporogenes; B — Pseudomonas sp.; *C — Bacillus megaterium;*
D — Salmonella typhi (original photographs: Nowak, 1927; after Pelczar and
Reid, 1965)

The flagellum is several times longer than the cell, and much more slender. Many
rod-like bacteria, but very few cocci have flagella.

Different types of bacterium have a characteristic number of flagella and
distribution of flagella on the surface. Thus the flagella may be located at, or near,
one or both ends of the cell (polar flagellation), or they may project from all sides of
the cell (peritrichous flagellation). Members of the order *Pseudomonales* possess
polar flagella, whereas those of the order *Eubacteriales* have peritrichous flagellation. Monotrichous cells have a single flagellum at one end, amphitrichous cells
have two flagella, one at each end, and lophotrichous cells have more than one
flagellum located at one or both ends.

Flagella are too slender to be observed by light microscopy in living preparations.
They can, however, be made visible with special stains *(Fig. 27)*.

Some bacteria secrete a viscous substance which forms a mucous layer (the
capsule) around the cell. Capsule formation is considerably influenced by the
conditions of cultivation, in particular by the properties of the culture medium.
Zoogloea refers to a group of bacterial cells embedded in a single mucous capsule.

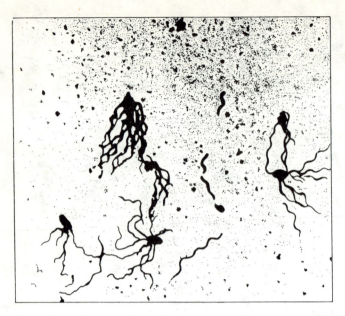

Fig. 27. Peritrichous bacterium *(Proteus vulgaris)*. Flagell staining, × 1660 (Pelczar and Reid, 1965)

Sporogenesis. Bacterial spores, more exactly termed endospores, are highly resistant oval or spherical bodies formed within the cells of some bacteria, especially species of the genera *Bacillus* and *Clostridium*. Endospore formation has recently been demonstrated in other species including *Thermoactinomyces vulgaris, Actinobifida dichotomica, Sarcina ureae, Sporolactobacillus inulinus* and *Methanobacterium omelianski* (Cross, 1970). In the sporangium of *Bacillus* and *Clostridium* species a single endospore is formed, i.e. sporogenesis in these genera is not accompanied by cell division. Sporogenesis, too, is strongly influenced by the composition and physical conditions of the culture medium, including the temperature of incubation.

Figure 28 shows the phase of endospore formation in *Bacillus cereus,* as observed in the phase-contrast microscope.

The position and size of the endospore are variable characteristics according to the bacterial species. The position of the spore in the sporangium may be central, terminal (at the end of the cell) or subterminal (between the centre and one end of the cell). The endospore may be larger or smaller in diameter than the vegetative cell, this being a reliable criterion which is used in the identification of different endospore-forming species.

When characterizing bacteria morphologically, it should be borne in mind that cells are usually larger and less uniform in the lag phase than in subsequent phases. The cultures used when bacteria are being characterized are most frequently in the logarithmic phase. Cells from old cultures may show unusual phenomena such as

76

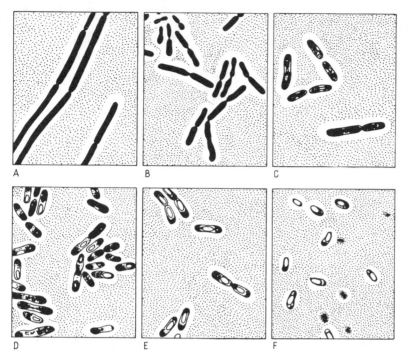

Fig. 28. Stages of sporogenesis in *Bacillus cereus*. Phase-contrast photomicrographs (Hashimoto et al., 1960)

A — long cells in the phase of rapid reproduction; *B* — vegetative cells that have become granular before spore formation; prospore stage; the spore has already developed at one end of the cells, but it is not yet refractive; *D* — spore-forming cells with endospores of increasing refractivity; *E* — spore-bearing cells after the autolysis of the sporangia

intercellular granule formation, which gives rise to irregular staining and pleomorphism.

L forms of bacteria. Under specific conditions, some bacterial species are capable of transforming into L forms which are minute bodies 0·3 to 0·5 μm in diameter. The formation of the L form is preceded by a considerable enlargement of the bacterial cell together with anabolic changes. Under some conditions, L forms can revert to the original vegetative form.

2.7.2.2   Characteristics of colony formation

In the identification of bacteria, the following types of growth are usually examined:

(1) colonies formed from bacteria inoculated on a solid medium;

(2) colonies formed on the surface of an agar slope after spreading with a loop or streaking with a needle;

(3) growth in liquid nutrient solution;

(4) growth in stab cultures in high strength agar and gelatin media.

Characteristics of colonies growing on agar plates:

(a) *Size*. Many bacteria develop moderately sized colonies irrespective of the length of the incubation period. Other species such as those of the genera *Pseudomonas* and *Proteus,* tend to spread over the entire surface of the plate (swarming growth).

(b) *Outline and margin*. Colonies may have indentations around the margin or filamentous protrusions.

(c) *Shape and texture*. The elevation, the convexity, and the texture of the colonies' surface are further important characteristics.

(d) *Colour*. Colonies may be colourless or of various colours, some species even staining the surrounding medium.

(e) *Optical properties*. Transparent, translucent, opaque, bright, or dull colonies are encountered.

Some typical colonies are shown in *Figure 29*.

The morphology of colonies is strongly influenced by the composition of the medium and the conditions of cultivation. On plates of high colony density, bacterial metabolites may inhibit colony growth. This type of inhibition can be

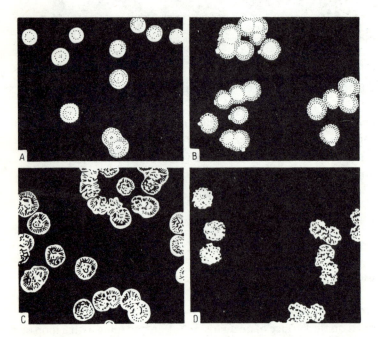

Fig. 29. Characteristic forms of bacterial colonies (Pelczar and Reid, 1965)

78

eliminated by preparing special cultures; for example, a single colony may be developed in an Erlenmeyer flask of 100–150 ml volume containing 30–35 ml of solid medium. Several cells are transferred to the centre of the medium's surface on the tip of an inoculating needle and the culture is incubated under the most favourable conditions for bacterial growth. After several days or weeks, a large colony develops which shows the characteristics of the species in the absence of self-inhibition. The giant-colony method is applied in the study of moulds and yeasts as well.

In some bacterial and yeast species, the structure of the colony is variable even under constant conditions. Some colonies (S-type colonies) have a smooth surface, and others (R-type colonies) have rough surface.

S and R colonies also differ with respect to cell structure and biochemical properties. Cells from S-type colonies disperse in a liquid medium, they are relatively virulent, and their flagellate forms are motile. R-type cells tend to aggregate, and are of low motility and low virulence.

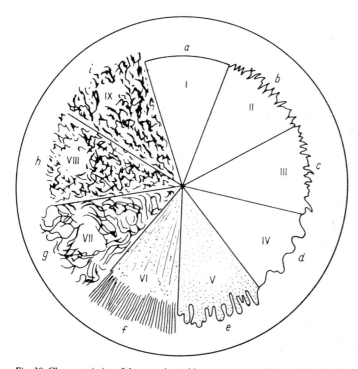

Fig. 30. Characteristics of the margin and inner structure of bacterial colonies (Görög, 1961)

a — entire; *b* — ragged; *c* — serrated; *d* — undulating; *e* — lobular; *f* — fringed; *g* — hair-like; *h* — crenate; *i* — arborescent; I — transparent; II — translucent; III — opaque; IV — smooth; V — finely granular; VI — coarsely granular; VII — wavy; VIII — filamentous; IX — arborescent

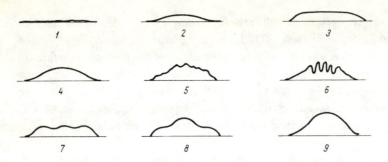

Fig. 31. Characteristic elevations of the colonies of different bacteria (Görög, 1961)

1 — flat; *2* — slightly convex; *3* — raised; *4* — convex; *5* — papillary; *6* — furrowed;
7 — crateriform; *8* — cumulous; *9* — strongly convex

Older colonies may show an annular structure, a phenomenon which may be attributed to some regularly fluctuating factor such as the temperature, humidity, intensity of illumination, etc.

The morphological characteristics of colonies and the corresponding abbreviations used to refer to them are presented in *Figure 30* and *31* (Görög, 1961).

As regards the distribution of different cells in a colony, valuable information can be obtained from impression preparation (Klatsch-Preparate). A thoroughly degreased coverslip lowered with forceps on to the colony, is gently pressed down so that the cells on the surface of the colony become attached to the coverslip. The preparation is fixed in the usual way, stained, and examined under the microscope.

Streak cultures on agar slopes are characterized by
a) the density of the growth (weak, moderate or abundant),
b) the outline and margin of the streak culture (regular or irregular),
c) the consistency of the bacterial mass (buttery, readily removable with an inoculating wire, viscous, tensile, dry, easily dispersable in liquid, etc.),
d) the colouring (i.e. pigment formation).

Some of the pigment-forming species retain the pigment in the cells, thus developing more or less intensely coloured colonies. Characteristic colours of the colonies of some species are: *Staphylococcus aureus:* golden-yellow, *Sarcina lutea:* lemon-yellow, *Micrococcus flavus:* yellow, *Micrococcus niger:* brownish-black. Other pigment-forming species (e.g. *Pseudomonas fluorescens, P. aeruginosa, P. syncyanea*) secrete pigment into the culture medium. The intensity of pigment formation is also influenced by the composition of the medium and the conditions of incubation; e.g., cultures of *Serratia marcescens* grown at room temperature have an intense brick-red colour, whereas those grown at 37 °C are almost colourless.

The shapes and nomenclature of streak cultures are shown in *Figure 32*.

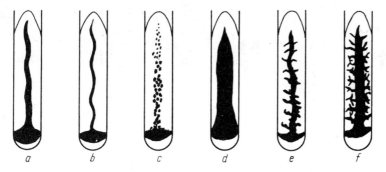

Fig. 32. Characteristic shapes of streak cultures (Görög, 1961)

a — filiform; *b* — vermiform; *c* — beaded; *d* — effuse; *e* — rhizoid; *f* — arborescent

Liquid cultures are characterized by

a) the rate of multiplication (slow, moderate or rapid),

b) the evenness of dispersion of the cells in the líquid medium (uniform turbidity or localized growth, either on the surface in the form of a membrane, fragments, a thick layer, and sometimes with the formation of foam, or in the sediment which may be particulate or viscous),

c) the odour (putrid, aromatic or odourless).

The characteristic appearance and nomenclature of deep-agar stab cultures are presented in *Figure 33*.

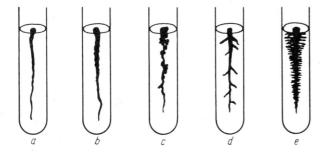

Fig. 33. Characteristics of stab cultures in deep agar (Görög, 1961)

a — filamentous; *b* — vermiform; *c* — beaded; *d* — rhizoid; *e* — arborescent

The characteristics of gelatin stab cultures [**167**] are the presence of

a) gelatin liquefaction caused by the bacterium; the liquefaction either extends throughout the medium or is confined to the upper region. In the latter case the part of the medium that has remained solid may exhibit various shapes, as shown in *Figure 34*.

6 Kiss

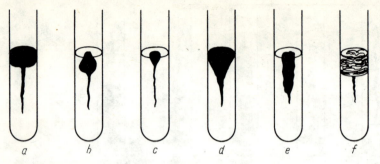

Fig. 34. Characteristic forms of liquefaction in gelatin stab cultures (dark areas represent lique-faction (Görög, 1961)

a — crateriform; *b* — napiform; *c–d* — infundibuliform; *e* — saccate; *f* — stratiform

b) the absence of liquefaction; bacterial growth appears only along the stab, sometimes "radiating" outward.

Bacterial motility is established not only by direct observation under the microscope, but also by the appearance of radial growth in a semi-solid medium. Non-motile bacteria form colonies along the stab without radial projections.

2.7.3   The morphology of yeasts

The two terms "mould" and "yeast" are colloquial, and have no taxonomic significance.

2.7.3.1   The main morphological properties of yeasts

The vegetative thallus of yeasts most commonly consists of single cells, which may be spherical or sausage-like in shape, sometimes tapering, or lemon-shaped *(Fig. 35)*. Yeast cells are much larger than bacteria, i.e. from 1 to 5 µm in width and 5 to 30 µm in length. The cells may be highly variable in size and shape even within the same strain, depending on the conditions of cultivation.

In the cytoplasm of yeasts, microscopic granules, inclusions, and vacuoles can be observed, the vacuoles occurring mainly in older cultures. The cell wall is thin and colourless in young cells, and becomes thicker later. Some species secrete a gelatinous substance in which the cells appear to be embedded. Yeasts have neither flagella nor cilia, and consequently are incapable of active movement.

According to the mode of vegetative reproduction, yeasts multiplying the cell fission (i.e. the genus *Schizosaccharomyces*) and yeasts multiplying by budding (all other yeasts, including the genus *Saccharomyces*) are distinguished. In budding

82

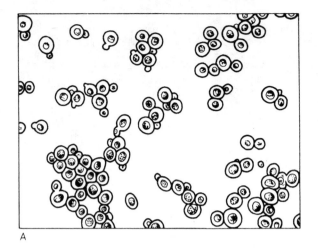

A

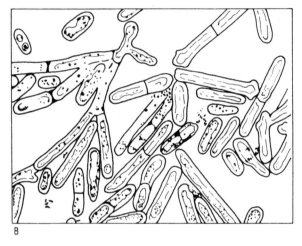

B

Fig. 35. Morphological characteristics of yeast cells
(Pelczar and Reid, 1965)

A — Saccharomyces cerevisiae; B — Endomyces magnusii; C — Nadsonia richteri; D — Schizosaccharomyces octosporus

yeasts, elongated buds may not become detached and thus may develop into pseudohypha or pseudomycelium (as in *Candida*).

Budding is an asexual process of reproduction *(Fig. 36, a)*. A small bulge is formed usually at a specific site on the cell wall, and groups until it reaches about one third of the original cell volume. Meanwhile the cell nucleus divides into two daughter nuclei. One of these remains in the present cell, and the other migrates into the daughter cell. The latter, in the process of budding off, may separate completely from the parent cell. Alternatively, the bud remains attached to the parent cell and may produce a bud itself; in this way chains of buds may be formed. Catepae of

6*

buds are an indication of an abundance of nutrients and generally favourable conditions for reproduction.

In the family *Schizosaccharomycetes,* vegetative reproduction occurs by cell fission, which is characterized by the formation of a cross-wall (septum) *(Fig. 36, b).*

Spore formation in yeasts comes about by condensation of the vegetative cells into 1, 4, 6, 8, or rarely 12 strongly refractive spheres, the so-called diploid yeast spores, or ascospores. These are located within the original vegetative cells, in asci *(Fig. 36, c).* At maturity the spores are liberated and develop into vegetative cells.

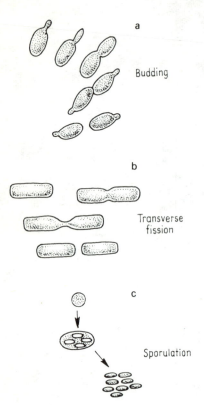

a

Budding

b

Transverse fission

c

Sporulation

Fig. 36. Modes of reproduction in yeasts (Pelczar and Reid; original publication: Ingold, 1961)

Sexual reproduction in yeasts involves the fusion of gametes that have developed from haploid ascospores. The shape of the ascospores in yeasts is characteristic of the genus.

Some species of the *Saccharomycetaceae* form ascospores readily in different types of environment, whereas others require special culture media and well-defined conditions for the formation of ascospores. Some yeasts tend to lose their spore-forming capacity after prolonged culture in the laboratory.

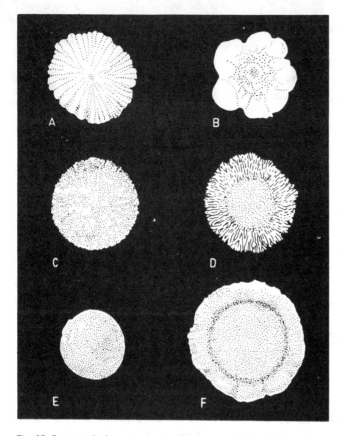

Fig. 37. Some typical yeast colonies (Pelczar and Reid, 1965)

A — Saccharomyces cerevisiae; B — Saccharomyces rouxii; C — Candida krusei; D — Saccharomyces melbis; E — Torulopsis magnoliae; F — Hansenula anomala

Attempts have been made to utilize the characteristics of colonies as a means of identifying yeast strains (e.g. S or R colonies, flat or convex colonies, regularly or irregularly shaped colonies, etc.) *(Fig. 37)*. Young colonies are ointment-like in consistency, while older ones are denser, tending to dry out, and may form pigments which give them a characteristic colour (yellow, orange, pink, brown, or black). Also the appearance of yeasts in liquid culture may be informative; some species grow on the bottom of the culture vessel forming a sediment, whereas others give rise to a uniform turbidity and a third group grows only on the surface, forming a skin or thicker layer.

2.7.3.2   Microscopic examination of yeasts

In order to make visible the cell components of yeasts, special methods of staining are required. The nucleus is usually examined in preparations stained with iron haematoxylin or Feulgen's stain. Fat droplets are stained with Sudan III. Metachromatic granules and vacuoles stain pale red with methyl red. Cellulose when stained with zinc chloride and iodine appears blue. Phase-contrast microscopy (see 2.4.4.4) makes visible some details of the subcellular structure which are not observable either by simple transmitted light microscopy or with any of the usual range of staining methods.

2.7.3.3   Examination of the sporogenesis of yeasts

The so-called wild yeasts sporulate rapidly, while cultured strains sporulate at a slower rate or not at all.

The sporulation of yeasts can be stimulated by inoculating plaster blocks with actively reproducing yeast cultures. The truncated cone-shaped blocks are immersed in an enamelled pot for 5–10 minutes before use. A quantity of water (the amount depending on the size of the block) is poured into a sporulation dish so that the water covers the bottom of the dish in a layer 2–3 mm deep. The block is taken out of the boiling water, placed in the sporulation dish with its larger surface face down, and the dish is covered and allowed to cool.

1–2 ml of a 24-h malt-broth culture [90] of the yeast is pipetted under aseptic conditions on the surface of the plaster block, drop by drop so that the liquid is absorbed into the block.

During the incubation two-times per day, some of the growth covering the surface of the block is scraped off for microscopical investigations. The time taken for the appearance of spores, the numbers of spores produced, and the shape of the spores are recorded.

2.7.4   The morphology of mould fungi and methods of examination

2.7.4.1   The morphological structure of mould fungi

The thalli of the small multicellular fungi referred to as moulds consist of filaments known as hyphae; these may branch and sometimes anastomose to form a tangled mass, any large portion of which is referred to as mycelium. Part of the mycelium is usually beneath the surface of the culture medium (submersed mycelium), the remainder (aerial mycelium) being just above or projecting up from the surface. Colonies may reach a considerable size, and generally, the morphological features of moulds are very variable. The so-called true fungi of importance in the food industry belong to the following classes: *Phycomycetes* (unicellular fungi), *Ascomycetes* (fungi forming ascospores) and *Fungi imperfecti (Deuteromycetes)*.

The fungi belonging to the class *Phycomycetes* generally do not contain septa, and, therefore, may be referred to as unicellular fungi. The fungal thallus has two parts distinguished on the basis of function, namely the vegetative structure and the reproductive organs. Of the fungi belonging to this class, some species of the genus *Pytophthora* (family *Phytiaceae*, order *Peronosporales*, subclass *Oömycetes*) and some of the genus *Peronospora* (family *Peronosporaceae*) play a part in the spoiling of raw materials of plant origin used in the food industry. For example, *Phytophthora infestans* parasitizes the leaves, stem and tubers of the potato plant and more rarely those of other members of the *Solanaceae* as well; *Peronospora destructor* parasitizes the common onion *(Allium cepa)*. The mycelium of *Phytophthora infestans* consists of hyphae 3–4·5 μm in diameter. Asexual sporogenesis in this fungus is indicated by the development of sporangiospores enclosed in a zoösporangium. The latter is formed on a sporangiophore that projects above the surface of the medium. The zoösporangium contains 6–16 zoöspores, each having two flagella.

The sporangiophore of *Peronospora* branches in tree-like manner, the branches stand two by two, with non-ciliated non-motile spores (conidia) on each.

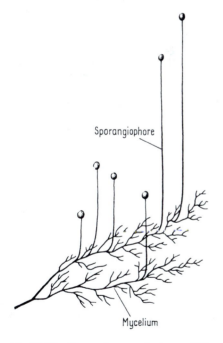

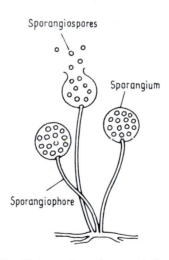

Fig. 38. Mycelium and sporangiophores of *Mucor hiemnalis*

Fig. 39. Formation of sporangia in the genus *Rhizopus*

87

Mucor and *Rhizopus* are important genera of the *Mucoraceae* family (order *Mucorales,* subclass *Zygomycetes*). The mycelia of these microscopic fungi are well-developed, showing a very pronounced branching. These fungi also form sporangiophores which grow out above the surface of the culture medium. The ends of the sporangiophores swell and form small spherical structures—the sporangia *(Fig. 38).*

Sporangia are 20–200 μm in diameter, and, therefore, may be visible to the naked eye. The wall of the mature sporangium breaks open to liberate the spores. The columella is a part of the sporangiophore which projects upward into the interior of the sporangium; its shape (ovoid, oblong, piriform, etc.) is characteristic of the species.

The sporangiophores of species of *Rhizopus* arise at the nodes of stolons from which rhizoids grow *(Fig. 39).* Sexual reproduction involves the fusion of cells (gametes) developing from neighbouring hyphae (zygophores). The zygote (zygospore arising from this process forms mycelium or produces a sporangium *(Fig. 40).*

In addition to yeasts also a number of mould fungi of importance to the food industry belong to the class *Ascomycetes.* The hyphae of these fungi are septated. The genera *Aspergillus* and *Penicillium* of the order *Plestascales* are the most

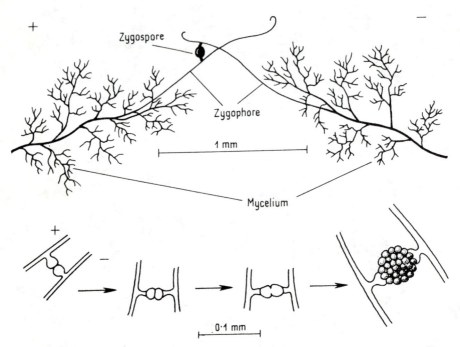

Fig. 40. Sexual reproduction in the genus *Rhizopus* (according to Pelczar and Reid; original publication: Ingold, 1961)

88

important as far as food microbiology is concerned. In the majority of species of these genera sexual reproduction is unknown. Therefore, these species used to be grouped to the so-called conidium-bearing fungi *(Fungi imperfecti, Deutero-mycetes)*, because producing is the commonest mode of reproduction, thin-walled exospores or conidia, which bud from the ends of hyphae.

In the order *Plectascales* as in the yeasts, sexual reproduction is characterized by ascospore formation. Asci develop in the enclosed perithecium in a more or less irregular arrangement. The perithecium of aspergilli shows some resemblance to a watering-can. In the genus *Aspergillus*, the mycelium ramifies abundantly, forming a rapidly developing mat above which the sporangia grow out. The tips of the sporangia form vesicles from which rod-like sterigmata are produced in every direction bearing chains of conidia. The whole structure resembles a dishevelled head *(Fig. 41)*.

The external appearance of colonies of *Penicillium* is very similar to that of *Aspergillus* colonies; they are indistinguishable to the naked eye, but microscopic differentiation is straightforward, on account of the brush-like ramicfation of the conidiophores of *Penicillium (Fig. 42)*. The brush is composed of flask-shaped sterigmata from which chains of conidia are cut off together with the metulae beneath them. Below the latter are branches, and below the branches is the conidiophore *stricto sensu*.

Some species develop a coremium, a column-like formation consisting of compact bundles of conidiophores attached to one another.

Several types of fungal spore are illustrated in *Figure 43*.

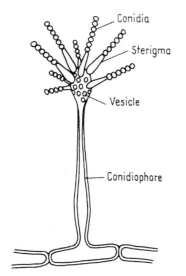

Fig. 41. Formation of conidia in the genus *Aspergillus*

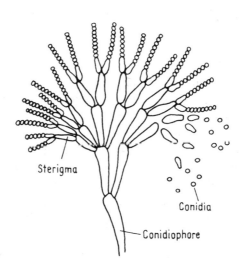

Fig. 42. Formation of conidia in the genus *Penicillium*

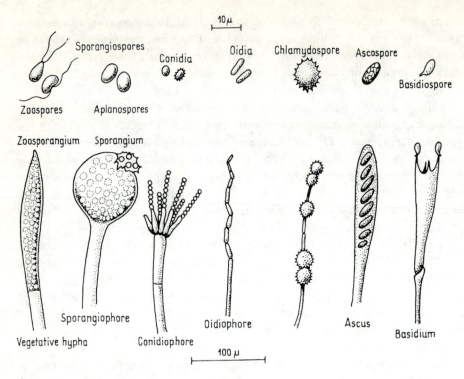

Fig. 43. Types of fungal spores

2.7.4.2 Methods of morphological examination of mould fungi

Mould cultures developing on solid media can generally be identified by morphological criteria. Colonies growing on agar plates are examined with a magnifying lens or under a low-power stereomicroscope, either from above or below the plate. Sufficient information may be obtained to determine the genus to which a fungus belongs.

Slide culture makes possible a thorough microscopic examination of intact mould colonies *(Fig. 44)*. Slide cultures are usually prepared as follows.

On a slide, three sides of a square are built with dental wax to a height of 1–2 mm depending on the size of the slide. A warm coverslip is gently pressed onto the wax to form a slide-chamber, which is then filled to two-thirds of three-quarters of its volume with mould inoculum mixed with melted agar at a temperature of 50 °C. The slide is placed in a sterile Petri dish containing moistened filter paper, and incubated for 24–72 h. The slide-chamber can be examined directly under the microscope.

90

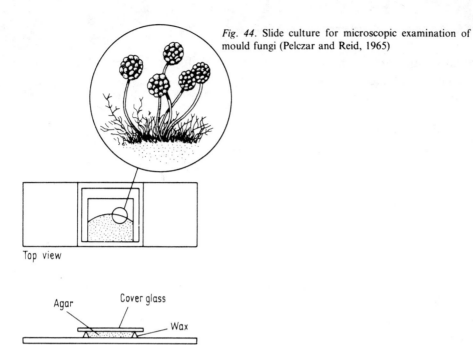

Fig. 44. Slide culture for microscopic examination of mould fungi (Pelczar and Reid, 1965)

Top view

Agar   Cover glass

Wax

Side view

2.8   Biochemical procedures

2.8.1   Liquefaction of gelatin

2.8.1.1   Nutrient gelatin

10–15% gelatin is added to nutrient broth [**32**], the final pH of which should be 7·2. The medium is sterilized by autoclaving for 20 minutes at 115 °C. Stab cultures are prepared with the nutrient gelatin and incubated at the temperature for the optimum for 30 days.

The shape and extent of the liquified portion of the gelatin are recorded. Where cultures have been incubated at temperatures above 25 °C, the characteristics of liquefaction are recorded only when the culture shows the presence of fluid following immersion in ice-cold water.

2.8.1.2   Frazier's gelatin agar (modified)

0·4% gelatin is added to nutrient agar [166] with a final pH of 7·2. The medium is sterilized by autoclaving for 20 minutes at 115 °C.

Poured, dried plates of this medium are inoculated across the whole surface and incubated at the optimum growth temperature for 2–14 days.

8–10 ml mercuric chloride solution (composition: mercuric chloride 15·0 g; concentrated hydrochloric acid 20·0 ml; distilled water 100·0 ml) are poured on to the culture. The unhydrolysed gelatin forms an opaque precipitate with the reagent and the hydrolysed gelatin thus shows up as a clear zone. The width of the clear zone between its outer limit and the edge of the colony is recorded.

Gelatin-charcoal disc

In this method results can usually be recorded after less than 24 h. The gelatin-charcoal discs consist of formalin-denatured gelatin containing charcoal. Commercially produced discs can be used or they can be prepared in the laboratory.

Peptone water [111] is dispensed in aliquots of 1 ml into 75 × 10 mm test-tubes, and sterilized by autoclaving for 15 minutes at 121 °C. One gelatin-charcoal disc is added aseptically to each tube immediately before use.

The medium is pre-warmed to the incubation temperature and heavily inoculated from young nutrient agar slope cultures. Incubation is carried out at the optimum growth temperature in a water-bath. The cultures are inspected at short intervals, e.g., hourly for the first 6 h and then at longer intervals. The incubation time of cultures which show negative results at successive inspections should be extended for as long as 24 h.

The hydrolysis of gelatin causes particles of charcoal to be liberated, so that they settle at the bottom of the tube. On shaking, the particles are resuspended and become visible.

It should be noted that formalin-denatured gelatin does not melt at the same temperature as normal gelatin and, therefore, this test may be performed at the optimum temperature of the micro-organism under investigation.

2.8.2   Hydrolysis of casein

Milk agar, i.e. nutrient agar [32] with 10% skim-milk added to it, is sterilized by autoclaving for 20 minutes at 115 °C. A more opaque milk-agar medium can be made by mixing 10 ml hot, sterile, 2·5% water agar with 5 ml of hot, sterile skim-milk immediately before the plates are poured. The milk agar formed in this way is of 30% strength.

A dried plate of this medium is inoculated by making a single streak across its surface, and then incubated at the optimum growth temperature for 2–14 days.

The test reagent is a mercuric chloride solution (2.8.1.2) containing either 1% hydrochloric acid, or 1% tannic acid.

Clear zones which become visible after incubation of the plates indicate that casein hydrolysis has taken place. This is then confirmed by flooding the plate with the test reagent which coagulates the remaining protein.

2.8.3 Hydrolysis of coagulated blood serum

Loeffler's serum consists of three parts of blood serum to one part of glucose nutrient broth [50]. Since low temperatures are maintained in the preparation of this medium, the glucose nutrient broth should be sterilized by autoclaving before it is mixed with the serum in a sterile flask. Aliquots of the medium are then transformed to small sterile screw-capped bottles, and coagulated in a sloping position by heating slowly to 85 °C. The medium is sterilized in inspissator at 85 °C for 10 minutes on each of three consecutive days.

The serum slopes are inoculated by streaking the surface in the usual way, and they are incubated at the optimum growth temperature for 2–14 days.

Visual examination will indicate whether hydrolysis, i.e. liquefaction of the coagulated serum, has occurred.

2.8.4 Production of indole from tryptophane

Material from a young agar culture is used to inoculate nutrient broth containing tryptophane [153]. The medium is incubated at the optimum growth temperature for 2–7 days. 0·5 ml of Kovács's reagent is added, stirred thoroughly in, and the mixture is allowed to stand. A deep red colour in the alcoholic layer indicates the presence of indole.

2.8.5 Production of ammonia from peptone or arginine

2.8.5.1 Peptone water

A tube of peptone water (peptone 1%, sodium chloride 0·5%; autoclaving at 121 °C for 15 minutes; pH = 7·2) is inoculated and incubated together with a sterile control at the optimum growth temperature for 2–7 days. A loopful of culture is mixed with a loopful of Nessler's reagent [99] on a slide or glazed porcelain tile, or 1 ml of culture is added to 1 ml of Nessler's reagent in a clean tube. An orange to brown reaction colour indicates the presence of ammonia. The same mixture made with the sterile control should turn pale yellow or show no colour reaction.

2.8.5.2   Arginine broth [7]

This culture medium is mainly used in the differentiation of streptococci.

A bottle of arginine broth is inoculated and then incubated together with a sterile control at the optimum growth temperature for 2–7 days. Testing is carried out as described in 2.8.5.1.

2.8.5.3   Thornley's semi-solid arginine medium [140]

This medium is mainly used in the differentiation of Gram-negative micro-organisms.

A test-tube containing 15 ml of medium is stab-inoculated and its surface sealed with vaseline or vaspar. The tube is incubated together with a sterile control at the optimum growth temperature for 2–7 days.

Hydrolysis of arginine with the formation of ammonia results in alkalinity, and is indicated by a change in the colour of the medium to red. *Pseudomonas* and *Aeromonas* produce ammonia from arginine in both sealed and unsealed media, whereas *Flavobacterium*, for example, produces a red colouration in an unsealed medium, but produces no reaction in a sealed medium.

2.8.6   Production of hydrogen sulphide

Hydrogen sulphide may be formed from organic sulphur compounds (e.g. cysteine and cystine), or from the reduction of inorganic sulphur compounds (e.g. sulphite).

2.8.6.1   Cystine or cysteine broth

Peptone water or nutrient broth is used as the basic medium with an addition of 0·01% cystine or cysteine. Filter paper is soaked in saturated lead acetate solution, dried, cut into strips and sterilized; this is used as indicator paper.

The medium is inoculated, and a strip of indicator paper is inserted between the plug and the glass so that the lower end of the paper is above the medium. The culture is incubated at the optimum growth temperature for 2–7 days, together with an uninoculated control.

Production and liberation of hydrogen sulphide causes blackening of the indicator paper. If no blackening has appeared by the end of the incubation period, 0·5 ml of 2N hydrochloric acid is added, immediately replacing the plug and the indicator paper. If sulphide has been produced but has remained in solution, the addition of acid will result in the liberation of hydrogen sulphide.

94

2.8.6.2   Ferrous chloride gelatin

The medium is prepared by adding freshly-prepared 10% ferrous chloride solution to boiling nutrient gelatin to give a final concentration of 0·05%. The medium is dispensed into sterile, narrow test-tubes, which are quickly cooled and sealed with sterile rubber stoppers.

The tubes are inoculated by stabbing, and incubated at 20–25 °C for 7 days. Production of hydrogen sulphide is indicated by blackening of the medium.

N.B. This medium will also indicate the liquefaction of gelatin. This is the recommended method for the differentiation of bacteria of the *Enterobacteriaceae* family.

2.8.6.3   Nógrády and Ródler's modified double-layer polytropic medium [107]

Slopes of the medium are streak-inoculated and butt-stabbed.

Incubation is carried out at 37 °C for 1–2 days. The various phenomena that may be observed in the medium are interpreted as follows:

Lower region (bottom layer):

blackening	hydrogen sulphide formation
blackening, bubbling and/or disruption	hydrogen sulphide and gas formation

Upper layer butt region:

yellow colour	decomposition of glucose with acid formation
purple colour	decomposition of urea
no colour change	no decomposition of glucose or urea
bubbles and/or disruption	decomposition of glucose with gas formation

Upper layer slope region:

yellow colour	decomposition of lactose and/or glucose, sucrose
purple colour confined to slope surface	alkaline pH produced by metabolites
deep purple colour	metabolites produced on the slope surface and butt: decomposition of urea

no colour change

no decomposition of lactose or sucrose or glucose; no alkaline metabolites produced

2.8.7 Reduction of nitrate

Nutrient solution prepared for the demonstration of nitrate reduction [103, 104] is inoculated, and incubated at the optimum growth temperature for 2–7 days. 0·1 ml of Griess–Ilosvay reagent [56] is added. The presence of nitrite is indicated by the development of a red colour within 1 minute; the intensity of the colour depends on the quantity of the nitrite formed.

2.8.8 Litmus milk test [76]

Some bacteria ferment lactose with the formation of acid or acid plus gas. In an acid medium milk proteins coagulate to form an acid clot, whereas a soft clot is that produced by rennin—an enzyme present in the milk. These two kinds of clot can be distinguished by means of an indicator, or by the harder consistency of the acid form. The soft clot settles out relatively slowly.

The proteolytic enzymes of bacteria which do not ferment lactose, hydrolyse the protein clot, and the peptonization causes the pH to become alkaline. If peptonization is strong, hydrolysis is complete, and the fluid will be straw-yellow with a putrid odour.

2.8.9 Lysine-decarboxylase test

Medium [82] is inoculated from agar slope culture and incubated at 37 °C for 18–24 h. 1 ml of 4 N NaOH is added to each culture and mixed well. 2 ml of chloroform are measured into each tube and the mixture shaken thoroughly. Centrifugation at 3000 rpm is carried out for 10 minutes so that the chloroform separates in a clear layer. (The cadaverine is extracted by the chloroform). 0·5 ml of the clear chloroform layer is transferred with a sterile pipette into an empty sterile test-tube, and 0·5 ml of 1% ninhydrin-chloroform solution is added. The mixture is warmed for 4 minutes at 37 °C and the result recorded immediately.

A violet colour indicates a positive result, and no colour reaction indicates a negative result. Incubation at a higher temperature or for more than 4 minutes may produce a false positive result because other components present in small quantities may also produce the same colouration.

The composition of the medium is different from that given in *Carlquist*'s original description, in that it contains trypticase or "casitone" instead of acid hydrolysate of casein.

2.8.10   Phenylalanine-deaminase test for the demonstration of phenylpyruvic acid

Medium [39] is inoculated with a large amount of culture obtained from an agar slope, and incubated for 4 h (or for 18–24 h if necessary) at 37 °C. 10% ferric chloride solution is dropped (4 or 5 drops) on the surface of the medium and is allowed to run down over the slope. A green colour indicates a positive reaction, and no colour change indicates a negative result.

The medium differs from the original composition in that it contains 6 ml of yeast extract prepared as described by Eörsi (1947) instead of 3 g yeast extract, and 20 g agar instead of 12 g.

2.8.11   Demonstration of oxidizing and fermenting activity

0·5 ml of steril 10% glucose or other carbohydrate solution is added to 5–5 ml melted Hugh–Leifson medium [59]. The media are mixed well. Then culture to be tested is inoculated in duplicate tubes by stabbing with a straight wire. To one of the tubes a layer of sterilised melted liquid paraffin (to a depth of about 1 cm) is added to make a cover layer. Both test-tubes are incubated at the optimum temperature for 1–14 days with periodic obse. vation.

Results	*Open tube*	*Sealed tube*
oxidation	yellow	green or blue
fermentation	yellow	yellow
no action on carbohydrate	blue or green	blue or green

Note: The medium can also be used for detecting gas production and motility.

2.8.12   The glucose test (Andrade's nutrient broth)

Some bacteria produce acid and/or gas when they metabolize glucose. In medium [6], the development of a red colour indicates acid formation; the appearance of gas in the Durham tube indicates gas formation.

2.8.13   Lactose fermentation

Three test-tubes with Durham tubes containing lauryl-sulphate-tryptose medium [79] are inoculated with 1 ml each, respectively, of three consecutive dilutions of the microbial suspension. The tubes are incubated at 30 ± 1 °C. Those tubes showing gas formation after 24 h of incubation are noted, and incubation is continued for another 24 h.

The result can be confirmed by transferring a loopful of the gas-forming cultures into Durham tubes containing brilliant green–lactose–bile medium [14] and incubating as above.

2.8.14   Production of carbon dioxide from glucose

The medium used in this test is Gibson's semi-solid tomato-juice medium, containing a mixture of 4 parts skim-milk to 1 part nutrient agar [138], to which 0·25% yeast extract, 5% glucose and 10% tomato-juice have been added. The final pH should be 6·5. The medium is dispensed into tubes to give a depth of 5–6 cm. Stamer et al. (1964) have shown that the tomato-juice on which lactic acid-forming bacteria grow can be replaced by manganese. This may be conveniently added as a solution of manganese sulphate (final concentration of Mn^{2+} 1–10ppm), instead of tomato-juice.

The medium is melted and then cooled to 45 °C. It is inoculated with ~ 0.5 ml of young broth culture, mixed and cooled in tap water. When the medium has solidified, nutrient agar at ~ 50 °C is added to give an overlayer 2–3 cm deep, and the cultures are incubated at the optimum growth temperature for 14 days.

The semi-solid medium and the agar overlayer trap any carbon dioxide gas that is produced in the medium. This is shown by disruption of the agar seal and by the presence of gas bubbles in the medium.

2.8.15   Methyl red test

The isolate to be tested is inoculated into buffered glucose medium [53] and incubated at 35–37 °C for 5 days. 5 ml of the culture are pipetted into an empty test-tube, and shaken with 5 drops of methyl red indicator [95]. A red colour indicates acid formation (positive test), and a yellow colour indicates a negative result; an intermediate colouration cannot be evaluated.

2.8.16   Voges–Proskauer test

The isolate to be tested is incubated in buffered glucose medium [53] at 35–37 °C for
48 h. 1 ml of culture is pipetted into an empty test-tube, and 0·6 ml α-naphthol and
0·2 ml potassium hydroxide solution [163] are added. A cherry-red colour
developing within 2–4 hours indicates a positive reaction, i.e., acetyl-methyl-
carbinol formation has been formed from glucose.

2.8.17   Hydrolysis of starch

Microbes can be tested for starch hydrolysis either in solid or in liquid medium. The
use of starch agar is somewhat more convenient; this consists of nutrient agar [138]
with an addition of 0·2–1·0% soluble starch. The best results are obtained by
pouring 10 ml of nutrient agar into a Petri dish, allowing the agar to solidify, and
overlaying this with 5 ml of starch agar.

Plates are inoculated by streaking once across the surface followed by incubation
at the optimum growth temperature for 2–14 days. 5–10 ml iodine solution [55] are
then added over the surface of the medium.

The presence of unhydrolysed starch is indicated by a blue colour. The
hydrolysed starch of the clear zone is produced by beta-amylase activity. The width
of the clear zone from the edge of the growth to the outer borderline of the zone
should be recorded. A reddish-brown zone around the colony is suggestive of starch
hydrolysis (dextrin formation) resulting from alpha-amylase activity.

2.8.18   Utilization of citrate as the sole carbon source

Koser's [70] or Simmon's [120] medium is inoculated with the aid of a needle from
pure culture. The use of the needle and the small size of the inoculum (practically a
6-hours broth culture) are both necessary to obtain reliable results. In the case of
Simon's agar, a stab culture is prepared in a short slope of the medium. Cultures are
incubated for 72–96 h when Koser's citrate medium is used, or 48 h when Simmon's
medium is used, in either case at a temperature of 35–37 °C. Visible growth is taken
as a positive result, the colour of the culture medium changing from green to blue at
the same time.

2.8.19   Production of polysaccharide from sucrose

The medium used in nutrient agar [32] containing 5–10% sucrose. It is sterilized by autoclaving at 121 °C for 15 minutes.

Plates are inoculated so as to obtain solitary colonies. Incubation is carried out at 20–25 °C (or optimum growth temperature) for 1–14 days. The synthesis of dextran or laeven is indicated by the development of mucoid cultures.

2.8.20   Test for acid tolerance

This test is described in 2.6.4.2.

2.8.21   Detection of butyric acid, propionic acid and acetic acid in culture media

The medium that has been used for the cultivation of a given microbe is acidified and steam-distilled. The organic acids are extracted with ether, transformed into ammonia salts and separated by chromatography.

2.8.22   Decomposition of organic acid substrates

1) 1% potassium-sodium tartarate (abbrev. D-tartarate)
2) 0·5% L-tartaric acid (L-tartarate)
3) 0·5% I-tartaric acid (I-tartarate)
4) 1% sodium citrate
5) 1% mucic acid

(The concentrations refer to the complete medium in each case.)

The pH of solutions 1–4 is adjusted with 5N NaOH to 7·4.

The culture medium is inoculated with a loopful of 20 hours old nutrient broth [132] culture and incubated at 37 °C for 14 days. Results are read daily on the basis of the colour reaction. At the final reading 0·5 ml of saturated lead acetate solution are added to the cultures; the medium containing mucic acid is judged only on the basis of colour change. Partial decolourization of the original blue colour to greenish-yellow or total decolorization, indicates that the substrate has been decomposed, i.e., the test is positive for the particular organic acid. The test is also positive for substrates 1–4 if a minimal amount of precipitate is formed after the addition of lead acetate. The test is negative if the blue colour remains unchanged and a massive precipitate is formed with lead acetate. (After standing for 24 h the precipitate may occupy two thirds of the liquid volume).

Sometimes a positive reaction is obtained on addition of the lead acetate in spite of persistence of the blue colour. For this reason, the addition of lead acetate should never be omitted. If it is of interest to establish how long complete decomposition takes, several replicate inoculations are carried out, and the lead acetate is added to one culture on each consecutive day from day 4 to day 8. This is particularly important in the case of L-tartarate, which takes more than 5 days to be decomposed by some species of *Salmonella*.

2.8.23   Hydrolysis of tributyrin

The medium is nutrient agar (pH 7·5) containing yeast extract [168] but no milk; 10 ml of tributyrin are emulsified in 100 ml of medium, using (preferably) an electric stirrer. Sterilization is carried out by steaming for 30 minutes on three consecutive days.

Poured, dried plates are inoculated by streaking once across the surface and incubated at the optimum growth temperature for 2–14 days.

Where hydrolysis of tributyrin has occurred a clear zone is visible in the medium. The width of the zone from the edge of the colony to the outer limit of the clear area is recorded. This reaction is usually regarded as being mediated only by lipase, although Sierra (1964) has reported the existence of a bacterial proteolytic enzyme which is capable of hydrolysing tributyrin (but no other complex fats).

2.8.24   Hydrolysis (lipolysis) of butter-fat, olive oil and margarine

2.8.24.1   Butter-fat agar and olive-oil agar

The medium used is yeast-extract agar, pH 7·8 [168] with an addition of 5% butter-fat or olive oil (no milk is added). The mixture is emulsified by vigorous manual shaking.

A poured, dried plate of the medium is inoculated by streaking once across the surface, and incubated at the optimum growth temperature for 2–14 days.

The plate is flooded with 8–12 ml of saturated copper sulphate solution and allowed to stand for 10–15 minutes. The reagent is poured off and the plates washed gently in running water for 1 h to remove excess copper sulphate. Where lipolysis has occurred, a bluish-green zone appears due to the formation of insoluble copper salts of the fatty acids released in the lipolysis.

2.8.24.2 Victoria-blue (or olive-oil) agar

The medium used is the same as that described in 2.8.24.1, except that it contains 0·006% Victoria blue as an indicator of lipolysis. The dye is added to the medium aseptically as a sterile solution, and the medium is then emulsified with the fat by vigorous hand shaking to produce an emulsion of a pinkish-mauve colour.

The medium is inoculated and incubated as described in 2.8.24.1.

Where lipolysis has occurred, the free acids combine with the Victoria blue to form deep blue salts. Deep blue zones observed around the microbial growth thus indicate lipolytic activity.

2.8.24.3 Victoria-blue margarine agar

Difficulties may be encountered with the medium described in 2.8.24.2, owing to fading of the Victoria-blue medium. This can be prevented by the use of partially hydrogenated, and, therefore, more stable fats such as margarine. These may be used as described in 2.8.24.2 or they may be saturated with Victoria blue and then added to the nutrient agar. When required for use, the medium is heated until it liquefies, allowed to cool to 45 °C, and emulsified by vigorous hand shaking.

The test is carried out as described in 2.8.24.2.

2.8.25 Hydrolysis of Tween derivatives

Any one of the *Tween derivates* — Tween 20 (a lauric acid ester), Tween 40 (a palmitic acid ester), Tween 60 (a stearic acid ester) or Tween 80 (an oleic acid ester), can be used, but Tween 80 is the one usually selected for this test.

Tween agar in a Petri dish is inoculated by streaking once across the centre, and is incubated at the optimum growth temperature for 1–7 days.

2.8.26 Hydrolysis of lecithin

2.8.26.1 Egg-yolk agar

Egg-yolk agar is based on nutrient agar [147] with an addition of 1% sodium chloride and 10% v/v egg-yolk emulsion. A plate of the medium is inoculated by streaking once across the surface, and is incubated at the optimum growth temperature for 1–4 days. Lecithinase activity (i.e. the hydrolysis of the lecithin in the egg-yolk agar) is indicated by the appearance of an opaque zone around areas of microbiological growth. The width of the zone is measured.

102

2.8.26.2 Egg-yolk broth

The medium used is nutrient broth [147] containing 1% sodium chloride and 10% v/v egg-yolk emulsion.

The inoculated medium is incubated at the optimum growth temperature for 1–4 days. Lecithinase activity brings about opacity in the medium, usually with a thick curd.

2.8.27 Urease test

Christensen's medium [20] changes in colour from yellow to pink or red on account of ammonia being formed from urea.

2.8.28 Catalase test

1 ml of 10% v/v hydrogen peroxide solution is poured on to an agar-plate culture or is added to 1 ml of bacterial suspension. Alternatively, a loopful of the culture is added to a drop of hydrogen hyperoxide on a slide. In the presence of catalase, gas formation is observed.

2.8.29 Oxidase test

A few drops of a 1% solution of tetramethyl-p-phenylenediamine-dihydrochloride are observed on to a piece of No. 1 Whatman filter paper in a Petri dish. The filter paper is then smeared with some bacterial culture taken in a platinum loop.

A purple colour developed within 5–10 minutes is usually recoded as a positive reaction, although colour development may take 10–15 minutes in the case of a few species (weak positive reaction).

The test reagent may be kept in a dark bottle for up to 2 weeks in a refrigerator. Filter paper soaked in the test reagent and dried may be stored at room temperature for a year.

2.8.30   Coagulase test

2.8.30.1   Tube method

Centrifuge 0·5 ml rabbit blood mixed with 0·5 ml 2% sodium citrate in a small tube. The supernatant is inoculated from a 24-h broth culture and incubated at 37 °C. The result is recorded after an incubation period of 1–7 h. If coagulation is not observed, incubation is continued up to a total of 24 h.

A centrifuged mixture of 0·5 ml rabbit blood and 0·5 ml 2% sodium citrate solution serves as a control.

2.8.30.2   Slide test

One or more for *S. aureus* characteristic or suspected colonies are selected from plate cultures. As control cultures, a positively reacting *Staphylococcus aureus* culture and a negatively reacting *Staphylococcus epidermidis* strain are used.

Each colony is mixed in one drop of water on a slide to give a dense suspension. One loopful of rabbit plasma is mixed with the suspension.

Microscopically visible clotting within 5 s indicates that the strain under test is positive for coagulase.

The plasma should be stored in a refrigerator and warmed to 37 °C immediately before use so that coagulation is not delayed.

2.8.31   Phosphatase test

The coagulase-positive staphylococci also show phosphatase activity. Whereas the coagulase test is carried out on individual colonies, the phosphatase activity can be detected by cultivating the organisms on a nutrient agar medium containing 0·01% phenolphthalein phosphate [40]. Since phosphatase-positive organisms belonging to genera other than *Staphylococcus* may interfere with the test, 8% sodium chloride may be incorporated in the medium to make it more selective for the growth of staphylococci. This medium can then be used to identify potentially pathogenic staphylococci in foodstuffs by using the *Miles and Misra surface-count technique*. However, the addition of salt to phenolphthalein phosphate agar may interfere with the phosphatase reaction, and, therefore, the use of a medium such as *Baird–Parker's medium* (egg-yolk–tellurite agar) [10] is preferable in these circumstances.

Nutrient agar is melted and cooled to 50 °C. Sterile 1% phenolphthalein phosphate solution is added to give a final concentration of 0·01%. The medium is mixed and poured into plates.

The plates are inoculated and incubated overnight at 37 °C. Each plate is then exposed to ammonia vapour by adding a few drops of concentrated ammonia solution to a filter paper inserted under the lid of the dish.

Pinkness or redness of the colonies indicates the presence of phenolphthalein released by phosphatase activity.

2.8.32 ONGP (beta-galactosidase) test

0·2 ml sterile distilled water is measured into a sterile, short-length test-tube and a loopful of agar slope culture is suspended in the water. A reagent paper is inserted in the tube and the result is read after incubation periods of 1, 2 and 4 h, in a water bath of 37 °C. A lemon-yellow colour, resulting from the release of nitrophenol by β-D-galactosidase activity, indicates a positive reaction. A colourless medium appearing (whitish from the suspended bacteria) indicates a negative reaction. Positive (*Citrobacter, E. coli*) and negative (*Salmonella*) control tubes are used in the test.

2.8.33 Test for haemolysis

The medium used is nutrient agar [32] containing 0·85% sodium chloride and 5% (v/v) defibrinated or oxalated blood. Horse blood is suitable for streptococci, but for other organisms such as staphylococci, the blood of other animals (e.g. the sheep, rabbit, or ox) may be more suitable.

The nutrient agar medium is liquified, cooled to 50 °C, and the sterile blood (0·5 ml to 10 ml agar) is added aseptically. The medium is mixed and aliquots of 5–10 ml are poured into Petri dishes. Alternatively, 5 ml of blood agar can be poured over a thin layer of solidified nutrient agar and allowed to solidify.

For the examination of staphylococci, a dried plate is streaked so as to produce separate colonies. *Streptococci* may be incubated either in pour-plates or on streak-inoculated plates, at the optimum temperature for growth.

Clear zones around colonies indicate haemolytic activity. Beta-haemolysis refers to complete clearing of the blood agar, with destruction of the blood cells and transformation of the haemoglobin into colourless products. The edges of the beta-haemolytic zone are usually sharply defined. Alpha-haemolysis is characterized by a greenish, slightly transparent zone resulting from partial cell destruction and haemoglobin break-down. This greenish zone may be hazy and indistinct in outline. Among those microbes which are of importance in food microbiology, numbers of the genus *Staphylococcus* generally cause beta-haemolysis, whereas in *Group D of Streptococcus*, alpha-haemolysis is observed more frequently. Complete or partial lysis of blood cells may be caused by various toxic substances, including metabolites. The species-specificity of erythrocytes and the environmental temperature are important factors governing the course of haemolysis.

2.8.34 Test for motility

Medium [97] is stab-inoculated with a wire needle on which the colony to be examined or some growth from an agar slope subculture has been picked up. The stab culture is incubated at 37 °C for 24–48 h.

Acid-forming bacteria give rise to a yellow colouration of the whole medium, or if the fermentation is weak, the colour develops only along the stab channel.

Acid plus gas formation shows the same discoloration together with gas bubbles at the surface of the medium.

Motile strains not only grow in the neighbourhood of the stab channel, but also spread diffusely throughout the medium.

2.8.35 Growth in a medium containing potassium cyanide (KCN)

Medium [64] is inoculated with a loopful (loop diameter 3 mm) from a broth culture that has been incubated for 24 h at 37 °C. As a control, a medium without KCN is inoculated. Further cultures with the KCN-containing medium are set up to include a known positive control strain and a known negative control strain.

The inoculated media are incubated for 4 days and results are read daily. The result is taken to be positive if growth is observed in the presence of KCN, and negative if there is no growth in the medium containing KCN, the medium in the latter case remaining clear.

Ewing and Edwers (1960) used Bacto peptone (Difco) instead of the special Danish "Orthama" peptone of Moller's (1954) original medium.

2.8.36 Gram stain

This has been described in 2.6.4.1.

2.8.37 Demonstration of pigments

Pigment production and the nature of the pigment produced, are characteristics that can be utilized in the identification of microbes. Micro-organisms are incubated on nutrient agar at a temperature between 25 and 30 °C for 3 days, after which pigment production is examined. Water-soluble pigments diffuse into, and colour the medium e.g. many *Pseudomonas* species produce diffusible yellow, greenish-yellow or green pigments. Water-insoluble pigments can be extracted with organic solvents such as ethanol or chloroform. (The pigment produced by *Chromobacterium* is violet or purple, those produced by *Flavobacterium* and *Xanthomonas* are yellow or orange-yellow, and *Pseudomonas* and *Serratia* produce red pigments.) Some pigments are fluorescent in UV light, e.g. *Pseudomonas* appears yellow, greenish-yellow or green. For further details, see 6.3.6.

106

2.8.38 Dalmau's plate for the examination of mycelium, pseudomycelium and chlamydospore formation by yeasts

A freshly poured potato-glucose agar plate [17], or a corn-flour agar plate [73] is spot-inoculated or streak-inoculated after any condensation has been allowed to evaporate. The site of inoculation is covered with a coverslip and the plate is incubated at 25 °C for 24–48 h before examination under the microscope.

2.8.39 Examination of spore formation in yeasts

Inoculated Adams' [2] or McClary's [87] agar slope media are incubated at 25 °C for 2–6 weeks. Spore formation is examined in stained preparations.

2.8.40 Yeast spore stain

A smear is fixed by heat or with methanol, and stained in 5% malachite green for 3–5 minutes, the preparation is heated up to about 80 °C three or four times. The preparation is then rinsed in running water for 30 minutes and stained for 10–20 minutes in 1% safranin solution. Spores stain green, contrasting with the red staining of the cells.

2.8.41 Test for sugar fermentation by yeasts

E.g. any of the following sugars may be used: glucose (D), galactose (G), sucrose (S); maltose (M), lactose (L) or raffinose (R).] 4% sugar and bromcresol-purple are added to $0 \cdot 5\%$ peptone water and the medium is dispensed into small test-tubes, 3 ml per tube; the tubes are sterilized by steaming for 30 minutes on three consecutive days. The medium is then inoculated with the yeast suspension which is to be tested, and one drop of yeast extract per tube is added at the same time. A vaseline-paraffin layer (equal volumes of vaseline and solid paraffin) 2 cm deep is formed on the surface of the medium by adding the mixture in the molten state. The culture is incubated at 26 °C for 4–5 days. Fermentation of the sugar is indicated by a yellow colouration of the medium and lifting of the vaseline-paraffin layer. If the sugar present in the medium is not fermented by the yeast, neither of these changes occurs.

2.8.42   Test for sugar assimilation by yeasts (an auxanographic method)

One drop of yeast extract is added to a melted and cooled (40 °C) basal agar medium [131] containing no carbon source. A cell suspension of the yeast to be tested is thoroughly mixed with the medium. The mixture is poured into a Petri dish and any condensation is allowed to evaporate. Drops of sugar solutions (those listed in 2.8.41) are dispensed on to the agar surface directly over spots marked on the opposite side of the Petri dish, or filter paper strips each soaked in one of the sugar solutions are placed on the agar surface. The plates are incubated in the inverted position for 2–3 days at 25 °C. If the yeast assimilates a particular sugar, growth will be observed in the vicinity of the sugar spot.

2.8.43   Test for nitrogen assimilation by yeasts (an auxanographic method)

The principle of this procedure is analogous to that described in 2.8.42; the test in this case is performed with an agar medium containing no nitrogen source [106]. Potassium nitrate and other nitrogen sources are added to the surface of the dried medium. Ammonium sulphate is used as the control. Incubation is carried out at 25 °C for 2–3 days. Nitrogen assimilation is indicated by the growth of the yeast in the position of the nitrogen source.

3.   PRINCIPLES OF SAMPLING FOR MICROBIOLOGICAL GRADING

The qualitative characteristics of food samples, including their microbiological properties, are examined in order to obtain information about entire supplies of particular foodstuffs. This information may be instrumental in deciding the fate of a supply of food or in deciding how food processing is controlled. Making use of such information involves an inherent difficulty in that although only a small sample can be examined, the conclusions are applied to the whole supply of the foodstuff in question. The characteristics of foodstuffs may change from sample to sample, and may vary even within a sample. The question, therefore, arises as to which of the variable results should be regarded as being representative of the non-examined bulk of the foodstuff supply.

These problems are particularly valid with respect to microbiological characterization. In foods, micro-organisms tend to be unevenly distributed. Depending on the origin and reproductive characteristics of the contaminating micro-organisms, and the conditions of food processing and food storage, some microbes may be found distributed throughout the food, while others may be confined to the surface, or to foci or veins of contamination inside. The techniques of sampling and the number of the samples necessary for grading according to predetermined characteristics are governed by these factors.

The techniques of sampling will be described together with individual procedures. In this Chapter, the principles of collecting samples and grading are dealt with. These principles are equally valid for the grading of raw materials, semi-processed products, finished products and additives.

3.1   Mathematical–statistical principles of sampling

Any samples taken for microbiological testing must be regarded as a loss to the manufacturer. In many cases, the entire unit (box or flask) from which the sample has been taken is in danger of being spoiled; for this reason and for reason of laboratory hygiene, the unit must be disregarded after grading.

To minimize losses, only the smallest possible portions of the product are taken as samples. If a particular property of a food varies within the food as a whole, any conclusions with respect to that property are of limited value if they are based on small sample and there may be a considerable risk involved in applying such conclusions and making decisions therefrom.

Uncertainty and risk are unavoidable in the grading of food on the basis of sample tests. If grading is carried out consistently, but without regard for mathematical–statistical aspects of sampling, the risk or probable error involved can subsequently be computed. However, this exercise may only serve to demonstrate that the value of statistically uncontrolled grading is often greatly overestimated. Initially, acceptance of this was greeted with much reluctance and the widespread introduction of statistical methods has been slow. The different methods of sampling are closely related with corresponding methods of evaluating laboratory results, and since, therefore, sampling and grading are closely related activities, they are discussed together.

3.1.1　　Definitions

Population. The group of individual units (pieces, cans, etc.) forming an entire foodstuff, or the part of a bulk or liquid foodstuff which is to be graded.

In a homogeneous population, the random variation of a given characteristic among individual units results from many weak effects.

Classification (grading) of a population refers to the establishment of categories by which the population is judged to be good (acceptable) or worthy of rejection (unacceptable).

Sample. A part of a population, or a subset from a set of units, which is taken for examination in order to grade a population. A sample unit is any one of the units constituting a specified sample; each sample unit is collected and examined separately. In the computation, one set of results is designated as a sample result and each individual item of the data is a sample unit result.

Sampling. In the case of random sampling from a population, each individual of the population has the same chance of being sampled.

Operating characteristic curve (OC). A diagram or mathematical expression which gives the probability of a population being acceptable in relation to the mean value of the characteristic used as a criterion for acceptability.

Sample design. Sample design refers to the number of sample units required for a given grading operation and the methods of dealing with, for example, defective individuals and of defining the categories to be used in the grading scheme.

110

3.1.2   Grading categories, operating characteristic curve (OC)

The risks involved in sampling can be illustrated by the simplest type of categorization, e.g. that in which the result of the examination is expressed as "good" or "defective". In this procedure, at least two sets of data must be obtained, namely the number of sample units, and the acceptability number, i.e. the number of defective sample units which can be accepted before the lot represented by the sample is rejected.

While the defective units (e.g. cans with visible swelling) are unevenly distributed in the population, so the proportion of swollen cans in a sample is not identical with that in the population. If several samples are taken, the proportion may vary from sample to sample.

How than can the result from a sample nevertheless serve as a basis for dealing with the entire population?

Theoretically, it is assumed that a given proportion of defective sample units is obtainable (e.g. 1%). Having examined a very large number of samples, it is possible to calculate, in terms of probability, the proportion of samples that is expected to contain a greater or smaller number of defective units than some predetermined number, and the number of samples that are expected to contain the same proportion of defective units as that specified. If these values are calculated for several populations of various theoretical proportions of defective units, values can be obtained for the acceptability rates corresponding to stated numbers of defective units. On the basis of these values, the OC curve can be drawn. This curve shows the probable errors involved in a sampling procedure; its calculation is simplified by using a computer.

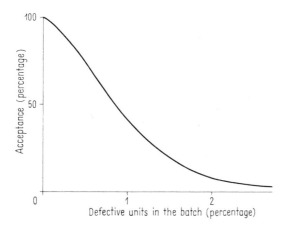

Fig. 44/a. Operating characteristic curve, if $n = 200$; $c = 1$

Example. On receipt of a type of canned food, a procedure is put into operation, which includes incubation of a sample of 200 units at constant temperature. A batch is to be pronounced acceptable if the number of swollen cans per 200 is 0 or 1.

If, in fact, the proportion of cans showing swelling is 1% of the whole population, then, assuming an even distribution, it might be expected that each sample would contain 2 swollen cans, so that every batch would have to be rejected. However, the swollen cans are randomly distributed in the population. Thus, then in 40% of the samples less then 2 cans are found swollen, i.e. 40% of the batches are qualified as acceptable. On the other hand, of those batches in which 0·5% of the units are defective, i.e. the batches are in fact acceptable, 2% of the lots will be rejected *(Fig. 44/a)*.

The detailed form of the OC is usually of no special interest because slightly differing curves have the same grading capacity. For this reason, only a few points of the OC are generally determined. As the OC has an inverse S-shape ($c \neq 0$) or an inverse J-shape ($c = 0$), an approximate description of the curve requires relatively few data.

In several cases two points of the OC are fixed: one at a high (e.g. 95%) acceptability value and the other at a low (e.g. 5%) value (points *A* and *R*) respectively, in *Figure 45*. Batches of the former degree of acceptability are rarely rejected, and thus the manufacturer can work relatively safely at this level. The rejection rate at point *R* is referred to as the manufacturer's risk. Batches with an acceptability corresponding to point *R* are almost always rejected. Thus the buyer can be reassured that such material will probably not reach him. The acceptance rate at point *R* represents the buyer's risk. The fate of batches falling between points *A* and *R* is uncertain. The threat of rejection compels manufacturers to produce material at a level near point *A*.

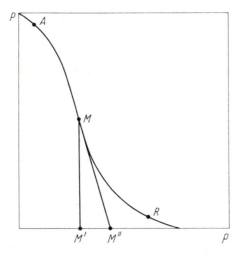

Fig. 45. Operating curve for sampling and its characteristic points

$A : P = 95\%; \; p = 0·18\%$ — acceptability point
$M : P = + 50\% \; p; \; = 0·84\%$ — median
$R : P = 5\% \; p; \; = 2·35\%$ — rejection point

Slope $= \dfrac{MM}{M'M'} = 62·5$ a 1%

increase in the number of defective elements is accompanied by a 62·5% decrease in the probability of acceptance

According to another interpretation, the point corresponding to a 50% acceptance rate *(M* in *Fig. 45)* and the slope of the curve (tan β in *Fig. 45*) are fixed. This, like other simplified methods for the determination of OCs, is not applied in microbiological grading.

It is a common property of OCs that the larger the number of data on the basis of which a batch is graded (i.e. the greater the value of *n*), the steeper is the OC at the point of control (i.e. the stricter is the control); also the higher the c/n ratio, the more the control point is shifted to the right (i.e. the less severe is the control).

3.1.3 Grading based on attributes and variable measurements

The abscissa in *Figure 45* shows the theoretical proportion of defective units in population. In grading based on qualitative attributes this is the only way of representing the characteristic under study.

Other characteristics, e.g. the total count and the numerical results of enzyme tests, may be transformed into attributes or represented by averages. Qualitative data may be transformed into attributes by fixing a limit value on some practical basis. If the value of a sample unit is on the "good" side of the limit, the unit (batch) can be graded as good. If it is on the unacceptable side, the unit is to be graded as being for rejection. Thus, the number of unacceptable units is determined according to an attribute-sampling scheme.

The mean values of measured quantities can be used for grading by comparing them with the limit value of another measurement-type sampling scheme. If the mean falls to the "rejection" side of the limit value, the population is to be rejected, etc.

A variable sampling plan yields more information than an attribute system. This has the advantage that the OC of the grading based on this is steeper than that of a grading based on an attribute sampling scheme with the same value of *n*.

Another advantage of a variable sampling plan is that further statistical parameters (e.g. standard deviation, type of the distribution) are determined in the procedure. In this way, some intrinsic traits of the fluctuations occurring in manufacture can be revealed. The OC of a measurement sampling scheme, however, depends on the above-mentioned statistical data and, therefore, correct grading from averages requires a distribution (characterized by the shape of the curve and the standard deviation) consistent with the fixed levels. At the expense of the steepness of the OC, the standard deviation can be substituted by a value estimated from the sample, but the parameters representing the form of the distribution must be predetermined by making a preliminary survey of the data. Such a preliminary survey, including the control of the form of the distribution and computation of dispersion, make for a lot of extra work, and, therefore, measurement sampling schemes are applied to the quality control system of the manufacturer (e.g. control of intermediary products, intake of raw material for manufacturing, internal control of finished products), but not to the control exercised by authorities or by commerce.

In the following example, the efficiency of an attribute sampling scheme is compared with that of a measurement sampling scheme.

Example. The product to be graded is a dried vegetable. Batches of this are graded according to the following attribute scheme. Five individual packages are randomly selected from each batch. The type of micro-organism specified in the control requirements is counted (number of cells/g) in each package. The grading of batches is carried out according to the following:
— a batch is graded as good if the number of units containing 100 or more germs/g is 0 or 1;
— it is graded as unacceptable in all other cases.

The OC pertaining to this example is shown by line *A* in *Figure 46*. The grading system applied to, and recommended for real practice is more complex (see three-class grading).

OCs similar in steepness to that of line *A* in *Figure 46* can be obtained if the grading is based on the means of the logarithms of the microbial counts from three sample units. By using logarithmic values, the shape of the distribution is more

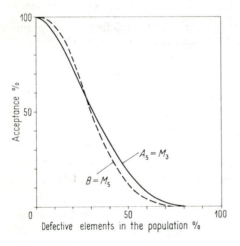

Fig. 46. Comparison of operating characteristic curves for attribute and measurement sampling schemes

A_s: attribute sampling scheme; $n = 5$, $c = 5$; a unit is defective if the logarithm of the viable count is > 2.

M_3: Measurement sampling scheme: $n = 3$; a population is defective if the mean of the logarithms of the 3 viable counts is > 1.65.

M_5: Measurement sampling scheme: $n = 5$; a population is defective if the mean of the logarithm of the 5 viable counts is > 1.65. The standard deviation of the log viable count is 0.7

favourable for grading purposes. In this system the grading limits should be fixed in terms of the logarithmic mean. If we wish to bring an OC into correspondence with OC *A,* then in fixing the limits for the logarithmic means, the standard deviation of the microbial counts must be taken into account. If, for example, the standard deviation is 0·7, then the log limit is 1·65 (i.e. the average cell count is 45 cells/g). If the standard deviation is larger, the limits must be fixed at lower and higher levels.

The larger the number of separately packed sample units, the steeper is the OC. If the logarithms of the total counts of 5 packages are averaged and the resulting mean is set at the above limit (1·65), the OC *B* will be obtained.

3.1.4   Three-class grading

The shortcomings of measurement procedures and the paucity of information in sample attribute procedures have prompted some authors (Clark and Thatcher, 1974) to develop three-class procedures. These attribute procedures involve two categorization steps. One of the limits determines a total count which approaches $1/100$ of that concentration of micro-organisms which causes spoiling of food or poisoning. The batch is accepted if this relatively high limit (M) is not exceeded in any of the sample units.

The other limit is determined by the Good Manufacturing Practices guidelines for any given micro-organism. This is designated as m. The M/m quotient ranges between 2 and 100, depending on the type of food and the properties of the micro-organism. The actual microbial count may exceed the m value in a limited number of sample units.

The OC for three-class grading can be derived by combining the OCs of the two separate categorization steps. The resulting curve represents a more stringent grading than either of the two-class curves. In this case every decision may have four outcomes from which only one decision gives acceptance of the lot.

In a three-class procedure, the recommended number of sample units *(n)* is 5 or 10, depending on the micro-organism and the food. The number of defective sample units *(c)* may be 0, 1, 2 or 3, judging at the m level, and exclusively 0 judging at the M level.

The possible OCs for the second subdivision of sample units are shown in *Figure 47.* The resulting OC derives from the products of the acceptance rate corresponding to $c = 0$ and that corresponding to the other c value fixed in the sampling scheme. Where $c = 0$, the proportion of defective units is given by the probability of units exceeding the $M = 0$ value; for other values of c, this proportion is given by the probability of units exceeding the m value. The OC must be prepared for each case because it depends on the ratio of the $M–m$ distance to the standard deviation (and on the shape of the distribution curve). This task can easily be carried out by programmable calculators.

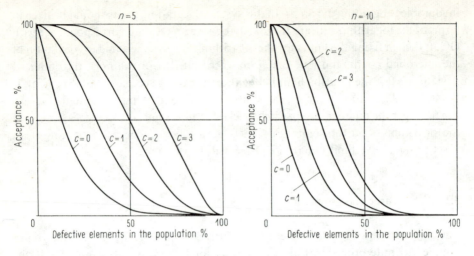

Fig. 47. Operating characteristic curves for second-step categorization in 3-class grading

n = number of sample elements; c = maximum number of defective sample elements for acceptance of a batch. The condition $c = 0$ applies in the case of elements exceeding the M limit, and any other value of c applies to elements exceeding the m limit. The c values pertaining to M, and n are fixed according to the type of foodstuff involved and the microorganism causing the problem

The resulting OC is not any steeper than the OC of the simple attribute procedure for the same number of sample units, but it is responsive to fluctuations in the total counts. If in any population this fluctuation is low, the acceptance of items with high average total counts will be determined by the lower limit of total count; if the fluctuation is high, the upper limit will be decisive.

3.1.5   Selection of samples

The actual OC of a grading operation will approximate the computed curve if each member of the population stands the same chance of being selected, i.e. if the sampling is random. However, random sampling must observe strict rules of procedure, otherwise distortion of the results may occur. Bias in the results arises partly from the fact that the inspector may subconsciously pick up sample units on the basis of their appearance or spatial distribution, and partly from the fact that not all sample units are equally accessible to the subject making the selection (e.g. it is difficult to take samples from the inside of a pile without going to a lot of extra trouble). Factors influencing microbiological characteristics, such as the rate of cooling, may vary from the centre of a bulk of material to its surface.

Subjective errors may be reduced by random sampling. In the case of foodstuffs stored in bulk quantities and divided into sub-populations of different packagings, sampling required division of the population into parts before it.

116

In the procedure of stratified sampling, the population is divided into strata, sometimes applying theoretical boundaries, and samples are selected from each stratum. The properties measured separately for each stratum are weighted, taking into account the size of each stratum as a proportion of the population. Deviations may be observed between individual strata, and if these are consistently observed, their distorting effect may be removed.

If large populations are arranged and packed into sub-populations, and if these are further subdivided, multi-stage sampling may be carried out. First, groups to be sampled are selected from the largest units of storage. From these groups smaller units, e.g. cans, are selected for sampling. The results obtained from such a sampling design are of equal weight.

3.2   The aim of grading

In practice, there are two groups of methods of microbiological grading. Those belonging to the first group are used where it is necessary to decide whether a population of the raw material, or a population (batch) of the finished product is suitable for use (i.e. they are used in the grading of a population or batch). If the outcome is favourable, the material will be used for production or the finished product will be sold. Otherwise, both the raw material and the finished product must be processed again or in the most severely affected cases, destroyed.

The aim of the other group of grading methods is to examine both the hygienic conditions of individual phases of the processing and the influence of each phase on the quality of the product (control of production). A favourable decision indicates that the processing is good and that no intervention is necessary. In the case of an unfavourable outcome, the technology must be changed. The OCs for, and procedures followed in the two kinds of grading are different, and, therefore, in the following each will be dealt with separately.

In one respect, both groups of procedures are the same; first of all, the OC of best fit for the given foodstuff must be established and the grading characteristic must be selected. This is the most difficult step in drawing up a sampling scheme. The value of the individual points of an OC depends on:
— the type of micro-organism and its properties (pathogenicity, etc.),
— the type of foodstuff and the usual mode of consumption,
— the subjects who will consume the food.

The more dangerous a micro-organism, the more strict is the OC that must be selected; if the microbe is highly pathogenic, only the lowest part of the curve can apply, i.e. the part which reflects a low probability of acceptance.

As to the mode of consumption, if the substance of a food and the mode of culinary preparation are favourable to the growth of a given micro-organism, then a stricter OC should be chosen.

Similarly, a stricter OC should be applied if the consumers are children, sick or old people, or patients on a diet. Selection of the most appropriate OC is made difficult by the fact that the principles of risk-taking with respect to grading are just beginning to be developed.

Once an OC has been fixed, the adequate sampling system, including the number of sample units and the limits, can be established by calculations.

The computation of OCs, especially those of variable plans, requires a great many data on the microbiological conditions of the finished products, and the conditions of processing. So, many of these basic data are still to be determined.

3.2.1 Grading of populations

The OCs that are most appropriate for the grading of foodstuff populations vary according to the properties of the micro-organisms under study. In this respect micro-organisms can be grouped as follows:
— pathogenic micro-organisms,
— micro-organisms causing spoilage,
— indicator micro-organisms.

3.2.1.1 Pathogenic micro-organisms

Pathogenic micro-organisms must not be present in foodstuffs in numbers that would result in disease. Therefore, evidence must be sought for the absence of pathogenic microbes.

Table 4
Relationship between volume of sample and probability of identification of micro-organisms

$P (\%)$	"a"
63	100
90	43
95	33
99	22
99·9	14

P is the probability (%) of there being an average of less than are micro-organisms in "a" % of a given volume of a sample in which no micro-organisms has actually been found.

118

If the distribution of a micro-organism in the whole batch is random, examination of one sufficiently large sample may provide a reliable indication of the count for the given microbe is below the fixed limit *(Table 4)*.

The difficulties are more pronounced if only some units of a product are contaminated by the pathogen. In such cases an attribute procedure must be applied and it should be stipulated that no pathogen must be present in the sample. With this requirement, the OC will take the shape of a reversed J *(Fig. 48)*. In the case of such OCs, it is sufficient to fix a single point, *viz.*, that corresponding with a low probability of acceptance (e.g. 1%). For low probabilities of acceptance, *Table 5* shows all the relationships between the total number of sample units and the theoretically possible proportions of samples containing a pathogen, for a random distribution of pathogen-containing units.

As shown in *Figure 48,* it is very difficult to establish the absence of pathogenic micro-organisms in all units of a foodstuff. Uncertainty arises if only a single unit is examined. Detection of a pathogen in a single unit justifies rejection of the batch, while failure to detect a pathogen may be a matter of chance, and the probability of error of this kind is high. If, on average, every second unit (e.g. a dry sausage) is contaminated by the pathogen, the probability of failure to detect the contaminant is 50% when only one unit is examined.

Obviously, the more dangerous a pathogen, the more sensitive must be the method specified for its detection. As shown in *Table 5,* a method of high sensitivity for the detection of highly pathogenic microbes needs a large number of sample

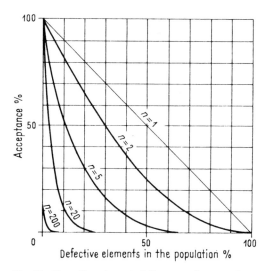

Fig. 48. Operating characteristic curves for sampling schemes in which a sample is rejected if even a single unit is defective, expressed as a function of the number of sample elements

Table 5

Contamination of excluded batches as a function of the number of sample units and the statistical confidence

Number of sample units	Probability of acceptance		
	5%	1%	0.1%
	frequency of error (%) which still can be expected in the population		
1	95.0	99.0	99.9
2	77.6	90.0	96.8
5	45.1	60.2	75.0
10	25.9	36.9	49.9
20	13.9	20.6	29.2
50	5.8	8.8	12.9
100	3.0	4.5	6.7
200	1.5	2.3	3.4
500	0.6	0.9	1.4
1000	0.3	0.5	0.7

Note: If the given proportion of the units in a batch is contaminated, it can be stated with probability (given at the head of the Table) that none of the n units of the sample taken (according to the Table) will be contaminated.

units. However, the cost of control grows proportionately with the quantity of the foodstuff that goes into sampling.

Thus the phase of the manufacturing process in which contamination by pathogenic micro-organisms arises cannot be screened out by quality control of the finished product alone. The risk of contamination must be reduced by rigorous control of the raw material and the stages of processing. In the control of finished products, on the other hand, it is promising to examine not the pathogen, but the characteristic concomitant microflora instead (see 3.2.1.2). This flora comprises various micro-organisms, some of which are relatively easily controlled. Thus control becomes simpler. The success of this approach depends on the constancy of the ε-factor, i.e. the numerical relationship between the indicator microflora and the pathogen. Efforts have been made to clarify the details of this relationship (Mossel, 1975; Drion and Mossel, 1977).

3.2.1.2   Micro-organisms causing spoilage

Spoilage micro-organisms usually occur in large numbers in the raw material of foodstuffs. Although they decline in number and activity during processing, such microbes are present in varying numbers in almost all finished products. Their activity, which depends on conditions within the foodstuffs itself and the mode of storage, eventually leads to spoilage.

In addition to the intrinsic contamination, unpacked foods may become contaminated from outside during storage and transport.

Spoilage micro-organisms are assessed in terms of their numbers and some of their measurable characteristics, not simply by their presence alone. Procedures based on mean microbial counts have been used more and more widely by manufacturers in finished product control. In setting out appropriate procedures, it should be taken into account that microbial counts may show variations of several orders of magnitude. For this reason, indices (e.g. the mean) may be used based on the logarithms of microbial counts.

For the extrinsic control of finished products, the three-class attribute grading is recommended.

The use of logarithmic values in the calculations is an appropriate method of handling microbial counts for micro-organisms influencing the shelf life of food. In the case of pathogenic microbes, the risk of illness occurring increases with the size of the microbial count, and thus the risk of infection cannot be assessed on the basis of logarithmic means, but rather on the basis of arithmetic means. Since the arithmetic mean increases faster than the logarithmic mean as the count increases, the risk of infection is higher than the risk of spoilage for microbial populations of identical distribution.

3.2.1.3   Indicator micro-organisms and indices

Owing to their small numbers, and other of their properties, pathogenic micro-organisms are difficult to detect in food. An alternative approach is, therefore, to examine the concomitant microflora, i.e. indicator micro-organisms, which are characteristically present along with a particular pathogen. Usually, the indicator micro-organisms are present in much higher numbers than the pathogen itself.

The indicator micro-organisms are assessed under less strict conditions by attribute procedures, and three-class procedures are used ever more widely. In the control applied by the manufacturer, variable plans are the more effective.

Micro-organisms that cause spoilage, and the general hygienic state of the food, are often indicated by the indicator microflora, usually by some kind of total microbial count or by enzyme reactions. The results are obtained as continuous random variables. In this case variable plans can be used for control.

121

3.2.2   Problems of sampling for production control

In the microbiological control of foodstuffs by the batch, the risk of erroneous assessment cannot be reduced without raising the number of sample units, and consequently the cost of the control operation as well.

The situation is more favourable if in the grading of a finished product, the population can be ragarded as more or less known as a result of some preliminary information being available. Such preliminary information may be obtained from the control results of earlier batches or the control results of processing. The latter are particularly useful because they are detailed and the requirements demanded of production control, thereby preventing the appearance of objectionable material among the finished product.

The microbiological control of processing can be divided into the following phases:
— control of the raw material;
— control of personal hygiene;
— control of the equipment used;
— control of processing.

In up-to-date, comprehensive control of foodstuffs, the control of processing does not end when the foodstuff has been sold, but follows the food to the point at which it is placed on the consumer's table. Thus the manufacturer endeavours to follow any changes occurring in the microbiological state of his product during its entire shelf life, even as far as the preparation of the food before consumption.

3.2.2.1   Control-chart control

Grading carried out during processing differs in some respect from the grading of batches. When finished products are graded, usually more than one characteristics are examined, whereas in the control of processing, one (occasionally a few) characteristics are chosen for each phase. During the control procedure, some of the characteristics of distribution are examined based on the preliminary information available. From these, the means of the critical microbiological data and working limits are established. An upper and a lower limit are fixed; any variation between these limits requires no intervention. If data are found with the monitory zone (i.e. outside one of the inner limits but inside the outer ones), then a warning is issued to the manufacturer, and usually more detailed tests are demanded. A monitory zone might be fixed for batch control as well, but in practice, second sampling after a finished product has been delivered is difficult to carry out.

Example. It is desired to prevent microbiologically objectionable products from reaching the quick-freezer operation. For this purpose, samples are taken at four sites from each batch of the material, and by examining these using one of the more

122

rapid microbiological tests (e.g. a reduction test), the microbiological state of the product is determined. The values relating to the four sample units are introduced on the control chart *(Fig. 49)* and they are averaged, then it is noted into which of the fixed zones the average value falls. For each zone there are operational instructions which are carried out according to the actual value of the sample average.

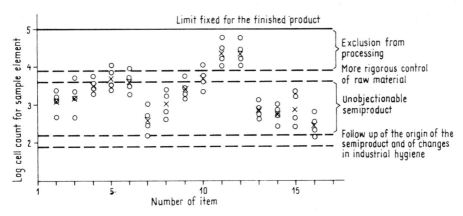

Fig. 49. Control chart for cell counts in semi-finished products

When the control-chart system has been introduced, the limits should be adjusted to the actual conditions of production. Control charts used in this way indicate, without extra cost, any shift in the quality level and any change in the stability of production. A change in quality level is followed by an unidirectional shift of the characteristic values without any change in the dispersion. A change in stability of production, on the other hand, is followed by an increase of dispersion, i.e. the control limits will be passed in any direction.

In microbiological grading, the lower limit has a lower weighting factor than the upper limit because the former corresponds to an absence of heavy microbiological contamination. Nevertheless, even favourable microbiological changes must be examined because an understanding of such changes may help in finding ways to improve the microbiological quality of the product.

3.2.2.2   Special traits of the different phases of processing

In the control of production, the control of raw materials is a similar operation to that of controlling batches. For any raw material, critical indices can be fixed, and these may vary from one manufacturer to another.

Raw materials consistently purchased from the same source, which is also a manufacturer producing large amounts of the same foodstuff, may be controlled by the control-chart system. In the control of raw materials, there is a possibility of being able to work with a large number of sample units by using contact plates. In the control of the technical part of the manufacturing process, it is advantageous to take small samples of the material undergoing processing, before it is packaged; the samples can be properly homogenized or concentrated, and in this way a lot of information can be obtained without loosing considerable amounts of the material. On this basis technological parameters (e.g. time of heat treatment) can be adjusted according to the mean value obtained for the initial microbial counts.

In critical areas of the equipments (e.g. sites difficult to clean) the equipment can be inspected by means of contact plates. If suspicion is aroused during this inspection, the tests can be performed as a regular target inspection; otherwise it is sufficient to screen apparatus at random intervals.

In public health control, personal hygiene should be checked as comprehensively as possible. Sampling procedures play no direct role in this respect.

4. DESCRIPTION AND IDENTIFICATION OF MICRO-ORGANISMS OCCURRING IN FOODSTUFFS

4.1 The taxonomy of bacteria (Sirockin and Cullimore, 1969)

Bacteria belong to the class *Schizomycetes* which includes the following orders:

I. *Pseudomonadales*	VI. *Actinomycetales*
II. *Chlamydobacteriales*	VII. *Beggiatoales*
III. *Hyphomicrobiales*	VIII. *Myxobacteriales*
IV. *Eubacteriales*	IX. *Spirochaetales*
V. *Caryophanales*	X. *Mycoplasmatales*

Bacteria belonging to the orders *Pseudomonales, Eubacteriales* and *Actinomycetales* are encountered with considerable frequency in the food industry. These bacteria, and the methods of studying them, are, therefore, matters of prime concern in this volume.

The systematics and identification of bacteria are based on external characteristics, namely shape, colour and size, as well as on biochemical properties.

In addition, the source of the carbon utilized by the bacterium is an aid to classification. Those micro-organisms utilizing carbon dioxide directly are designated as autotrophs, and those satisfying their carbon demand entirely from organic substances are called heterotrophs. Further classification is based on the nature of the carbon source *(Fig. 50)*. The shape of the bacterial cell as observed under the microscope proves a further criterion for classification (Chapter 2.7), as shown in *Figures 24* and *25*.

By studying these characteristics and biochemical responses (Chapter 2.8), it is possible to undertake the identification of bacterial isolates as far as the genus. For further identification, excellent handbooks are available (Buchanan–Gibbons, 1975; Harrigan and McCance, 1976; Skerman, 1959; Árpai, 1967).

A simplified identification key (after Sirockin and Cullimore, 1969) is presented here. The bacteria of interest to the food microbiologist are heterotrophs, and, therefore, only those genera containing heterotrophs are presented *(Tables 6–11)*.

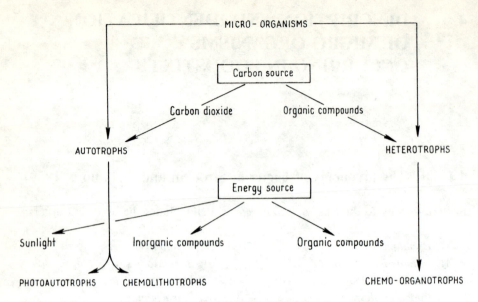

Fig. 50. Classification of micro-organisms according to utilizable carbon-energy source

126

Table 6
Key for the identification of bacteria (Sirockin and Cullimore, 1969)

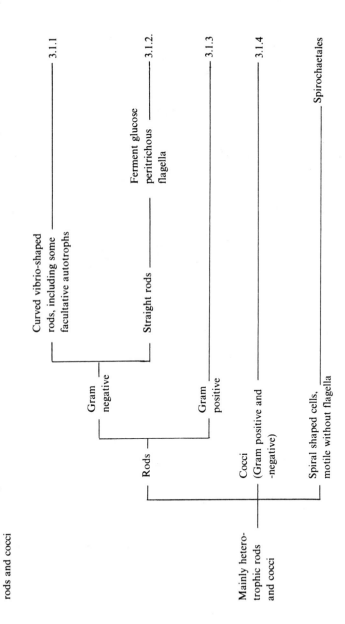

Mainly obligate autotrophic rods and cocci

Mainly hetero-trophic rods and cocci

Rods — Gram negative — Curved vibrio-shaped rods, including some facultative autotrophs ————— 3.1.1

Straight rods — Ferment glucose peritrichous flagella ————— 3.1.2.

Gram positive ————— 3.1.3

Cocci (Gram positive and -negative) ————— 3.1.4

Spiral shaped cells, motile without flagella ————— Spirochaetales

Table 7
Key for the identification of the Gram-negative curved rods (Sirockin and Cullimore, 1969)

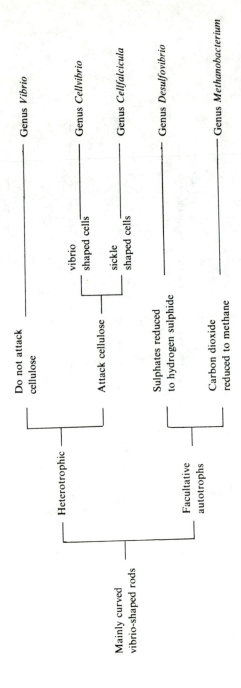

128

Table 8

Key for the identification of straight Gram-negative, heterotrophic rods (Sirockin and Cullimore, 1969)

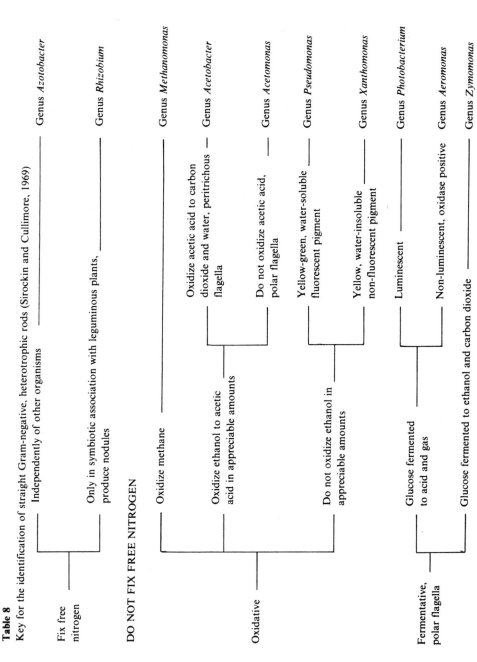

Fix free nitrogen
- Independently of other organisms ────────── Genus *Azotobacter*
- Only in symbiotic association with leguminous plants, produce nodules ────────── Genus *Rhizobium*

DO NOT FIX FREE NITROGEN

- Oxidize methane ────────── Genus *Methanomonas*
- Oxidative
  - Oxidize ethanol to acetic acid in appreciable amounts
    - Oxidize acetic acid to carbon dioxide and water, peritrichous flagella ────────── Genus *Acetobacter*
    - Do not oxidize acetic acid, polar flagella ────────── Genus *Acetomonas*
  - Do not oxidize ethanol in appreciable amounts
    - Yellow-green, water-soluble fluorescent pigment ────────── Genus *Pseudomonas*
    - Yellow, water-insoluble non-fluorescent pigment ────────── Genus *Xanthomonas*
- Fermentative, polar flagella
  - Glucose fermented to acid and gas
    - Luminescent ────────── Genus *Photobacterium*
    - Non-luminescent, oxidase positive ────────── Genus *Aeromonas*
  - Glucose fermented to ethanol and carbon dioxide ────────── Genus *Zymomonas*

9 Kiss

Fermentative, when motile peritrichous flagella

Plant pathogens, producing galls —————— Genus *Agrobacterium*

Yellow pigment —————— Genus *Flavobacterium*

Violet pigment, insoluble in water and chloroform, soluble in ethanol —————— Genus *Chromobacterium*

Litmus milk turned alkaline, no acid produced on glucose —————— Genus *Alcaligenes*

Non-pigmented using hexose sugars —————— Genus *Achromobacter*

Ferments glucose to acid and gas —————— Family **Enterobacteriaceae**

Small rods, parasites, good growth only on enriched media. No gas on glucose —————— Family **Brucellaceae**

Table 9

Key for the identification of the genera of the Enterobacteriaceae. (Sirockin and Cullimore, 1969)

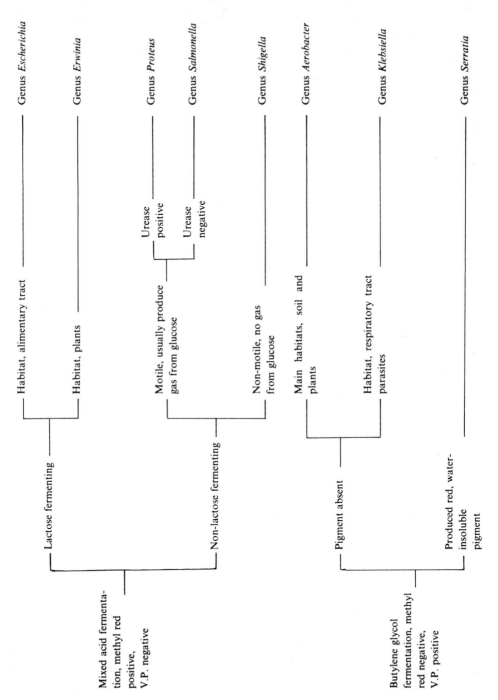

132

Table 10
Key for the identification of the Gram-positive rods (Sirockin and Cullimore, 1969)

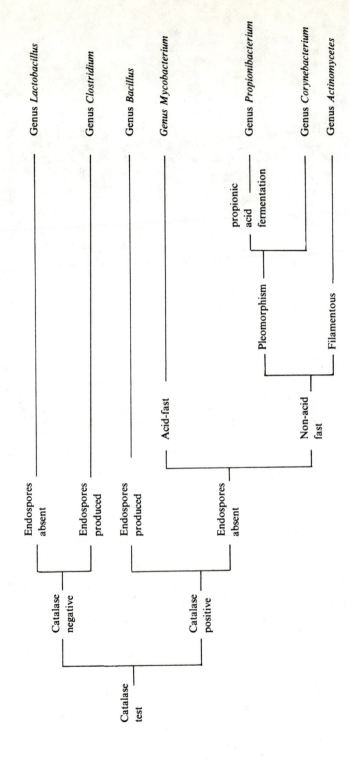

Table 11

Key for the identification of the heterotrophic cocci. (Sirockin and Cullimore, 1969)

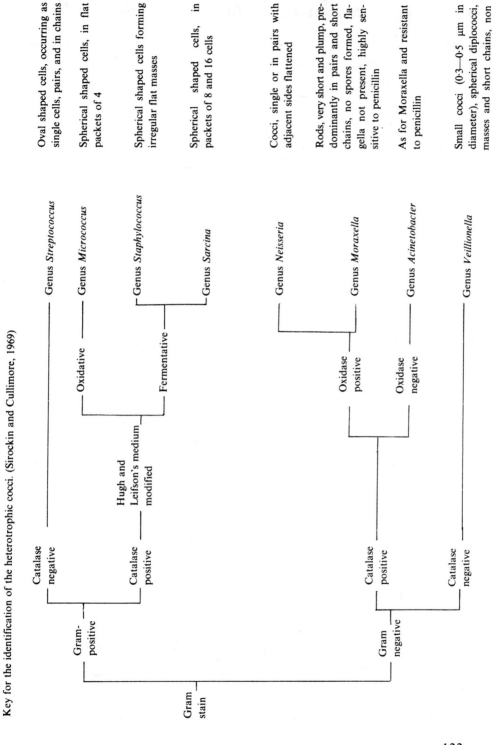

					Description
			Catalase negative	Genus *Streptococcus*	Oval shaped cells, occurring as single cells, pairs, and in chains
	Gram-positive	Catalase positive	Hugh and Leifson's medium modified — Oxidative	Genus *Micrococcus*	Spherical shaped cells, in flat packets of 4
Gram stain			Fermentative	Genus *Staphylococcus*	Spherical shaped cells forming irregular flat masses
				Genus *Sarcina*	Spherical shaped cells, in packets of 8 and 16 cells
	Gram negative	Catalase positive	Oxidase positive	Genus *Neisseria*	Cocci, single or in pairs with adjacent sides flattened
			Oxidase negative	Genus *Moraxella*	Rods, very short and plump, predominantly in pairs and short chains, no spores formed, flagella not present, highly sensitive to penicillin
				Genus *Acinetobacter*	As for *Moraxella* and resistant to penicillin
		Catalase negative		Genus *Veillionella*	Small cocci (0·3—0·5 μm in diameter), spherical diplococci, masses and short chains, non sporulating, anaerobic

133

4.2 Key to the identification of yeasts (after Lodder, 1970)

(The morphological and biochemical characteristics of the yeasts are described in Chapters 2.7 and 2.8, respectively)

1 a	Forming ascospores.................................	2
b	Not forming ascospores	12
2 a	Ascospores hat-shaped or Saturn-shaped...............	3
b	Ascospores not hat-shaped..........................	4
3 a	Assimilating nitrate *(Hansenula)*	
	Hansenula anomala (Hansen) H. et P. Sydow	
b	Not assimilating Nitrate *(Pichia)*	
	Pichia membranaefaciens Hansen	
4 a	Ascospores kidney-shaped *(Kluyveromyces)*	5
b	Ascospores not kidney-shaped........................	6
5 a	Cells 10–11 µm or less in length; not assimilating nor fermenting maltose	
	Kluyveromyces fragilis (Jörgensen) Van der Walt (= *Saccharomyces fragilis* Jörgensen)	
b	Cells 7–8 µm or less in length; usually assimilating and sometimes also fermenting maltose	
	Kluyveromyces lactis (Dombrowski) Van der Walt (= *Saccharomyces lactis* Dombrowski)	
6 a	Ascospores needle or awl-shaped *(Metchnikowia)*	
	Metchnikowia pulcherrima Pitt et Miller (= *Candida pulcherrima* (Linder) Windish)	
b	Ascospores spherical or oval	7
7 a	Fermentation intense and rapid *(Saccharomyces)*	8
b	Fermentation slow, weak, or nil *(Debaryomyces)*	
	Debaryomyces hanseni (Zopf) Lodder et Kreger-van Rij (= *Debaryomyces kloeckeri* Guilliermond et Péju)	
8 a	Assimilating sugar: R –*	9
b	Assimilating sugar: R+	10
9 a	Assimilating sugar: M+	
	Saccharomyces rouxii Boutroux	
b	Assimilating sugar: M –	
	Saccharomyces bisporus (Naganishi) Lodder et Kreger-van Rij var. *mellis* (Fabian et Quinet) Van der Walt	

* R = raffinose, M = maltose, D = glucose, G = galactose, S = sucrose, L = lactose.

134

(= *Saccharomyces mellis* [Fabian et Quinet] Lodder et Kreger-van Rij)

10 a	Assimilating sugar: G−	
	Saccharomyces bayanus Saccardo (= *Saccharomyces pastorianus* Hansen)	
b	Assimilating sugar: G+	11
11 a	Fermenting sugar: R+(1/3), cells longer then 21 μm not occurring	
	Saccharomyces cerevisiae Hansen (= *Saccharomyces ellipsoideus* Hansen)	
b	Fermenting sugar: R+ (complete), cells not limited to 21 μm in length	
	Saccharomyces uvarum Beijerinch (= *Saccharomyces carlsbergensis* Hansen)	
12 a	Not forming pseudomycelium or true mycelium..........	13
b	Both pseudomycelium and true mycelium sometimes formed	15
13 a	Cells budding bipolarly *(Kloeckera)*	
	Kloeckera apiculata (Rees emend. Klöcker) Janke	
b	Cells budding multilaterally	14
14 a	Colonies pink (carotenoid pigment formed) *(Rhodotorula)*	
	Rhodotorula glutinis (Fresenius) Harrison	
b	Colonies colourless (no pigment formation)	
	Torulopsis spp.	
15 a	Cells ogival; acetic acid copiously formed from glucose	
	Brettanomyces spp.	
b	Cells not ogival; acetic acid not formed................	16
16 a	Arthrospores developing in the colony by fragmentation of hyphae	
	Trichosporon spp.	
b	No arthrospores developing in the colony *(Candida)*	17
17 a	Assimilating nitrogen *Candida utilis* (Henneberg) Lodder et Kreger-van Rij (= *Cryptococcus utilis*/ Henneberg/Anderson and Skinner)	
b	Not assimilating nitrogen	18
18 a	Not fermenting sugars (except for a weak fermentation of glucose)	19
b	Fermenting sugars with ease, especially glucose	20
19 a	Cells 2–4 × 4–10 μm; showing tolerance to 2–9% NaCl	
	Candida valida (Leberle) van Uden et Buckley (= *Candida mycoderma*/Rees/Lodder et Kreger-van Rij)	

135

b   Cells 3–5 × 5–11 (or up to –20) μm; NaCl tolerance 10–14%
Candida lipolytica (Harrison) Diddens et Lodder

20 a   Fermenting sugars: D +, G, S, M, L, R −; max. temperature tolerance 43–45 °C
Candida krusei (Cast.) Berkhout

b   Fermenting sugars: D, G, S, L +, R ±, M −; max. temperature tolerance 37–42 °C
Candida kefyr (Beijerinck) van Uden et Buckley
(= *Cryptococcus kefyr* / Beijerinck/Skinner)

4.3   Key to the identification of moulds (partly after Vörös and Léránth, 1974a, b)

(The morphological characteristics and biochemical reactions of moulds are described in Chapter 2.7 and 2.8, respectively)

1 a   Thallus consisting of coenocytic mycelium without septa (septa may form beneath the reproductive organs). Sexual reproduction by zygogamy; asexual reproduction occurring by formation of sporangia (or rarely by formation of monosporous sporangiola functioning as conidia) *(Zygomycetes)*   2

b   Thallus consisting of septate mycelium. Sexual reproduction absent or occurring by ascogamy; modes of asexual reproduction including conidium formation, but excluding sporangium formation   7

2 a   Sporangium with columella *(Mucoraceae)*   3

b   Besides sporangia, sporangiola formed, each containing a few spores but without a columella *(Thamnidiaceae)*
Thamnidium elegans Link ex Wallr.

3 a   Sporangium without apophysis; rhizoids absent *(Mucor)* ...   4

b   Sporangium with apophysis; groups of sporangiophores developing from rhizoids *(Rhizopus)*   6

4 a   Colonies white, yellowish or light gray; wall of sporangium soon disintegrating; sporangiophore long, not branching along a pseudoaxis. Sporangium, yellowish–grayish–blue, large 100–300 μm in diameter. Gemmae absent. Spores 8–12 μm long
Mucor mucedo L.

b   Young colonies white, then grayish or brown in later development. Wall of sporangium fragile, or slowly disintegrating. Sporangiophores much-branched. Gemmae numerous   5

136

5 a	Growing rapidly at 37 °C	
	Mucor rouxianus (Calmette) Wehmer	
b	Not growing at 37 °C	
	Mucor racemosus Fresenius	
6 a	Spores 7–9 μm; growing rapidly at 37 °C; gemmae formed	
	Rhizopus oryzae Went et Prinsen	
b	Spores 10–20 μm; not growing at 37 °C; gemmae not formed	
	Rhizopus stolonifer (Ehrenberg ex Fries) Lind. (= *Rhizopus nigricans* Ehrenberg)	
7 a	Asci are formed. Asci and asexual reproductive cells (conidia) may occur together in some colonies *(Ascomycetes)*	8
b	Ascus formation unknown. Only asexual reproductive structures (conidia) occurring in colonies *(Deuteromycetes, Fungi imperfecti)*	10
8 a	Free asci developing in groups without peridium-surrounded fruiting body. Conidial form resembling *Penicillium* (but conidiophore not arranged in regular penicillus); phialoconidia formed	
	Byssochlamys fulva Olliver et Smith (stat. conid.: *Paecilomyces varioti* Bainier)	
b	Asci formed in a peridium-surrounded fruiting body; conidial form not *Paecilomyces*	9
9 a	Asci rounded, irregularly arranged in thin-walled yellow or orange-yellow cleistothecia. Apex of conidiophore terminated with a vesicle; conidia produced by sterigmata (phialides) arranged in a single layer	
	Aspergillus repens (Corda) de Barry	
b	Asci elongated, club-shaped, regularly arranged in a hymenium in true, thick-walled dark perithecia. Conidia are as blastospores in chains, orange- or flesh-coloured en masse	
	Neurospora sitophila Shear et Dodge (stat. conid.: *Monilia sitophila* [Mont.] Sacc.)	
10 a	Conidia inequally two-celled, pear-shaped, pink en masse. There is difficulty in observing the mode of conidium development *Trichothecium roseum* Link	
b	Conidia not inequally two-celled. The mode of conidium development is easily observed	11
11 a	Conidia not newly-formed cells, arising rather as a result of the fragmentation of hyphae (arthroconidia). Colony white, diffuse, flat, often resembling yeast colonies	
	Geotrichum candidum Link	
b	Conidia arising as newly-formed cells, not as arthroconidia	12

12 a   Asexual reproductive cells forming individually articulating on a wide base, not tending to become detached (aleurioconidia). Conidiophore undifferentiated; conidia one-celled, developing both terminally and laterally; conidial wall rough
Sprotrichum spp.

b   Asexual reproductive cells arising as newly-formed cells, detaching easily   13

13 a   Both inner and outer cell walls of the conidiophore (conidiogenic cell) participating in the process of conidium formation (blastoconidia)   14

b   Internal wall only of the conidiophore participating in the process of conidium formation. (The internal wall protrudes through a pore in the external wall—poroconidium formation.) Conidiophore and conidia dark; conidia divided by transverse and longitudinal walls, the apex ending in a short beak; long conidial chains may be formed
Alternaria tenuis Nees

c   Neither of the conidiophore walls taking part in conidium formation (phialoconidia)............................   16

14 a   Conidia arising at either a single site or a few sites on the terminal fertile branch (conidiogenic cell) of the conidiophore; conidia forming in chains (blastoconidia)   15

b   Conidia one-celled, arising simultaneously over entire surface of the conidiogenic cell, not forming chains (botryoblastoconidia). Conidiophore branching; colony mouse-gray.
Botrytis cinerea Pers. ex Fries

15 a   Colonies resembling yeast colonies, with mucous consistency, consisting of dark hyphae and mass of colourless, unicellular blastoconidia
Aureobasidium pullulans (de Bary) Arnaud (= *Dematium pullulans* de Bary)

b   Colonies mould-like, dark, non-mucous, often consisting of greenish-brown hyphae with conidiophores of the same colour, and chains of conidia. Earliest formed conidia 2- or 3-cellular, later conidia unicellular
Cladosporium herbarum (Persoon) Link

16 a   Conidia spherical or rounded, all unicellular, uniform in shape   17

b   Conidia elongated and of two kinds: Microconidia, unicellular (sometimes two-celled), forming chains; macroconidia, curved, fusoid, divided by 3–7 cross walls. The 4-day-

old colony is 2·5 cm in diameter and peach or violet in colour
Fusarium moniliforme Sheld.

17 a	Conidiophore simple, with dilated apex which forms a vesicle. Sterigmata (phialides) located in one or two layers on the vesicles *(Aspergillus)*	18
b	Conidiophore branching like a penicillus; sterigmata (phialides) located on conidiophore branches *(Penicillium)*	23
18 a	All sterigmata arranged in a single layer	19
b	Sterigmata usually arranged in two layers	20

19 a Yellow or orange cleistothecia, and grayish-blue conidial heads arising in the colonies (cf. **9 a**)
Aspergillus repens (Corda) de Bary

b Cleistothecia not forming in the colonies; oil-green or smoky-gray heads forming
Aspergillus fumigatus Fresenius

20 a Colonies and conidial heads black; wall of conidiophore smooth
Aspergillus niger van Thieghem

b Neither colonies nor conidial heads black; wall of conidiophore usually rough, granular   21

21 a Colonies and conidial heads ochre
Aspergillus ochraceus Wilhelm

b Colonies and the conidial heads yellowish-green   22

22 a Conidiophores > 1000 µm long; sterigmata arranged in one layer on the majority of the vesicles
Aspergillus oryzae (Ahlburg) Cohn

b Conidiophores < 1000 µm long; sterigmata arranged in two layers on most of the vesicles
Aspergillus flavus Link

23 a Conidiophores < 500 µm long, forming bundles (coremia) (Fasciculata subsection)
Penicillium expansum (Link) Thom

b Conidiophores not forming coremia   24

24 a Colonies velvety; conidiophores arising from the substratum (*Velutina* subsection)   25

b Colonies lanose or floccose; conidiophores arising from the aerial mycelium (*Lanata* subsection). Colonies remaining white for a long time, then becoming pale greyish-green after 10–14 days
Penicillium camemberti Thom

25 a   Conidiophore wall smooth; edge of colony not arachnoid; conidium spherical, 3–3·5 µm in diameter
Penicillium notatum Westling

b   Conidiophore wall rough; edge of colony arachnoid; conidium spherical, 3·5–5 (or up to 8) µm in diameter
Penicillium roqueforti Thom

5. TECHNIQUES FOR QUANTITATIVE DETERMINATION OF MICRO-ORGANISMS

5.1   Total count of cell mass

Methods include (a) counting of the viable and non-viable micro-organisms by optical or other physical techniques, (b) measuring the mass of microbial cells, and (c) determining the quantity of some component(s) of the microbial cell. The total numbers of viable and non-viable microbes are calculated from the results.

5.1.1   Microscopic methods

The majority of microscopic methods give a direct cell count per unit volume (e.g. Breed's counting method), whilst in some methods the result is expressed as a standard index, such as the Howard count.

5.1.1.1   Counting-chamber methods

Counting chambers are specially designed slides which allow exact volumes of microbial suspensions to be prepared in which the micro-organisms can be counted under the microscope.

5.1.1.1.A   The Thoma chamber

The main component of the *Thoma chamber* is a piece of glass 0·2 mm thick with a circular hole cut in its centre, fixed to the surface of a slide. Located centrally in the well thus formed is a smaller glass plate 0·1 mm thick *(Fig. 51)*. On this glass plate is a quadratic calibration with squares of 1/20 mm side (0·0025 mm² in area). The space around and above the glass plate is filled by means of a pipette with the microbe's suspension to be examined, and the suspension is tightly covered with a coverslip, taking care to avoid trapping bubbles. If the chamber is clean and the filling is done correctly, the depth of the chamber will be 0·1 mm. Mild pressure on the coverslip brings about the appearance of Newton's rings; if these do not appear,

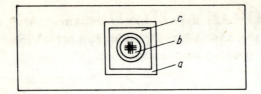

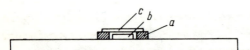

Fig. 51. The Thoma chamber
a — ring; b — sample; c — cover slide

the chamber must be cleaned and refilled, and the coverslip must be changed. The volume represented by each square is $0.0025 \, mm^2 \times 0.1 \, mm = 0.00025 \, mm^2$ (i.e. $1/4000 \, \mu l$). A dilution of the suspension is required which gives between 500 and 1000 cells in the 100 to 200 squares selected for counting. The squares that are to be counted should be selected from all parts of the chamber at random. Cells straddling the upper and right-hand borderlines of the squares should either be consistently included in the count or consistently disregarded. The total for the 100–200 squares usually represent the whole chamber. This total count is divided by the number of squares to give the mean number of cells per square. To obtain the cell count per ml, this number is multiplied by 4×10^6. Yeast cells are very suitably counted in the Thoma chamber.

5.1.1.1.B   The Buerker chamber

In the Buerker chamber, there are two horizontal surfaces on each of which a mesh of squares is engraved *(Fig. 52)*. The mesh consists of squares of $1/20$ mm side and $1/5$ mm side. When the chamber is properly closed with a coverslip, the depth of the liquid is 0.1 mm. Thus, the volume represented by each small square is

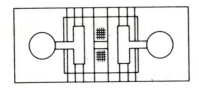

Fig. 52. The Buerker chamber

142

0·00025 mm³, and that represented by each larger square is 0·004 mm³. Multiplication by 4×10^6 and $2·5 \times 10^5$, respectively, gives the cell count per ml. The Buerker chamber is mainly used for counting yeast and mould cells.

5.1.1.1.C   The Helber–Glynn, Petroff–Hauser and Helber chambers

In construction the first and second of these chambers are identical with the Thoma chamber, whereas the Helber chamber has the same construction as the Buerker chamber. The squares are $1/20 \times 1/20$ mm and the depth of the liquid is 0·02 mm in the correctly closed chambers. The volume of microbial suspension per square is 0·00005 mm³. The mean cell count per square must be multiplied by 2×10^7 to give the cell count per ml. The 0·02 mm depth of the chamber makes it possible to count bacteria in these chambers with a high degree of accuracy.

5.1.1.1.D   Slide chamber

The counting chambers described above can be substituted by the ordinary slides and coverslips commonly used in laboratories (Vas, 1962). Of the microbial suspension to be examined, a drop (e.g. 0·004 ml) is measured on to the slide from a micropipette. Alternatively, it can be weighed on an analytical balance. The drop is then covered with a coverslip and the edges of the coverslip are sealed with vaseline using a heated inoculating loop. With a knowledge of the area of the field, the number of fields counted and the thickness of the fluid layer, it is possible to calculate the cell count per ml from the numbers of cells counted in several separate fields.

Using an objective micrometer slide, the diameter and area of the field can be determined. The depth of the fluid layer can be calculated by dividing the volume of the suspension dropped on the slide by the area of the coverslip; e.g.

diameter of the field: 0·12 mm
area of the field: 0·0113 mm²
volume of the suspension: 4 mm³
area of the coverslip: 400 mm²

$$\text{depth of the suspension, } d = \frac{4 \text{ mm}^3}{400 \text{ mm}^2} = 0·01 \text{ mm}$$

Thus the volume scanned in the field is 0·000113 mm³, and the mean microbial count per field must be multiplied by 8,840,000 ($\sim 8·8 \times 10^6$) to give the cell count per ml.

143

It should be borne in mind that owing to the imperfect surfaces of ordinary slides and coverslips, the variation of the cell count from one area to another is greater than that occurring in a counting-cnamber.

The migration of cells can be prevented by mixing the suspension with an equal volume of sterile 0·4% agar, and taking the dilution into account in the calculation of the final result.

A total of at least 500 cells must be counted for each cell suspension. Each cluster of cells and each budding yeast cell should be regarded as one cell.

Total cell count $= N \times F \times D$/ml, where $N =$ the mean number of cells per smallest unit of the grid of the chamber, $F =$ a coefficient which is characteristic of the chamber, and $D =$ the dilution factor.

The coefficients' characteristic of the different chambers are presented in *Table 12.*

Table 12

Coefficients for counting-chambers

Type of chamber	Coefficient
Thoma and Buerker	$4·0 \times 10^6$ (or $2·5 \times 10^5$)
Helber–Glynn, Petroff–Hausser, Helber	$2·0 \times 10^7$
Slide (20 × 20 mm coverslip, 0·12 mm field diameter)	$8·8 \times 10^5$

The applicability of couhting chambers is restricted to cell densities which are sufficiently high for counting in this way. This limitation can be overcome to some extent by examining a larger number of squares, but the procedure then becomes time-consuming. The lower limit of applicability is $1·0 \times 10^6$ cells per ml for the *Thoma* and *Buerker chambers,* and $1·0 \times 10^7$ cells per ml for the others. Cell suspensions that are too dense to be examined can be diluted before counting.

The ranges of error for the chamber methods vary from $\pm 10\%$ to $\pm 15\%$.

Example. The number of yeast cells in a culture medium is to be determined using a Buerker chamber; the medium is first diluted a hundredfold.

827 cells are counted in 95 large squares.

Mean cell count per square $= \dfrac{827}{95} = 8·6$;

Chamber coefficient $= 2·5 \times 10^5$;

Total cell count $= 8·6 \times 2·5 \times 10^5 \times 100 = 2·2 \times 10^8$ cells per ml.

5.1.1.2   Accuracy of counting

The calculated cell count is inevitably not 100% accurate. Since the actual error is not known, it is necessary to define an upper and lower limit called the *confidence limits* or *fiducial limits,* within which the actual cell count occurs with a well-defined degree of probability. The distance between the confidence limits is called the confidence interval. The range of safety for an estimated cell count may be set, according to the degree of safety required at the 90%, 95% or 99% confidence levels. For example, the 90% confidence interval indicates that in 90 out of 100 determinations the real mean is expected to lie within the confidence interval. If the limits are widened, the confidence probability will increase (i.e. the probability of an error will decrease). Conversely, narrowing of the interval means a lower confidence level.

Using the diagrams in *Figures 53–55,* we can very rapidly calculate the minimum total cell count and the number of fields, respectively, which are required for determining the cell count of a microbial suspension with a predetermined degree of confidence.

Example (with confidence limits set at 90%). The half width of the confidence interval is taken as 20% of the mean cell number per field (e.g. the confidence interval will be ± 2% if the mean cell number per field is 10%. Thus the probability of the real value occurring within the interval 8–12 cells per field is 90%. In the

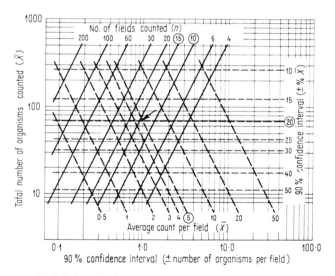

Fig. 53. Relationships between total number of cells counted, number of fields examined, and the 90% confidence intervals for cell counting (Cassel 1965)

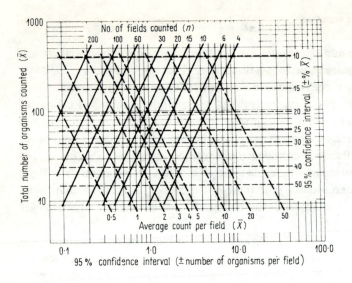

Fig. 54. Relationships between total number of cells counted, number of fields examined, and the 95% confidence intervals for cell counting (Cassel 1965)

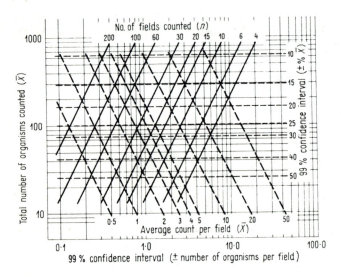

Fig. 55. Relationships between total number of cells counted, number of fields examined, and the 99% confidence intervals for cell counting (Cassel 1965)

diagram *(Fig. 53)* the value 20 is encircled and the corresponding level is indicated by a horizontal dotted line.

The experimental mean value is obtained by counting the cells in four fields. The total count is found to be 20, and the cell count per field $(\bar{x})$ is, therefore, 5. This is indicated by an oblique broken line in *Figure 53*. The number 5 is encircled.

An arrow points to the intersection of the two dotted lines. The co-ordinates of this point on the abscissa and the ordinate represent the confidence interval and the number of cells to be counted, respectively. The former is ± 1 cell, the latter, $x = 69$.

The steep continuous line on the left side of the point of intersection is designated by the number 15 on the top of the diagram, and the parallel line on the right is designated by the number 10. Relative to these, the point of the intersection can be interpolated as corresponding to the number 14. This means that under the conditions of the example, the cells in 14 fields must be counted to reach the 90% confidence level.

5.1.1.3   Stained preparations

A known quantity of a microbial suspension is smeared on a predetermined area of the surface of a slide. The smear is stained and the cell count is determined under the microscope.

Procedure. The material to be examined is smeared (usually 0·01 ml or 10 mg) on a well-defined area of the slide surface (e.g. 1 cm²) as evenly as possible. The smear is dried, fixed, degreased if necessary, and stained (2·6). The number of microbial cells (or clusters of cells) is then determined. From the diameter (or area) of the field, the cell count per unit of volume (or weight) of material can be calculated.

This method was originally used for the grading of milk (Breed, 1911). Since even today its main field of application is the grading of milk and dairy products, the method is described in detail in connection with the total microbial counts of milk (8.1.2.1).

5.1.2   Electronic particle counting

This method, originally developed for blood cell counting and the estimation of particle size distribution in powders, has recently been introduced in the microbiological laboratory. It finds application in the direct determination of total microbial counts in pure microbial suspensions, and the determination of the microbial count in mixed populations after some of the individual cells have developed into microcolonies. In addition, electronic particle counting is widely used in the determination of leucocyte counts in milk samples, a procedure which is important in the diagnosis of mastitis.

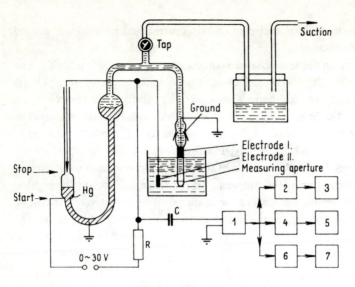

Fig. 56. Wiring diagram of an electronic particle counter

1 = low-noise impulse pre-amplifier; *2* = integral discriminator; *3* = counter; *4* = cathode-ray tube; *5* = auxiliary circuits; *6* = differential discriminator; *7* = rotamotor

   The particles that are to be counted electronically *(Fig. 56)* are suspended in an electrolyte solution, e.g. saline.

   A test-tube containing a piece of glass capillary is submersed in the suspension and electrodes are located at either opening of the capillary. The resistance between the electrodes remains constant until a particle, or particles, of low conductivity enter the capillary tube, whereupon the resistance increases until the particle leaves the capillary tube. The current impulses can be observed on an oscilloscope and their number can be totalled on a counter (Farkas, 1962; Kubitschek, 1969).

   The suspension under study is drawn through an aperture the diameter of which (50 — 100 μm) depends on the size of the particles; only one particle at a time should be able to pass through the capillary. The signal, i.e. the increase in resistance, is in direct proportion to the volume of the particle (cell), and thus the distribution of particle sizes can be registered automatically by means of a programmed impulse-size analyser equipped with a differential discriminator. The suspension is drawn through by a mercury. In the open arm of the U-shaped tube, there are two electrical contacts and the volume between these points equals the volume of the suspension sucked in through the measuring hole. The contacts, when touching the mercury level, switch on and off the counter, which is able to count 50,000 cells per minute. The accuracy of the method depends on (a) the proper choice of measuring aperture (on the basis of cell size), (b) the cell density, (c) the number of contaminant particles and (d) the signal-to-noise ratio deriving from the electric background

148

noise. If the cell density is low, filtration through a membrane filter of 0·5 μm average pore size is recommended. In this way, cell densities as low as a few hundreds of cells per ml can be determined with reasonable accuracy. Cell counts of between 10^3 and 10^6 cells per ml are easily carried out. The probable error of the method in an ideal case and assuming a Poisson distribution, is N (N = number of particles counted). If blood cells are counted, the total error is equal to, or less than, 2·5% with reproducibility ±1%.

5.1.3   Measurement of cell mass

5.1.3.1   Optical methods

The principle underlying these methods is that changes in light intensity occur when light passes through a microbial suspension. The intensity of the incident light tends to be reduced by the absorption and scattering effects of the microbial cells.

Optical procedures essentially give an indication of the cell mass rather than cell number, since different phases of growth and reproduction cell diameters vary considerably. In spite of this, the results obtained by using optical methods are often expressed in terms of cell numbers. The data are calibrated using the microscopically determined cell count for the logarithmic growth phase of the culture. The microbial counts obtained in this way relate to the number of cells of average size which would give those results (Monod, 1949) rather than the actual number of cells.

In suspensions of low cell count ($< 10^6$ cells per ml) the measurement of light scattering may give better results, methods on light absorption are more suitable for suspensions containing $\geqslant 10^6$ cells per ml. At such high concentrations, the relationship between cell density and light absorption in a nephelometer is not linear.

5.1.3.1.A   Turbidimetric methods

Turbidimetry involves the measurement of transmitted light (I_m). Photometers constructed for microbiological and chemical work, or modified forms of the latter, may be used for this purpose. The visual reading of earlier instruments has been replaced by the incorporation of photoelectric cells in up-to-date photometers. Extinction (optical density, OD or D) is calculated from the formula,

$$D = \log \frac{I_0}{I},$$

where I_0 is the intensity of the incident light and I is the intensity of the transmitted light. From the above,

$$D = \log I_0 - \log I$$
$$= \log I_0 - \log (I_m \cdot dI_s),$$

where I_m is the intensity of the light that is transmitted through the suspension without being scattered, I_s is the intensity of the back-scattered light passing in the direction of illumination, d is a coefficient which is a function of cell density.

For very dense cell suspensions, the value of d is higher and thus the computed value of D decreases. Thus the relationship between cell density and light absorption plus light scattering is not linear at high cell concentrations, i.e. that part of the calibration line which relates to high cell densities is curved. In the interests of accuracy, measurements are preferably made at cell concentrations corresponding to the linear part of the calibration curve.

The first step in making turbidimetric measurements is to obtain a calibration curve. The D values for dilutions of the stock suspension are measured, and plotted against the corresponding microscopically determined cell counts (with the D values on the ordinate).

The cell suspension which is to be assayed must be free of insoluble particles, and any effect of the colour of the suspension should be reduced to a minimum by using a colour filter. The suspension should be as homogeneous as possible. Turbidimetric methods meet the requirements specified for biological assays; the standard deviation expressed as a percentage of the mean value is within $\pm 2\%$.

5.1.3.2   Gravimetry

Gravimetric methods allow the cell mass per unit volume to be determined. Total cell counts can be obtained from the relationship between total cell count and cellular dry matter.

Methods based on weighing require pure cell suspensions.

Filtration. The mass of micro-organisms are separated from the suspending medium by passing the suspension through a sintered glass filter. A pore size is chosen which easily accommodates the size of the cells, e.g. a glass filter of average pore diameter $1.3 - 2.0\ \mu$ (G5 filter) is suitable for quantitative work. For yeasts, a G4 filter is satisfactory. A known volume (which is determined by the cell density) of the cell suspension is poured through the filter and the microbial mass is dried on the filter by drawing air through it, then the material is washed three times with distilled water, and dried to constant weight in an oven at 105 °C. The filter with the dried mass of cells is weighed, and the weight of the clean filter is subtracted to obtain the weight of the cells.

150

Yeast cell masses can be weighed after filtering through a Whatman No. 50 filter paper in a Büchner funnel. The funnel is dried at 105 °C before use. The cell mass that remains on the filter paper after a known volume of cell suspension is filtered, is washed by resuspending it three times in distilled water and removing the water under vacuum until the cell mass shows cracks. Then the cell mass is dried on the filter paper to constant weight at 105 °C. The predetermined weight of the filter paper is subtracted to obtain the weight of the cells.

Centrifugation. The suspension to be assayed is centrifuged, the supernatant is cautiously decanted, and the cell mass is washed by resuspension and centrifugation in water. After the third centrifugation, the cells are resuspended in a little water and transferred to a dry vessel. After evaporation in a water bath, the cell mass is dried to constant weight in an oven.

Example. Relationship between cell count and cell dry matter content. 20 g of yeast are suspended in sterile water and a dilution series is prepared by adding equal amounts of saline. The cell count is determined in a counting chamber and the dry matter weighed for each dilution. *Figure 57* shows the cell count (ordinate) plotted against the dry weight (abscissa) of the yeast in each dilution.

Given the volume and dry-matter content of a yeast suspension, the cell count can be read from the calibration curve.

Example. 8 g of yeast suspension are found to contain 1·324 g dry matter, or 0·1655 g dry matter per ml. From the calibration curve, this corresponds to $6·3 \times 10^9$ cells per 1 ml of the yeast cell suspension.

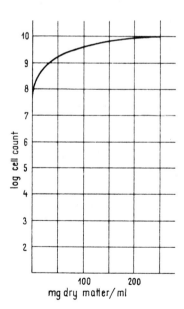

Fig. 57. Relationship between cell count and cell dry matter content

The lower cell density limit for reliability in this method ranges between 10^7 and 10^{10} cells per ml depending on the cell size. The standard error for three parallel measurements is about 1%.

In some laboratories the wet weight is measured. Known volumes of culture fluid or microbial suspension are centrifuged, the supernatant is decanted, and the tube containing the wet cell mass is weighed. Cell density is read from a calibrated curve of the wet-weight–cell count relationship.

5.1.3.3 Chemical methods

Components known to be present in cells in nearly constant proportions may serve as a basis for the determination of cell mass. The cell mass must be separated very thoroughly from the culture medium, so that components of the medium, e.g. nitrogen, phosphorus and carbon do not interfere with the assay.

5.1.3.3.A Protein assay

In the logarithmic phase of cell multiplication, the protein content of the cells is constant, provided that the conditions of cultivation are constant. The weight of the cells in a suspension can be calculated from the protein concentration present in the suspension.

The sensitivity of the biuret method carried out as described by Stickland is relatively low, but the method is quick to follow through.

A few millilitres of culture are centrifuged, the cell mass is resuspended in saline, and the optical density adjusted to 0·7–0·8. The *OD* of dilution prepared from the culture is measured in a photometer at 660 nm. 2 ml of each dilution are transferred to a test-tube, and 0·5 ml of a 0·3 N NaOH solution is added. The tubes are maintained in boiling water for 10 minutes, then cooled to room temperature. 0·1 ml of a 25% solution of $CuSO_4 \cdot 5 H_2O$ are added and the tube is immediately and vigorously shaken. After centrifugation at 5000 g, the supernatant is examined in a photometer at 578 nm. A calibration curve is prepared using albumin; thus the *OD* is empirically expressed as a function of protein concentration (mg protein per ml), the latter being linearly related to cell density.

Lowry's protein assay (Lowry et al., 1951) is more widely used, and although it is a sensitive method, only dissolved protein can be assayed. Insoluble protein must be solubilized first.

1 ml of the solution or suspension containing protein (e.g. a bacterium suspension) is adjusted so that the *OD* is between 0·1 and 0·3 at 660 nm (this corresponds to between 10 and 200 µg protein). 0·5 ml of a 0·3 N NaOH solution is

152

added and the mixture heated for 90 minutes in a capped centrifuge tube in a water bath at 60 °C. After cooling, 5 ml of the reagent are added with continuous shaking, and the tubes are allowed to stand in the dark for 10 minutes. (Reagent: 0.1 g $CuSO_4 \cdot 5\,H_2O$ is dissolved in 20 ml of a potassium-sodium tartarate solution. 1 ml of this is added to 50 ml of a sodium carbonate solution on the day of use.)

0.5 ml of Folin's reagent (1 part Folin's reagent [Merck] + 2 parts distilled water) is added and the mixture is immediately shaken and allowed to stand in the dark for 30 minutes. After centrifugation, the supernatant is examined in a photometer at 623 nm. The control contains water in place of the protein solution. A calibration curve is prepared using serum albumin fraction V at a concentration of 5–10 mg per ml.

5.1.3.3.B   Determination of dry material content of cells

The dry matter content of a mass of cells can be obtained from measurements of the organic matter content (Bailey and Meymandi-Nejad, 1961). The sample (contains 0.2–2.0 mg dry material) is mixed in at most 1 ml water or inorganic puffer solution, is mixed with 2 ml of a 2% dipotassium–dichromate solution and heated at 100 °C for 30 minutes. When the mixture has cooled, up to 5 ml water are added, and the colour intensity relative to that of the reagent solution is measured in a spectrophotometer at 580 nm. The dry matter content is obtained from a calibration curve.

5.2   Determination of total microbial count (cultivation methods)

The purpose of the cultivation methods is to determine the total number of viable cells, including bacteria, yeasts and moulds, present in a suspension.

First, a homogeneous microbial suspension is prepared. This is not easy to obtain if the material is pulpy, fibrous, or otherwise difficult to break up. If the material is readily miscible or easily dissolved, the ratio of material to diluent is usually 1 : 4 (20 g + 80 ml), otherwise greater dilutions are used, e.g. 10 g + 90 ml, 5 g + 95 ml, 1 g + 99 ml, or 10 g + 190 ml.

The material under investigation is weighed out accurately to the nearest 0.1 g, or if it is fluid, a precise volume is taken in a pipette; accordingly, the results are given per g or ml of the material. The test material is homogenized by one of the procedures described elsewhere, and the homogenate is retained as a stock suspension. Under aseptic conditions, the stock solution (and/or dilutions made from it) is used to inoculate a nutrient solution or the surface of an agar medium, (rarely a gelatin medium). The method of preparing the serial dilutions is the same whether the dilution method or the colony-counting method is applied. The two methods are different with respect to the inoculation technique and the calculation of the cell count.

5.2.1 The dilution technique and the dilution tube count

Dilutions of a stock solution are prepared until a dilution is obtained which can be expected to contain no cells (see "Most Probable Number", MPN). Test-tube cultures are prepared from the last few dilutions of the series, and after allowing some time for growth, the number of tubes which show microbial growth are noted. The number of viable cells is computed from the number of positive tubes by means of appropriate mathematical–statistical techniques.

A prerequisite of the use of the MPN method is that the cells are randomly distributed in the stock suspension, i.e. the cell concentration is the same throughout the suspension; cells should neither form aggregates nor exert any repulsive forces upon one another. In the subcultures, microbial growth should be apparent even if only one viable cell was present in the inoculum.

In the dilution procedure, sampling, preparation and inoculation must be performed aseptically. Special care should be taken to prevent microbial contamination.

From each of the series of tenfold dilutions, usually 1 ml is inoculated into each culture medium. In the preparation of the dilution series, it is advisable to change pipettes at each step. During inoculation of the culture media, on the other hand, changing the pipette may be omitted if the inoculations are carried out in order from the highest dilution to the least diluted of the series.

The aim of the dilution technique is to determine that dilution which contains one viable cell per ml. Thus it is absolutely necessary to test dilutions which, by rough estimation, would be expected to contain no viable cells per 1 ml volume. For the sake of thoroughness, inoculations are carried out from several dilutions, beginning with one that is clearly expected to produce microbial growth in each of the inoculated cultures, and ending the series with a dilution that is expected to give no growth at all. From the results, the MPN can be calculated by using a *probability table*. The number of cells contained in the original material can be calculated from a knowledge of the dilution factors and volumes used in the procedure.

The precision of the method can be increased by
— decreasing the dilution rate,
— increasing the number of replicate tubes, and
— increasing the number of repetitions.

Depending on the degree of accuracy desired, 2, 3, 4 or 5 replicate tubes should be inoculated with each dilution.

Table 13

Hoskin's table for determining the most probable number (MPN) of microbes per ml of sample, if 2 parallel inoculations have been made from each sample

Key numbers	Table values corresponding to the key numbers	Key numbers	Table values corresponding to the key numbers
0 0 1	0·45	1 1 2	2·8
0 0 2	0·90	1 2 0	2·1
0 1 0	0·46	1 2 1	2·9
0 1 1	0·92	1 2 2	3·7
0 1 2	1·4	2 0 0	2·3
0 2 0	0·94	2 0 1	5·0
0 2 1	1·4	2 0 2	9·5
0 2 2	1·9	2 1 0	6·2
1 0 0	0·6	2 1 1	13
1 0 1	1·2	2 1 2	21
1 0 2	1·9	2 2 0	24
1 1 0	1·3	2 2 1	70
1 1 1	2·9		

The dilution method is equally applicable, provided that a suitable culture medium and method of cultivation are available, for the determination of the total viable cell count and total cell number of special groups or species of micro-organisms. With this method, turbidity, gas formation, pH change, etc., as well as the development of mould colonies on solidified culture media can be taken as positive signs of growth. The MPN value can be calculated if the number of tube cultures regarded as positive (showing growth) and the dilution factors are known (*Tables 13–16*).

Calculation of the dilution count. First of all, the highest dilution must be found at which all the incubated tube cultures showed signs of microbial growth. At this level, the number of positive tubes is equal to the number of replicate tubes. In the second step, a three-digit key number is determined. The first digit of the key number is identical with the above replication number of the tubes; the next two digits are the numbers of positive tubes at the next two (higher) dilution levels (Example 1). If there is microbial growth at an even higher level of dilution, the number of positive tubes at this level should be added to the third digit of the key number (Example 2). If a negative tube is recorded from the lowest of the series of dilutions, it is advisable to repeat the test, beginning with a lower first dilution. If this is impossible, the number of positive tubes obtained from the first tested dilution is taken as the first digit of the key number (Example 3).

Example 1: a key number of 321 corresponds to a table value of 15. The dilution corresponding to the first digit of the key number is 10^1. Thus, the estimated dilution tube count is $15 \times 10^1 = 1.5 \times 10^2$.

Table 14

Hoskin's table for determining the most probable number (MPN) of microbes per ml of sample, if 3 parallel inoculations have been made from each sample

Key numbers	Table values corresponding to the key numbers	Key numbers	Table values corresponding to the key numbers
0 0 0	– –	2 0 0	0·91
0 0 1	0·3	2 0 1	1·4
0 0 2	0·6	2 0 2	2·0
0 0 3	0·9	2 0 3	2·6
0 1 0	0·3	2 1 0	1·5
0 1 1	0·61	2 1 1	2·0
0 1 2	0·92	2 1 2	2·7
0 1 3	1·2	2 1 3	3·4
0 2 0	0·62	2 2 0	2·1
0 2 1	0·93	2 2 1	2·8
0 2 2	1·2	2 2 2	3·5
0 2 3	1·6	2 2 3	4·2
0 3 0	0·94	2 3 0	2·9
0 3 1	1·3	2 3 1	3·6
0 3 2	1·6	2 3 2	4·4
0 3 3	1·9	2 3 3	5·3
1 0 0	0·36	3 0 0	2·3
1 0 1	0·72	3 0 1	3·9
1 0 2	1·1	3 0 2	6·4
1 0 3	1·5	3 0 3	9·5
1 1 0	0·73	3 1 0	4·3
1 1 1	1·1	3 1 1	7·5
1 1 2	1·5	3 1 2	12
1 1 3	1·9	3 1 3	16
1 2 0	1·1	3 2 0	9·8
1 2 1	1·5	3 2 1	15
1 2 2	2·0	3 2 2	21
1 2 3	2·4	3 2 3	29
1 3 0	1·6	3 3 0	24
1 3 1	2·0	3 3 1	46
1 3 2	2·4	3 3 2	110
1 3 3	2·9	3 3 3	– – –

Table 15

Hoskin's table for determining the most probable number (MPN) of microbes per ml of sample, if 4 parallel inoculations have been made from each sample

Key numbers	Table values corresponding to the key numbers	Key numbers	Table values corresponding to the key numbers	Key numbers	Table values corresponding to the key numbers
0 0 0	–	1 0 0	0·26	2 0 0	0·6
0 0 1	0·23	1 0 1	0·51	2 0 1	0·91
0 0 2	0·45	1 0 2	0·78	2 0 2	1·2
0 0 3	0·68	1 0 3	1·0	2 0 3	1·6
0 0 4	0·9	1 0 4	1·3	2 0 4	1·9
0 1 0	0·23	1 1 0	0·52	2 1 0	0·93
0 1 1	0·46	1 1 1	0·79	2 1 1	1·3
0 1 2	0·68	1 1 2	1·1	2 1 2	1·6
0 1 3	0·91	1 1 3	1·3	2 1 3	2·0
0 1 4	1·1	1 1 4	1·6	2 1 4	2·3
0 2 0	0·46	1 2 0	0·8	2 2 0	1·3
0 2 1	0·69	1 2 1	1·1	2 2 1	1·6
0 2 2	0·92	1 2 2	1·3	2 2 2	2·0
0 2 3	1·2	1 2 3	1·6	2 2 3	2·4
0 2 4	1·4	1 2 4	1·9	2 2 4	2·8
0 3 0	0·7	1 3 0	1·1	2 3 0	1·7
0 3 1	0·93	1 3 1	1·4	2 3 1	2·0
0 3 2	1·2	1 3 2	1·6	2 3 2	2·4
0 3 3	1·4	1 3 3	1·9	2 3 3	2·8
0 3 4	1·6	1 3 4	2·2	2 3 4	3·2
0 4 0	0·94	1 4 0	1·4	2 4 0	2·1
0 4 1	1·2	1 4 1	1·7	2 4 1	2·5
0 4 2	1·4	1 4 2	2·0	2 4 2	2·9
0 4 3	1·7	1 4 3	2·3	2 4 3	3·3
0 4 4	1·9	1 4 4	2·6	2 4 4	3·7

Table 15 (cont.)

Key numbers	Table values corresponding to the key numbers	Key numbers	Table values corresponding to the key numbers
3 0 0	1·1	4 0 0	2·3
3 0 1	1·6	4 0 1	3·4
3 0 2	2·0	4 0 2	5·0
3 0 3	2·6	4 0 3	7·1
3 0 4	3·1	4 0 4	9·5
3 1 0	1·6	4 1 0	3·6
3 1 1	2·1	4 1 1	5·5
3 1 2	2·6	4 1 2	8·1
3 1 3	3·2	4 1 3	11·0
3 1 4	3·8	4 1 4	14·0
3 2 0	2·1	4 2 0	8·2
3 2 1	2·7	4 2 1	9·4
3 2 2	3·3	4 2 2	13·0
3 2 3	4·0	4 2 3	17·0
3 2 4	4·7	4 2 4	21·0
3 3 0	2·8	4 3 0	11·0
3 3 1	3·4	4 3 1	16·0
3 3 2	4·1	4 3 2	22·0
3 3 3	4·8	4 3 3	28·0
3 3 4	5·6	4 3 4	36·0
3 4 0	3·5	4 4 0	24·0
3 4 1	4·3	4 4 1	39·0
3 4 2	5·0	4 4 2	70·0
3 4 3	5·9	4 4 3	140·0
3 4 4	6·7	4 4 4	—

158

Table 16

Hoskin's table for determining the most probable number (MPN) of microbes per ml of sample, if 5 parallel inoculations have been made from each sample

Key numbers	Table values corresponding to the key numbers	Key numbers	Table values corresponding to the key numbers	Key numbers	Table values corresponding to the key numbers
0 0 0		1 0 0	0·2	2 0 0	0·45
0 0 1	0·18	1 0 1	0·4	2 0 1	0·68
0 0 2	0·36	1 0 2	0·6	2 0 2	0·91
0 0 3	0·54	1 0 3	0·8	2 0 3	1·2
0 0 4	0·72	1 0 4	1·0	2 0 4	1·4
0 0 5	0·9	1 0 5	1·2	2 0 5	1·6
0 1 0	0·18	1 1 0	0·4	2 1 0	0·68
0 1 1	0·36	1 1 1	0·61	2 1 1	0·92
0 1 2	0·55	1 1 2	0·81	2 1 2	1·2
0 1 3	0·73	1 1 3	1·0	2 1 3	1·4
0 1 4	0·91	1 1 4	1·2	2 1 4	1·7
0 1 5	1·1	1 1 5	1·4	2 1 5	1·9
0 2 0	0·37	1 2 0	0·61	2 2 0	0·93
0 2 1	0·55	1 2 1	0·82	2 2 1	1·2
0 2 2	0·72	1 2 2	1·0	2 2 2	1·4
0 2 3	0·92	1 2 3	1·2	2 2 3	1·7
0 2 4	1·1	1 2 4	1·5	2 2 4	1·9
0 2 5	1·3	1 2 5	1·7	2 2 5	2·2
0 3 0	0·56	1 3 0	0·83	2 3 0	1·2
0 3 1	0·74	1 3 1	1·0	2 3 1	1·4
0 3 2	0·93	1 3 2	1·3	2 3 2	1·7
0 3 3	1·1	1 3 3	1·5	2 3 3	2·0
0 3 4	1·3	1 3 4	1·7	2 3 4	2·2
0 3 5	1·5	1 3 5	1·9	2 3 5	2·5
0 4 0	0·75	1 4 0	1·1	2 4 0	1·5
0 4 1	0·94	1 4 1	1·3	2 4 1	1·7
0 4 2	1·1	1 4 2	1·5	2 4 2	2·0
0 4 3	1·3	1 4 3	1·7	2 4 3	2·3
0 4 4	1·5	1 4 4	1·9	2 4 4	2·5
0 4 5	1·7	1 4 5	2·2	2 4 5	2·8
0 5 0	0·94	1 5 0	1·3	2 5 0	1·7
0 5 1	1·1	1 5 1	1·5	2 5 1	2·0
0 5 2	1·3	1 5 2	1·7	2 5 2	2·3
0 5 3	1·5	1 5 3	1·9	2 5 3	2·6
0 5 4	1·7	1 5 4	2·2	2 5 4	2·9
0 5 5	1·9	1 5 5	2·4	2 5 5	3·2

Table 16 (cont.)

Key numbers	Table values corresponding to the key numbers	Key numbers	Table values corresponding to the key numbers	Key numbers	Table values corresponding to the key numbers
3 0 0	0·78	4 0 0	1·4	5 0 0	2·3
3 0 1	1·1	4 0 1	1·7	5 0 1	3·1
3 0 2	1·3	4 0 2	2·1	5 0 2	4·3
3 0 3	1·6	4 0 3	2·5	5 0 3	5·8
3 0 4	2·0	4 0 4	3·0	5 0 4	7·6
3 0 5	2·3	4 0 5	3·6	5 0 5	9·5
3 1 0	1·1	4 1 0	1·7	5 1 0	3·3
3 1 1	1·4	4 1 1	2·1	5 1 1	4·6
3 1 2	1·7	4 1 2	2·6	5 1 2	6·4
3 1 3	2·0	4 1 3	3·1	5 1 3	8·4
3 1 4	2·3	4 1 4	3·6	5 1 4	11·0
3 1 5	2·7	4 1 5	4·2	5 1 5	13·0
3 2 0	1·4	4 2 0	2·2	5 2 0	4·9
3 2 1	1·7	4 2 1	2·6	5 2 1	7·0
3 2 2	2·0	4 2 2	3·2	5 2 2	9·5
3 2 3	2·4	4 2 3	3·8	5 2 3	12·0
3 2 4	2·7	4 2 4	4·4	5 2 4	15·0
3 2 5	3·1	4 2 5	5·0	5 2 5	18·0
3 3 0	1·7	4 3 0	2·7	5 3 0	7·9
3 3 1	2·1	4 3 1	3·3	5 3 1	11·0
3 3 2	2·4	4 3 2	3·9	5 3 2	14·0
3 3 3	2·8	4 3 3	4·5	5 3 3	18·0
3 3 4	3·1	4 3 4	5·2	5 3 4	21·0
3 3 5	3·5	4 3 5	5·9	5 3 5	25·0
3 4 0	2·1	4 4 0	3·4	5 4 0	13·0
3 4 1	2·4	4 4 1	4·0	5 4 1	17·0
3 4 2	2·8	4 4 2	4·7	5 4 2	22·0
3 4 3	3·2	4 4 3	5·4	5 4 3	28·0
3 4 4	3·6	4 4 4	6·2	5 4 4	35·0
3 4 5	4·0	4 4 5	6·9	5 4 5	43·0
3 5 0	2·5	4 5 0	4·1	5 5 0	24·0
3 5 1	2·9	4 5 1	4·8	5 5 1	35·0
3 5 2	3·2	4 5 2	5·6	5 5 2	54·0
3 5 3	3·7	4 5 3	6·4	5 5 3	92·0
3 5 4	4·1	4 5 4	7·2	5 5 4	160·0
3 5 5	4·5	4 5 5	8·1	5 5 5	—

Example 2: a key number of 534 corresponds to a table value of 21. The dilution corresponding to the first digit of the key number is 10^2. Thus, the dilution tube count is 2.1×10^3.

Example 3: 221 may be taken as the key number if the test cannot be repeated, and the dilution cell count may be based on the 10^2 dilution.

The cell count thus obtained is present in 1 ml of the stock solution. The MPN per unit amount of the original material is calculated as follows:

$$\frac{\text{viable cell count}}{\text{for stock suspension}} \times \frac{\text{weighed of material (g)} + \text{diluent (g)}}{\text{weighted of material (g)}} = \text{MPN (1g material)}^{-1}$$

Example 1

	10^0	10^1	10^2	10^3	10^4	Dilution
	0	1	2	3	4	
	+	+	+	+	−	
	+	+	+	−	−	
	+	+	−	−	−	

Number of positive tubes   3   2   1

Key number: 321   *3*   *3*   *1*   *3*

Example 2

	10^2	10^3	10^4	10^5	10^6	10^7
	+	+	+	+	+	−
	+	+	+	+	+	−
	+	+	+	−	−	−
	+	+	−	−	−	−
	+	+	−	−	−	−

Number of positive tubes   5   3   2   2

Key number: 534

Example 3

	10^2	10^3	10^4	10^5	10^6
	+	+	+	−	−
	+	+	−	−	−
	−	−	−	−	−

Number of positives tubes   2   2   1

Key number: 211 (if possible, the procedure should be repeated with a lower dilution)

E.g., 10 g of material are diluted in 90 g diluent and the results of the inoculations are as follows:

	Dilution				
	10^2	10^3	10^4	10^5	10^6
	+	+	+	+	−
	+	+	+	+	−
	+	+	−	+	−

Key number: 323
Table value: 29
Base dilution: 10^3
Number of cells in the stock suspension: 2.9×10^4.
The MPN for viable cells in the stock suspension is

$$2.9 \times 10^4 \times \frac{10+90}{10} = 2.9 \times 10^5 \, g^{-1}$$

This method has no upper restriction on the cell density of the starting material. The lower limit is that arising from the requirement that the number of viable cells inoculated must be > 1. If the viable count is low, larger amounts of material must be used in the test. The factors determining the 95% confidence limits for the MPN values are presented in *Table 17*. Here, we have taken into account the dilution factor and the replication factor, but the error of dilution was neglected. The lower fiducial limit is the quotient of the key number and the factor in the Table, whereas the upper limit is the product of the same two values. Cell count data falling outside the 95% confidence interval represent the MPN value with a probability of only 5%.

Table 17

Reliability of MPN data (Cochran, 1950)

Number of inocula- tions per dilution	Factors required for the calculation of the 95% fiducial limits			
	Dilution factor			
	2	4	5	10
1	4·00	7·14	8·32	14·45
2	2·67	4·00	4·47	6·61
3	2·23	3·10	3·39	4·68
4	2·00	2·68	2·88	3·80
5	1·86	2·41	2·58	3·30
10	1·55	1·86	1·95	2·32

162

According to *Table 17*, the precision of the results may be approximately the same under different experimental arrangements, e.g. dilution factor 2, number of replicate tubes 3; dilution factor 10, number of replicate tubes 10. However, in the first case many more tubes are required than in the latter. On the basis of this consideration and an approximative estimation of the cell count, the time and work involved in carrying out the procedure can be calculated for any required level of precision.

Example. A stock suspension is prepared from 20 g of a sample of material that is to be tested, and 80 g of diluent. A 1 : 10 dilution series is prepared, and each of three tubes is inoculated with 1 ml from one of the dilutions. Results:

Dilutions of stock suspension	10^0	10^1	10^2	10^3	10^4	10^5
Number of positive tubes	3	3	2	1	0	0

Key number for calculation of MPN: 321. This corresponds to a table value (in the Hoskins table) of 15. Thus, the MPN for the viable cells per 1 g of the material is

$$MPN = 15 \times 10^1 \times 5 = 7 \cdot 5 \times 10^2.$$

According to the Hoskins table, the 95% confidence limits in the above example are

$$\frac{7 \cdot 5 \times 10^2}{4 \cdot 68} = 1 \cdot 6 \times 10^2,$$
$$\text{and } 7 \cdot 5 \times 10^2 \times 4 \cdot 68 = 3 \cdot 5 \times 10^3.$$

In other words, in a large number of identical repetitions of the procedure results less than 160 or more than 3500 would be obtained in not more than 5% of cases.

5.2.2 Colony-counting methods

5.2.2.1 Counting in Petri dish culture

A series of dilutions is prepared from the stock suspension, and Petri plates are inoculated with those of the dilutions which are expected to give rise to countable colonies. After inoculation and incubation, colonies are counted and the viable cell count per unit amount of the original material is calculated. By using special media and controlling the conditions of cultivation, the procedure can be adapted to the counting of particular groups of micro-organisms.

Petri-dish cell-counting methods have many variations. The preparation of the material and the dilutions is always carried out the same way, i.e. homogenization with the diluent is followed by serial dilution and inoculation.

5.2.2.1.A   Pour-plate method

This method is mainly used for the cultivation of relatively heat-tolerant, obligate aerobes.

1 ml of each of the last four dilutions prepared from the stock suspension is pipetted into a labelled Petri dish (the label should indicate the sample and degree of dilution), beginning with the highest dilution. About 15 ml of melted medium, usually containing agar as the solidifying agent are poured at a temperature of 45–50 °C into each Petri dish from a test tube or flask. The raised lid of the Petri dish is lowered again and the suspension is cautiously mixed with the medium by a combination of to-and-fro and circular movements over the horizontal surface of the bench.

The exact procedure consists of the following movements: five times to-and-fro, five circular turns clockwise, five times to-and-fro at right angles to the former movement, and five circular turns anticlockwise. Mixing should be continued until the inoculum appears evenly distributed in the medium. Care should be taken to ensure that the medium does not splash on the lid or the sides of the Petri dish.

The plates are allowed to set and are then inverted to prevent condensation, back-dripping, migration of colonies, and drying out of the medium.

Incubation is carried out at an appropriate temperature and for an appropriate period.

The medium must be transparent, otherwise the colonies growing within it are difficult to observe; for the same reason, the medium should be free of all kinds of particulate matter.

5.2.2.1.B   Surface counts

Sterile melted medium at 45–50 °C is poured into Petri dishes and allowed to set. The formation of condensate on the surface of the medium is prevented by employing one of the following methods of drying:

without lid, at 50 °C for 30 minutes;

with lid closed, at 50 °C for 2 h;

with lid closed, at 37 °C for 4 h;

with lid closed, at 20–22 °C for 16 h.

Plates dried by any of these methods can be stored in an air-tight box at +2–+4 °C for 7–10 days.

Surface cultivation can be carried out in the following ways.

164

(a) Surface cultivation on an undivided surface of medium. Inocula consisting of 0·1 ml of each of the last members of a dilution series are dropped on prepared plates. The drop of suspension is smeared over the surface of the medium with a sterile glass rod specially made for this purpose, or with a sterile test-tube or platinum wire. The aim is to achieve an even distribution of micro-organisms over the surface. The glass rod should be 3·5 mm in diameter, with a 20-cm-long handle and a 3-cm-long head part making an angle of 120° with the handle. It is advisable to keep several glass rods immersed in alcohol ready for use. These rods are flamed and cooled directly before they are used. The platinum wire fixed in a handle, is similarly shaped; it is somewhat simpler to use than the glass rod.

The inoculated plates are allowed to dry for 15 minutes before being placed inverted in the incubator.

(b) Surface cultivation on divided plates. An advantage of this method is that inocula from three dilutions can be examined on the one Petri dish. Each plate is divided into three sectors, and the amount of inoculum used is 0.02 ml per sector, dispersed from a commercial micropipette or from a laboratory pipette adapted for this purpose. The inocula are cautiously smeared over the surfaces of their respective sectors avoiding contamination of the neighbouring sectors. In this way, three Petri dishes can be used for three replicate inocula from each of three dilution levels (i.e. the amount of medium and number of Petri dishes required are reduced to one third). After 15 minutes during which the inoculum is absorbed by the medium, the plates are inverted and placed in an incubator.

(c) "Drop Plate Method". As with the use of divided Petri dishes, this method saves material, apparatus and time. It is suitable for the counting of any aerobic microbes (Thatcher and Clark, 1968).

The underside of a Petri dish is divided into three or more sectors, using plates with dry surfaces. Cells suspended in Ringer solution [115] of 1/4 strength serve as the inoculum. One drop (0·02 ml) per sector is dispensed from a special pipette, and dropped from a height of 3 cm with the pipette held vertically. Drops are released at the rate of one per second.

Individual colonies should develop separately from one another, the degree of dilution being adequate if 20–30 separate colonies develop from each drop.

The plates are allowed to stand at room temperature for 15 minutes, then they are inverted and incubated at 30 °C for 48 h.

In the calibration of the pipette, it should be noted that the volume of the drop formed at the tip of the pipette depends on the properties of the fluid, e.g. milk drops formed on the same pipette are 30% larger than water drops.

5.2.2.1.C   Enumeration of colonies

For this purpose, plates with between 30 and 300 colonies are most appropriate *(Table 18)*. Colony numbers are multiplied by the dilution factor and the viable count is expressed per unit volume or weight of the original material.

Table 18

Relationship between the error (at the 95% probability level) and the number of colonies counted in the pour-plate method (Cowell and Morisetti, 1969)

Number of colonies	Error ±%	Limits of error
500	9	455 – 545
400	10	360 – 440
320	11	285 – 355
200	14	172 – 228
80	20	80 – 120
50	28	36 – 64
30	37	19 – 41
20	47	11 – 29

The errors which may arise from the counting of colonies derive from two sources, namely from the dilution procedure and from the uneven distribution of colonies on the plate (Jennison and Wadsworth, 1940).

The error of dilution was discussed in 2.3.4.2.A. The error of colony distribution can be expressed by the standard deviation of replicate data, or by the standard deviation of the mean value (i.e. the standard error). The total error can be calculated as follows:

$$\text{total error (\%)} = \pm (\% \text{ error of dilution})^2 + \\ + (\% \text{ distribution error})^2.$$

Example. Let us suppose that the dilution is 1 : 10 and the volume of the inoculum 1 ml; the standard deviation of the pipetting volumes and inocula doses is 1%. Numbers of colonies developing on the plates:

Dilution	10^3	10^4	10^5	10^6
Number of colonies	> 300	290	35	4
per plate	> 300	250	20	2
(3 replicates)	> 300	280	33	0

166

For the 10^4 dilution, the error of the arithmetical mean of the colony numbers is calculated as follows. The error of dilution expressed as the standard deviation is $\pm 5.7\%$ (see *Table 2*); the error of distribution can be computed from the replicate data (in the formula, multiplication by 10^6 is omitted):

$$\pm \sqrt{\frac{2.9^2 + 2.5^2 + 2.8^2 - \dfrac{(2.9 + 2.5 + 2.8)^2}{3}}{3}} = \pm 0.173.$$

Expressed as a percentage of the mean value (2.73), the error of distribution is $\pm 6.3\%$. Thus, the approximate value of the total error, expressed as a standard deviation, is

$$\pm \sqrt{5.7^2 + 6.3^2} = \pm 8.5\%.$$

In the sense of the definition of standard deviation, this value tells us that the probability of the actual count being in the range limited by $\pm 8.5\%$ of the mean value, i.e. between 2.50 and 2.96, is 68%.

In taking the colony count, those plates on which the colony distribution is even and individual colonies are well separated are taken first of all. It should be noted whether at the lowest dilution the colony number is less than 30 per plate at the 100-fold dilution, e.g. less then $3 \cdot 10^3$ cells per ml in the stock suspension. Colony numbers greater than 300 at the highest dilution make possible counting of smaller areas of the plate; in such cases, if the colony number in an area as small as 1 cm² is less than 10, the colonies may be counted diagonally in 7 and 6 (total 13) areas of 1 cm². The arithmetic mean of the data thus obtained is then multiplied by the area of the plate (e.g. 65 cm²); this cell number then relates to the volume of the inoculum. If the colony number is greater than 10 per cm², counting of the colonies in 4 × 1 squares is satisfactory.

It must be borne in mind that the precision of the basic data from which the mean is derived may vary considerably among the data, because the data utilized in the calculation of each colony count are divided from plates inoculated with different dilutions, so that the numbers of colonies per plate also show considerable variation. It is thus reasonable to attach a weighting factor to the result for each dilution before taking the mean (Vas, 1972). Weighting may be done by means of the following formula:

$$m = \frac{d_1 (\Sigma c_1 + \Sigma c_2)}{n_1 + \dfrac{n_2 d_1}{d_2}}$$

where m = the wieghted mean, d_1 = dilution factor of the lowest countable dilution, d_2 = dilution factor of the next (higher) dilution, c_1 = the total number of colonies

167

counted from the lowest (d_1) dilution on n_1 plates, c_2 = the total number of colonies counted from the d_2 dilution on n_2 plates.

For aerobic microbes, the recommended period of incubation is shorter for surface cultures than for pour-plate cultures. The use of a magnifier lens or a microscope may shorten the required incubation time, e.g. from 3–4 days to 24 h. The counting of colonies may be facilitated by a calibrated mesh placed under the Petri dish, or by spotting the counted colonies with a waterproof marker. Counting may be further facilitated by a mechanical apparatus counting. An up-to-date method of colony-counting involves illumination of the Petri dish through an oblique glass plate. The user surveys the plate through a magnifier lens and labels each colony with a marker or touches each colony with a needle. The needle closes an electric circuit and the electric impulses are totalled on a register.

5.2.2.2 The membrane-filtration technique

Membrane filtration is used for the determination of microbe numbers in gases, fluids and solid substances. In the latter case, a homogeneous suspension must be prepared. This method makes possible the determination of total and/or viable in substances of very low cell count, since the microbial cells are concentrated on the filter. Cells collected on the membrane surface are stained and counted under the microscope. Alternatively, microbes can be cultivated on the surface of a solid medium where the colony number can be counted.

The pore size of the membrane filter is selected according to the size of the micro-organism. For microbiological purposes, membranes of 0·2–2·00 µm pore diameter are usually used. Filters that are not presterilized can be sterilized by autoclaving at 121 °C for 15–30 minutes. The most widely used types of membrane filters are listed in *Table 19*.

If the SM 125 00 membrane filter is used, contaminating particles can be removed without diminishing the count. This can also be achieved by sedimentation, or filtration through filter paper.

The filter head and the funnel in which the filter plate and cell suspension are placed, can be sealed under aseptic conditions. The funnel may be made of stainless steel, glass or plastic. All parts can be sterilized. The type of pump most widely used is the water-flow type. The volume of the filter is not critical, and the volume to be filtered may be introduced all at once or in portions. For microscopic counting, a cell density of between 1000 and 3000 per ml is most suitable, and, the filter plate should be 40–70 mm in diameter.

When all of the suspension has been filtered, about 10 ml sterile water or sterile diluent are added so as to wash the micro-organisms adhering to the wall of the funnel onto the surface of the filter.

168

Table 19

Types and characteristics of membrane filters (Millipore and Sartorius); Anon. (1969)

Millipore filters

Designation	Pore size, μm	Permeability	
		water*	air**
AA	0·8 ± 0·05	220	9·8
DA	0·65 ± 0·03	175	8·0
HA	0·45 ± 0·02	65	4·9
PH	0·30 ± 0·02	40	3·7
GS	0·22 ± 0·02	22	2·5

Sartorius filters

Designation	Pore size (a) μm	Permeability	
		water*	air***
SM 113 02	0·2	440	2·7
SM 114 05	0·8	150	0·65
SM 130 05	0·8	150	0·65
SM 114 06	0·6	65	0·4
SM 114 07	0·3	25	0·22
SM 125 00	10·0	4500	20·0

* $ml/min/cm^2$ with 700 Hg mm pressure difference.
** $litre/min/cm^2$ with 700 Hg mm pressure difference.
*** $litre/min/cm^2$ with 500 water column pressure difference. (a) Hagen–Poiseuille.

The microbes are then stained; first the stain (e.g. 0·01% aqueous methylene blue solution) and then the washing fluid are passed through the filter. The membrane filter is allowed to dry whereupon a small piece of it is placed on a slide, immersion oil is added, and the preparation is examined under the microscope.

Another staining method may be employed as follows. After filtration, the membrane filter is washed by passing through it 100 ml of a phosphate buffer, pH 5·2 (this is necessary if, for example, the suspension was prepared directly from a food sample), and a filtered gentian violet solution is pipetted onto the membrane. After 1 minute, the stain is drawn through with the pump, and then the filter is rinsed with 100 ml of phosphate buffer followed in sequence by 20 ml of propanol, 20 ml of a mixture of equal volumes of propanol and xylol, and 20 ml xylol.

In the colony-counting procedure, the membrane filter with the micro-organisms on it, is placed on the surface of an agar culture medium. No bubbles should be trapped between the membrane and the surface of the plate. To accelerate the diffusion of nutrients through the filter, it is advisable to use media of lower than normal agar concentration, or to wet the agar surface. Alternatively, a thick filter paper is placed in the Petri dish to take up about 2 ml of the medium, and the membrane filter is placed on this paper. The use of discs soaked in a nutrient medium further simplifies the method. 3 ml of sterilized distilled water are pipetted into a sterile Petri dish; this will be taken up by the disc that contains the nutrient. The membrane filter with the filtered microbes is then placed on the disc.

Of the Sartorius culture disc, SM 140 01 should be used in conjunction with the SM 117 07 filter for the counting of enterococci, SM 140 05 with the SM 114 07 filter for the total cell number, SM 140 05 with the SM 114 06 membrane for microbes only, SM 140 07 with the SM 114 07 filters for salmonellae, and the SM 140 08 dry nutrient pad should be used with the SM 130 05 filter for yeasts and moulds. The conditions of cultivation are set according to the special requirements of the microbes under investigation. Bacteria can be expected to form microscopically visible colonies after an incubation period of 4–5 h.

Cell and colony numbers are determined by the counting-chamber or colony-counting methods.

5.2.2.3   The roll-tube procedure

Owing to their simplicity, reproducibility and accuracy, colony-counting methods in Petri dishes are widely used in microbiological laboratories. However, in routine control studies carried out by manufacturers, still simpler procedures may suffice. These are sometimes less accurate, but are also less expensive than the Petri-dish methods. Media and cultures are less readily contaminated and are less disposed to drying out in these procedures. It is a further advantage of some of these methods that information about the microbiological state of the sample is obtained after a relatively short period of incubation.

Among methods of this type, the roll-tube procedure has gained the widest acceptance.

In essence, a layer of nutrient agar into which the sample is mixed, is incubated on the inner surface of a test-tube or a special vial closed with a rubber stopper. The procedure in its original form requires no special equipment. Recently, however, vials, rubber stoppers, rotators and colony-illuminators have been widely introduced as laboratory aids.

The special vials (Fig. 58) are 75 mm long, 25 mm in diameter at the bottom, and 13–14 mm at the neck. The inner surface area on which the medium is spread and solidified is about 2/3 of that of a standard Petri dish of 9 cm diameter.

The rubber stopper is ribbed, so that air and water vapour can escape from the vial during sterilization.

170

Rotation of the vial ensures that the melted culture medium, which has been mixed with the inoculum, coats the glass surface in an even layer. Seven vials can be placed in a rotator at the same time. While the vials are rotated, their outer surfaces are cooled with a water spray to bring about rapid solidification of the agar in a thin layer.

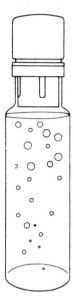

Fig. 58. Special vessel for the roll-tube technique of culturing

Procedure. The culture medium containing 2·5% agar is dispensed in aliquots of 4·0–4·5 ml into the special vials. The special rubber stoppers are placed loosely in the necks of the vials and the vials are sterilized in an autoclave. The vials are then closed tightly.

When the vials are used, the stoppers are loosened and the agar melted by placing the vials in boiling water and then cooling them in a water bath set at 45–47 °C.

A tenfold dilution series is prepared from the sample and 0·5 ml of each is transferred into two or three vials; by touching the tip of the pipette on the lower part of the neck of the vial, losses due to retention in the pipette can be prevented. The vials are then stoppered tightly.

An agar film is formed by rotating the vials, and the cultures are incubated with the vials inverted, i.e. with the stopper downwards, for 3 days at 30 °C (if the total count is to be determined). The colonies are counted using an illuminator, taking first of all those vials in which the colony number is between 10 and 200. The colony number on the agar surface is divided by two times the dilution factor to obtain the microbial count per ml.

The special vials may be substituted by ordinary test-tubes, in which case the agar film can be formed by holding the test-tubes under the tap and rotating them in the hands.

The roll-tube method gives somewhat lower counts than the Petri-dish method, nevertheless the overriding advantages of using the former method have been emphasized by the majority of authors. Compared with the pour-plate method, the roll-tube method is inexpensive (1/3–1/4 of the amount of solidified culture medium is required); it is more convenient because the culture medium can be prepared and dispensed in advance; the risks of contamination and drying out are negligible and post-infection in the inoculated vials is impossible. For the latter reason, the use of the roll-tube procedure finds particular favour in mobile laboratories.

5.2.2.4   The tube-plate method

The tube-plate method resembles the roll-tube method in both principle and performance. The only difference is that the thin layer of inoculated agar medium does not cover the entire inner surface of the vial, but forms a flat layer on one side, 7·5–10 cm long.

Aliquots of 1·5–2·0 ml of nutrient agar are dispensed into sterile test-tubes (16 × 160 mm). The liquid agar is maintained at 45 °C for 30–60 minutes, so that most of the condensate will run back into the agar. Then aliquots of the sample, and if necessary of dilutions prepared from it, are transferred by pipette into the still liquid nutrient agar, 0·1 ml to each vial. Smaller volumes are transferred in loops calibrated to carry a known volume (8.1.2.1. A). The agar is stirred and allowed to solidify in the tube in a nearly horizontal position. During incubation, the tubes are kept upright. The use of a magnifier lens greatly facilitates colony counting. According to the state of the original material, the optimal colony number is estimated to be between 30 and 300, but in our opinion, colony counts exceeding 100 or 150 are best avoided.

5.2.2.5   Deep agar culture

Deep agar cultures are suitable for obligate anaerobes and facultative anaerobes. Inocula from the sample or a dilution of the sample are transferred to test-tubes, each containing 10 ml of liquified agar precooled to 45 °C. The agar is thoroughly mixed by circular movements of the tubes in the vertical position; the agar is allowed to solidify in the vertical position. To ensure that conditions are anaerobic, a solution of reduced methylene blue is layered on the top of the medium. A blue colour in this layer indicates that conditions are not anaerobic.

The procedure is simple and quick to carry out, and the sample cannot be contaminated during cultivation. However, microbes with vigorous gas formation

172

cannot be cultivated in this way because the medium is disrupted by the gas. Another difficulty is that colonies growing deep inside the agar columns may go unrecognized. To eliminate these difficulties, some authors have introduced flat test-tubes, which are unfortunately difficult to clean. Alternatively, a sterile glass rod, possibly made of dark glass, is inserted aseptically with forceps. The glass rod should extend from the bottom of the tube to the methylene blue layer, so that thickness of the agar between the rod and the wall of the tube is 4–5 mm. The colonies growing in this zone are then easily observed.

5.2.2.6   The miniature plate method

This method requires a very small amount of culture medium and is less time-consuming than the Petri-dish method.

The inoculated preparations are incubated for a relatively short period and the colonies can be stained and counted before they are visible to the naked eye.

The procedure developed by Frost (1921) has been modified several times, but the principle of the test has remained the same.

Procedure. A sample of test material is mixed with an equal amount of sterile nutrient agar at 45 °C. The miniature plate is prepared on a slide maintained at 45 °C by pipetting a volume of the sample–agar mixture at this temperature on the slide and spreading it evenly over a well-defined area using a platinum needle.

The most widely used volumes and areas are 0·03 ml cm^{-2}, and 0·2 ml 10 cm^{-2}. The area over which the mixture is to be spread can be marked out by lines, or etched on the slide. (The diameter of a circle 1 cm^2 in area is 1·128 cm; an area of 10 cm^2 is usually formed by a rectangle 40×25 mm.)

An alternative procedure for the first two steps of the method is as follows. A quantity of the molten medium and a quantity of the suspension are placed on the slide, mixed with the platinum needle and spread. The slide is placed on a cooling surface to speed up solidification of the film. Then the slide is placed in a moist chamber with the agar layer facing upwards. The temperature and period of incubation are 28 °C and 16–20 h, respectively, or 32–35 °C and 12–20 h, respectively, where total counts are required. The number of psychrotrophs is determined after incubation at 7 °C for 4 days, or 17 °C for 4 h followed by further incubation at 7 °C for 44 h.

As a moist chamber, a sterile Petri dish (or other type of wide-mouth vessel) containing a little sterile water can be used. The slides are laid on glass rods to keep them above the water.

After incubation, the slides are dried on a hot-plate at 80–90 °C for 10–15 minutes and are then stained in acetic acid–thionine [30] for 1–5 minutes, immersed in water for 1–2 minutes, and dried again at 70–80 °C.

173

Colonies are counted at 90–100-fold magnification; they appear deep blue against a pale red background. It is advisable to examine a sufficient number of fields so that at least 400 colonies are counted, otherwise different parts of the agar film will be differently represented in the result. With samples spread over a surface of 2 cm², only the mid-part of the film is examined, since some drying out will have occurred near the edges.

The number of fields covering the entire area of the preparation can be calculated by dividing the area of the plate by the area of the field. To determine the cell count per ml of the sampled material, we multiply together the number of microcolonies per field, the number of fields equivalent to the entire area of the preparation, the degree of dilution of the suspension, and the reciprocal of the amount of the sample smeared on the plate.

Example

Area of miniature plate	1000 mm²
Area of field	3·08 mm²
Volume of sample spread over plate area	0·2 ml
Dilution factor	1·0
Average number of colonies per field	15

1000 : 3·08 = 324·7.

$$\text{Cell count per ml} = 15 \times 324·7 \times 1·0 \times \frac{1·0}{0·2} = 24{,}000.$$

It is sometimes possible to count unstained microbial colonies under the stereomicroscope at a magnification of × 20.

5.2.2.7   Other methods of microcolony counting

In spite of its advantages, the miniature plate method has not been widely introduced because microscopic colony counting is tedious, and injurious to the eyesight. Tolle et al. (1968) developed an electronic colony-counting method which was originally intended for grading milk, but which is finding application in other fields as well.

The method is based on the fact that micro-organisms growing in a solid culture medium under well-defined conditions of incubation develop microcolonies occupying a limited volume. The microcolonies are fixed, the culture medium is melted and suspended in an electrolyte, and the colonies can be enumerated as separate particles. In this method, Koch's classical pour-plate technique and the miniature-plate technique are combined with electronic particle-counting.

174

Procedure. 0·02 ml milk is added to 10 ml of nutrient gelatin [139] (i.e. the milk is diluted 500-fold). When thoroughly mixed, the gelatin is allowed to solidify in a test-tube. The surface is topped with 1 ml of nutrient gelatin and a plug inserted in the test-tube. Incubation is carried out at 21 °C for 20 h, or in the case of psychrotrophs, at 6 °C for 120 h.

After incubation, 2 ml of a formalin–hydrochloric acid mixture (4 parts 35% formalin to 1 part 5 N hydrochloric acid) are poured on the gelatin. The gelatin is melted in a water bath at 30 °C and 7 ml formalin–electrolyte (0·9% NaCl and 1% formaldehyde in aqueous solution) are added to the liquified medium. The test-tubes are removed from the water bath, and the microcolonies are shaken in the electrolyte and counted electronically in a Coulter Counter, a Partikelzähler ZG 1, or a Granulometer ZG 2, both of the latter being produced in the German Democratic Republic. The capillaries used are 200 μm in diameter, and the particle-counter is adjusted to count every particle larger than 6000 μm².

Tolle et al. (1968) and Johst (1970) claim the following advantages of electronic microcolony counting over the classical pour-plate method: (a) reliable results are obtained in considerably less time, (b) the error of dilution is eliminated or greatly diminished and (c) the procedure is less laborious. Even the counting of psychrotrophs, which are gaining increasing importance in the grading of milk, is considerably quicker because the growth of suitably sized microcolonies takes less time.

Electronic microcolony counting, however, becomes unreliable when milk samples of low microbial count (< 50,000 or 100,000 per ml) are examined. In such cases, it is necessary to mix a larger volume of milk with the gelatin, which then tends to coagulate. Milk samples of high microbial count are easily graded if appropriately diluted.

5.2.2.8   Burri's procedure

In many of the methods outlined above, materials of high microbial count are diluted with sterile liquid, and we use those dilutions in which the probable cell count per ml are between limits of determinable cell count of applied method. This, however, requires much in the way of work, materials, and expensive equipment. In order to simplify matters, methods have been evolved in which instead of using high dilutions, very small volumes, *viz.* 0·1, 0·01 or 0·001 ml, or 1-mg amounts of the undiluted material, are used for inoculating the culture medium. It is easy to measure portions of 0·1 ml by pipetting, but measuring doses of 0·01 ml or 0·001 ml requires specially calibrated loops.

The amount of material taken up in a loop depends on the composition and surface tension of the material. For this reason, the loop must be specially calibrated to take up a desired volume of a particular material. According to Burri, for example, a loop taking up 0·001 ml of milk should be made of wire 0·38 mm in thickness, with an inner loop diameter of ~1·50 mm; 0·01 ml of milk can be sampled by a loop with an inner diameter of 4 mm. The end of the wire forming the loop should just touch the neck of the loop, so that the loop is closed. The procedure first described by Burri (1928) is based on the handling of minute quantities.

Procedure. About 4 ml of nutrient agar are sterilized in a test-tube. The tube is layed in a nearly horizontal position to form an agar surface 8–10 cm long; the solidified agar is then dried by supporting the tube in an inverted position (with the cotton plug downward) for two days at 37 °C in an incubator.

Using the calibrated loop, 0·001 ml of liquid is transferred to the agar by touching the agar surface at three points and smearing the sample evenly by rapid movements. The culture is incubated at 30 °C for 48–72 h after which the colonies are counted.

The colony number must be less than 100 to be properly evaluated. Since 0·001 ml was spread, the result must be multiplied by 1000 to give the colony count per ml.

If the microbial count of the original material exceeds 100,000 per ml, it need not be diluted, whereas raw milk samples must be diluted 1 : 1000 or 1 : 100 before they are examined.

The procedure is simple, inexpensive and rapid. It can be applied equally to anaerobes and aerobes, and a further advantage is that the sample cannot be contaminated during incubation.

However, one disadvantage of the method is that the small amounts tested may not be representative of the whole sample. Thorough homogenization of the sample is, therefore, particularly important. Inaccurate calibration of the loop, unsatisfactory sampling, and uneven spreading of the inoculum may result in further error. It should be noted that the volume of the liquid taken up in the loop depends on the angle at which the loop is drawn out of the liquid; thus the loop should not be immersed beyond its neck and should be drawn out vertically. It should also be borne in mind that the demands of microbes growing on a dry agar surface are not the same as those of organisms growing in poured plates.

Obviously, Burri's method is inferior to the pour-plate method both in terms of reproducibility and accuracy. Nevertheless, it is an acceptable method for control tests, e.g. for testing the efficacy of heat treatment, for detecting contamination, for carrying out production phase examination, etc.

The main principle of the Burri procedure, i.e. using as sinoculum a minute amount of liquid held in a loop, has recently been applied to the pouring of plates. Two alternative methods have been described.

(a) 1 ml saline [45] is pipetted into each Petri dish. The inoculum taken with Burri's 0·001-ml loop (or a 1 mg sample) is distributed evenly in the saline. Liquid agar is poured into the dish.

Samples of low microbial count should not be diluted. If a high microbial count is

176

expected, the sample must be diluted, usually tenfold in the first instance. The colony count thus obtained is multiplied by the degree of dilution, and the product by 1000 to give the cell count.

(b) Thompson and Black (1967) washed the sample into a Petri dish with sterile liquid, using a syringe attached to the loop.

5.2.2.9   The Bacto strip method

This is a rapid, approximative method, which can be carried out without any microbiological laboratory equipment. Cells are counted by means of absorbent paper strips containing absorbed nutrient agar, which when immersed in a sample of the material being investigated, becomes inoculated. After an appropriate period of incubation, colonies are made visible with an indicator.

The method is used primarily for the detection of coliforms. The absorbency of the paper and the amount of the medium absorbed are predetermined according to the aim of the test.

The method requires no special equipment, except for the sterile strip of paper soaked in culture medium and ready prepared in a plastic bag. Each bag contains a strip divided into two unequal parts by perforations. The larger part serves for the test itself (the given absorbency refers to this part), and the smaller part facilitates sterile manipulation of the paper.

Procedure. The plastic bag is opened with scissors, and the paper strip is gripped over the perforations; it is taken out of the bag and immersed in the liquid under investigation. The strip is withdrawn and shaken gently to remove any droplets adhering to the paper. Teh bag is opened by pressing the edges inwards, and the strip is immediately replaced taking care to prevent contact with the bag until the entire length of the strip is inside the bag. The smaller portion is detached at the perforation and discarded. The bag is finally sealed by drawing the open end through a Bunsen flame, a spirit or a lighted match.

The conditions of cultivation (temperature, incubation period) depend on the microbes that are most likely to be present; in any case the incubation takes less time compared with pour-plates. The colonies are indicated by minute coloured points. Where strips with an absorption capacity of 1 ml have been used, colonies are counted on both sides; strips of 0·1 ml absorbing capacity are examined on only one side. If the latter type of strip is used, the result must be multiplied by 10.

An advantage of the method is that a quantitative result can be obtained in a relatively short time and without special equipment. Paper strips soaked in culture medium can be prepared in the microbiological laboratory, or commercial products can be used. The procedure itself is simple, and does not call for any special professional skill.

However, since this method is less reproducible, it is not equivalent in any way to the classical procedures. Its application is justified if the conditions necessary for carrying out more accurate procedures are lacking, and if a simple routine method without the need for high accuracy serves the purpose.

5.2.3 Comparative evaluation of the dilution and colony-counting methods

In order that the results of cell counting procedures are fully comparative, the results should be reported together with the method used, the medium of cultivation and the conditions of incubation.

Before choosing the method, the advantages and disadvantages of dilution methods and colony-counting methods should be considered as they would be applied in a particular case.

Usually, the Petri-dish method is preferable because it is more reproducible and the results are more reliable. In addition, a qualitative examination of colonies growing on Petri plates is easy to carry out.

Results obtained using the dilution techniques are less accurate, even if 5 replicate tubes are inoculated. Nevertheless, the use of this method is justified in many cases because:

— some micro-organisms grow poorly in solid media and cannot be examined with respect to production of metabolites, e.g., acid and gas;

— the cultivation and enumeration of anaerobes in nutrient broth may be simpler than in an agar medium.

The pour-plate method cannot be recommended for the detection of low count of micro-organisms; if the colony count per Petri dish is less than 30, the errors of the procedure are unacceptably high. In such cases, the sample volume pipetted into the Petri dish can be increased, but even by this means, the method cannot be taken beyond the lower limit of less than 1 microbe per ml of sample. By using the dilution method, on the other hand, samples as large as 100 ml can be examined, i.e. the presence of a single microbe in 100 ml sample can be detected (0·01 cell/ml).

For this reason, moderate faecal contamination of water is preferably examined by the dilution technique, and in such cases, double- or triple-strength media are used to counterbalance the diluting effect of adding the sample.

5.3 Determination of cell count by measuring biochemical activity

The determination of total cell counts and the detection and estimation of specific groups of micro-organisms may be based on metabolic tests. Since, however, the classical chemical tests are of poor sensitivity, quantitative changes in metabolites are not indicated in these tests unless the material is very heavily contaminated with

178

microbes producing or decomposing the metabolite under study. For example, a decrease in the pH of milk by 0.4–0.9 SH (i.e. the production of 0.01–0.02% lactic acid) indicates that at least 5–20 million lactobacilli are present per ml. It is thus desirable to have available highly sensitive methods for the examination of specific metabolites produced by microbes.

In some fields of microbiology, the nonspecific enzyme activities of microbes, or the activities characteristic of particular species, are examined. From the enzyme activity, the cell count can be estimated. Results based on these enzyme activities are less precise than results obtained by more complicated procedures, yet because they are simple, inexpensive and rapid, such enzyme tests are becoming more and more widely used. Recently, grading procedures based on these methods have been introduced in the meat industry and the canned food industry.

The activities of the nonspecific enzyme systems of microbes can be related to the total cell count. Measurement of the activity may be based on changes brought about by the micro-organisms in the redox potential of the medium. Using redox indicators, i.e. dyes indicating changes in the redox potential by a colour change, we can estimate the reducing effect of the microbes present in the sample, either from the degree of colour change observed after a given time, or from the time necessary for a given colour change to occur.

Methylene blue, resazurine, and 2, 3, 5-triphenyltetrazolium chloride (TTC) are the most commonly used redox indicators. Methylene blue is blue in its oxidized form and colourless in its reduced form. The blue resazurine is converted into pink resorufine as the result of an irreversible process, and the latter is reduced to dihydroresorufine in a reversible reaction. Thus the colour of the medium (blue, lilac, pink, colourless) reflects the value of the redox potential of the medium, TTC in its oxidised form is colourless, and red in its reduced form (formazan).

In general, the more time required for the reduction of a dye, the lower is the microbial count of the sample. However, the result may be influenced by a number of other factors connected with the properties of the material under test, the growth medium used, and the type of microbe present. The bacteria must be placed in an environment conducive to continued metabolic activity during the test, and, therefore, a bacterial culture medium may need to be used as a diluent.

Well-defined groups of micro-organisms can be examined by measuring the activities of enzyme systems that are exclusively characteristic of the organisms, e.g., coliforms are capable of reducing nitrate to nitrite; since lactobacilli do not possess nitroreductase, the reduction of nitrate added to milk indicates the presence of a contaminating flora. The rate of nitrite formation is positively correlated with the degree of contamination.

The details of the determination of cell counts from measurements of enzyme activity are described in Chapter 8.

6. DESCRIPTION AND IDENTIFICATION OF SOME IMPORTANT MICRO-ORGANISMS OCCURRING IN FOODSTUFFS

6.1 Pathogenic and toxin-producing microbes

The detection of pathogenic and toxin-producing microbes, including conditionally pathogenic microbes is of the greatest importance. The International Commission on Microbiological Specifications for Foods (ICMSF) has published, in agreement with the World Health Organization, a list of pathogenic toxin-producing microbes and parasites the presence of which should always be considered when foodstuffs are examined.

(1) Highly pathogenic organisms:
   Bacillus anthracis
   Clostridium botulinum
   Vibrio cholerae
   Salmonella typhi
   Shigella dysenteriae
   Brucella melitensis
   Hepatitis A virus
   Fungi producing mycotoxin
   Trichinella spiralis
   Echinococcus

(2) Organisms presenting a moderate health hazard:
   Clostridium perfringens
   Bacillus cereus
   Vibrio parahaemolyticus
   Salmonella typhimurium
   Other Salmonella spp.
   Shigella sonnei
   Yersinia enterocolitica
   Pseudomonas aeruginosa
   Staphylococci producing enterotoxin
   Taenia spp.
   Anisakis spp.

Table 20

The common types of fruit spoilage and the organisms causing them (Compiled from data published by Tomkins, 1951; Mossel and Ingram, 1955; Árokszállásy et al., 1968; Nickerson and Sinskey, 1972; Müller, 1974)

Type of spoilage	Micro-organisms causing the spoilage	Pome-fruits	Bananas	Citrus	Stone-fruits	Date	Fig	Raspberry	Strawberry	Vine
Anthracnose	*Colletotrichum lindemuthianum*		×	×						
Brown soft rot	*Monilia* sp. (*Sclerotinia* sp.)	×		×	×					
Brown rot	*Penicillium expansum*	×								
White mould	*Phytophthora cactorum*	×		×					×	
Black mould	*Aspergillus niger*		×	×	×	×				×
Black rot	*Alternaria* sp., *Physalospora* sp.	×		×				×		
Blue-mould rot	*Penicillium italicum*	×		×						
Bitter or yellowish-brown rot	*Gloeosporidium album*, *G. fructigenum*	×								
Core rot	mainly *Fusarium avenaceum*	×								
Moist or soft rot	*Rhizopus nigricans* in grapefruit: *Alternaria citri*			×	×				×	×
Grey-mould rot	*Botrytis cinerea*	×			×				×	×
Scabbing	*Endostigme* sp. (*Fusicladium* sp.)	×								
Green rot (green-mould rot)	lemon: *Penicillium digitatum* apple: *P. expansum*	×		×	×					

181

Table 21

Common types of spoilage of vegetables and the organisms causing them
Árokszállásy et al., 1968; Nickerson and Sinskey, 1972)

Type of spoilage	Micro-organisms causing the spoilage	Potato	Melon	Garlic
Anthracnose	*Colletotrichum lindemuthianum* lettuce: *Marssonina*			
Bacterial soft rot	*Erwinia carotovora, E. atroseptica,* *Pseudomonas* sp., *Xanthomonas* sp.	×	×	×
White-mould rot	potato, melon: *Phytophthora* sp. onion: *Peronospora* sp.	×	×	
White rot	*Sclerotium cepivorum*			
Black-mould rot	*Aspergillus niger*			
Black rot	*Alternaria* sp.; potato: *A. solani;* cabbage: *A. brassicae* cucumber: *Mycosphaerella citrillina*	×	×	
Fusarium rot	*Fusarium* sp.	×	×	
Cladosporium rot	*Cladosporium* sp.		×	
Blue-mould rot	*Penicillium* sp.			×
Powdery mildew	peronospora effusa (farinosa)			
Moist rot	*Pythium ultimum*	×		
Oidium rot	*Geotrichum* sp			
Rhizoctonia rot	*Rhizoctonia crocorum*			
Rhizopus soft rot	*Rhizopus* sp.		×	
Lettuce peronospora	*Bremia*			
Septoria rot	*Septoria apiicola*			
Sclerotinia rot	*Sclerotinia sclerotiorum*			
Grey-mould rot	*Botrytis cinerea, Mucor* sp.;onion: *B. allii*			
Rhizopus rot	*Rhizopus stolonifer* (*Rh. nigricans*)			

182

(Compiled from data published by Tomkins, 1951; Mossel and Ingram, 1955;

Root vegetables	Leguminous plants	Cruciferae	Leaf-vegetables	Green paprika	Tomato	Lettuce	Asparagus	Cucumber	Leaves of celery Leaves of parsley	Onion
	×				×	×		×		
×	×	×	×	×	×	×	×	×	×	×
					×					×
										×
										×
×		×		×	×					
					×		×			×
		×		×						
										×
			×							
					×					
×	×	×					×			
		×		×	×					
						×				
			×						×	
×	×	×				×		×	×	×
×		×		×	×				×	×
			×							

183

(3) Other pathogens:
   Haemolytic staphylococci and streptococci
   Streptococcus pneumoniae
   Genus Pasteurella
   Arizona spp. of Salmonella subgenus III
   Genus Escherichia (in newborns)
   Erysipelothrix rhusiopathiae
   Genus Listeria
   Corynebacterium pyogenes

(4) Conditionally pathogenic micro-organisms:
   the streptococcal spp. of the viridans groups.
   Faecal streptococci of the Lancefield D group (enterococci)
   Genus *Escherichia* (in adults)
   Genus *Proteus*
   Genus *Pseudomonas* (except *P. aeruginosa*)
   Genus *Clostridium* (except *C. botulinum* and *C. perfringens*)
   Non-haemolytic staphylococcal spp.
   Aerobic spore-formers (except *B. cereus*)

6.2   Micro-organisms causing food spoilage

Organisms responsible for the spoilage of raw foods are listed in *Tables 20–22*.

The spoilage of raw foods such as meat, poultry and fish kept in cold storage is caused mainly by cold-tolerant bacteria, fruits and vegetables spoil mainly by moulds and yeasts. Vegetables are often spoiled by bacteria also (*Pseudomonas*, *Xanthomonas* and *Erwinia* spp.). The typical spoilage of grains is moulding, most frequently caused by *Alternaria* spp., *Fusarium graminearum*, *Helminthosporium* spp., *Cladosporium* spp., *Aspergillus* spp., *Penicillium viridicatum*, *P. cyclopium*, *Chaetomium* spp. and *Sordaria* spp.

The spoilage of sliced or minced vacuum-packed meats may be caused by *Lactobacillaceae* spp. These, although they do not belong to the usual microflora of fresh meat, being facultatively anaerobic, may predominate over the aerobic *Pseudomonas (Acinetobacter) Alcaligenes* group under such circumstances.

The important micro-organisms causing spoilage of processed foods are listed in *Table 23*.

Owing to the heat resistance of their spores, spore-forming bacteria are usually the main organisms responsible for the spoilage of canned food. Yeasts, non-sporing bacteria *(Lactobacillus* and *Leuconostoc)* and moulds do not attack canned foods unless there is leakage or the heat treatment is inadequate.

184

Table 22

Micro-organisms causing the spoilage of raw foodsuffs of animal origin (Compiled from data published by Niven, 1951; Mossel and Ingram, 1955; Hammer and Babel, 1957; Ayres, 1960; American Meat Institute Foundation, 1960; Pelczar and Reid, 1965; Árokszállásy et al., 1968; Nickerson and Sinskey, 1972

Foodstuff	Type of spoilage	Micro-organism
Poultry	surface slime, bad smell	*Pseudomonas* spp. producing no pigment
		P. fragi
		P. putida
		P. ambigua
		Pseudomonas spp. producing pigment
		Pseudomonas fluorescens
		Xanthomonas spp.
Fish	putrid spoilage	*Pseudomonas putrefaciens*
		P. fragi
		Achromobacter spp.
	discolouration	*Flavobacterium* spp.
		Pseudomonas spp. producing pigment
Raw meat under aerobic conditions	surface slime and bad smell	*Pseudomonas* spp. producing no pigment
		P. fluorescens
		Acinetobacter (Achromobacter)
		Clostridium spp.
		Proteus vulgaris
	moulding after prolonged cold storage	*Mucor* spp.
		Clostridium herbarum
		Sporotrichum carnis
		Thamnidium elegans
		Rhizopus spp.
		Penicillium expansum
		Aspergillus glaucus
		Yeasts
Raw meats under anaerobic conditions	putrid sour fermentation rot	Proteolytic clostridia producing butyric acid
		Coliforms
		Clostridium spp.
		Facultative anaerobic bacteria
Raw milk	"sweet clotting" foamy (gaseous) fermentation	*Streptococcus lactis* var. *anoxyphilus*
		Coliforms
		Some clostridia
		Candida spp.
	bluing "ropiness"	*Pseudomonas syncyanea*
		Streptococcus lactis var. *hollandicus*
		Micrococcus spp.
		Alcaligenes viscolactis
	souring yellowing off-odour	*Streptococcus lactis*
		Pseudomonas fluorescens
		P. fluorescens (*Bac. fluorescens liquefaciens*)
		P. fragi
		"*Achromobacter*" spp.

Table 22 (cont.)

Sweet cream	yellowing	*Pseudomonas synxantha*
Egg	black rot	*Proteus* (*P. melanovogenes*)
		Alcaligenes spp. (rarely)
		Serratia spp. (rarely)
	bluing	*Pseudomonas syncyanea*
	reddening	*Serratia marcescens*
		Rhodotorula glutinis
	mouldy rot	*Penicillium* spp.
		Mucor spp.
		Cladosporium
		Sporotrichum
	pink rot	*Pseudomonas* spp. producing nonfluorescing pigment
	yellowing	*P. synxantha*
	colourless rot	*Pseudomonas* and *"Achromobacter"* spp. producing no pigment
	green rot	*Pseudomonas fluorescens*
		P. ovalis

186

Table 23

Micro-organisms causing the spoilage of commercial food products (Compiled from data published by Gray, 1959; Pelczar and Reid, 1965; Nickerson and Sinskey, 1972; Müller, 1974)

Foodstuff	Type of spoilage	Micro-organism
Wine	acetification	*Acetobacter orleanse* *A. ascendens, A. xylinum*
	cabbage-pickle taste, pungent taste	*Lactobacillus fermenti* (*Bact. intermedium*), (*Bact. gayonnii*) *Lactobacillus buchneri* (*Bact. mannitopecum*)
	ropiness	lactic acid bacteria causing mucosity *Aureobasidium pullulans* yeasts
	surface skin formation	*Candida mycoderma* *Pichia membranaefaciens*
	turbidity, sediment formation	*Saccharomyces cerevisiae* (alcohol-tolerant races of wine yeast)
Chocolate	swelling	yeasts (*Sacch. rouxii, Brettanomyces bruxellensis*) *Clostridium* spp.
	surface moulding	*Aspergillus* spp. *Penicillium* spp.
Fruit juices	fermentation, off-odour	*Byssochlamys* spp. *Penicillium* spp. *Phialophora* spp. *Botrytis* spp. *Monascus* spp. *Rhizopus* spp. *Saccharomyces* spp. *Hansenula* spp. *Pichia* spp. *Torulopsis* spp. *Acetobacter* spp. *Lactobacillus* spp.
Cocoa butter	off-odour	*Penicillium expansum* (*P. glaucum*) *Aspergillus oryzae*
Bread	moulding	*Rhizopus stolonifer* (*R. nigricans*) *Penicillium* spp. *Aspergillus glaucus* *Aspergillus niger* *Neurospora sitophila* *Mucor mucedo* *Mucor pusillus* *Aspergillus flavus* *Aspergillus flavus-oryzae*
	ropiness, off-taste	*Bacillus subtilis* (mucus-producing variants: *B. mesentericus fucus, B. panis*)
	chalk illness	*Trichosporon variabile* (*Monilia variabilis*) *Endomycopsis fibuliger*

Table 23 (cont.)

Foodstuff	Type of spoilage	Micro-organism
Condensed milk	lump formation	*Aspergillus repens*
Flour	moulding	*Rhizopus stolonifer* (*R. nigricans*) *Fusarium* spp. *Aspergillus repens* *Aspergillus flavus* *Aspergillus tamarii* *Aspergillus niger* *Penicillium nigricans* *Mucor* spp. *Syncephalastrum* spp.
Pasteurized milk	sweet clotting after prolonged storage souring	*Pseudomonas fragi* lactic-acid streptococci
Cured cooked meats, meat products	slime formation souring greening moulding	*Leuconostoc* spp. *Lactobacillus* spp. *Streptococcus* spp. *Microbacterium thermosphactum* yeasts *Streptococcus faecium* *S. faecalis* *Microbacterium thermosphactum* *Leuconostoc* spp. *Lactobacillus viridescens* *Pediococcus* spp. streptococci (enterococci) *Aspergillus* spp. *Rhizopus* spp. *Penicillium* spp.
Cheese	surface slime	*Alcaligenes metalcaligenes*
Ale	turbidity and off-taste ropiness souring	*Pediococcus cerevisiae* (*Sarcina cerevisiae*) coliforms *Zymomonas anaerobia* *Saccharomyces pastorianus* *Lactobacillus brevis* (*L. pentoaceticus*) *Acetomonas oxydans* var. *viscosum,* *A. oxydans* var. *capsulatum*
Dried fruit	white spot formation on the surface, alcoholic or yeasty taste, sometimes sour or bitter taste	*Zygosaccharomyces* spp. *Hansenula* spp. *Saccharomyces* spp. *Debaryomyces* spp. *Zygopichia* spp. *Aspergillus glaucus* group

188

Table 23 (cont.)

Carbonated soft drinks	fermentation	yeasts lactic-acid bacteria
Butter	hydrolytic rancidity	*Pseudomonas fragi* *P. fluorescens* *Serratia marcescens* *Achromobacter lipidis* *Geotrichum* spp. *Cladosporium butyris*
	blackening, gray-green spots, decomposition, fishy taste, yeasty taste	*Pseudomonas nigrifaciens* *Penicillium* spp. *Alternaria* spp. *Oidium lactis* *Torula cremoris* *Torula sperica*

Table 24

Spore-forming bacteria mainly responsible for the spoilage of canned food, and the heat-tolerance of their spores (Ingram, 1969)

pH of the food	Spore-formers causing spoilage	Heat tolerance of spores (D value, min) at 120 °C	Heat tolerance of spores (D value, min) at 100 °C	Temperature range of intensive vegetative reproduction °C

>4·5	*Clostridium*
	  thermosaccharolyticum
	C. nigrificans
	Bacillus
	  stearothermophilus
		3–4
		2–3
		
		4–5
			
			3000
				35–55

	Clostridium
	  sporogenes
	C. botulinum
	  A and B
	Bacillus
	  *licheniformis**
	*B. subtilis**
		0·1–1.5
		
		0·1–0·2
			
			13
			11
				10–40

4·0–4·5	*Bacillus coagulans* (thermacidurans)	0·1		35–55
	Clostridium butyricum *C. pasteurianum* *Bacillus macerans* *B. polymixia*		0·1–0·5 0·1–0·5 0·1–0·5 0·1–0·5	10–40

* Mainly in cured meat products.

The spore-forming bacteria principally responsible for the spoilage of canned food, and their main properties are listed in *Table 24*.

Canned fruits are sometimes spoiled by surviving ascospores of *Byssochlamys fulva* or by the sclerotia of *Penicillium* spp.

6.3   Bacteria

6.3.1   Enterobacteriaceae

6.3.1.1   Criteria of classification

The members of the family *Enterobacteriaceae* are Gram-negative, non-sporing, rod-shaped bacteria. Those having no flagella are non-motile, while the peritrichous groups show vigorous movement. They ferment glucose rapidly with gas formation, reduce nitrate to nitrite, give a negative response to the oxidase test, and grow on the commonly used culture media.

Since all members of the family are closely related, the differentiation of genera and species requires the use of biochemical and serological tests. Moreover, there are intermediary strains which are not readily identifiable on the basis of the usual tests. The tribes within the family have been distinguished according to biochemical properties common to all members of the tribe, whereas the genera within each main group have been classified on the basis of biochemical reactions specific to the genus. Within genera, 0 antigen groups can be differentiated, and the species can be identified according to further biochemical reactions, and serological and phage typing.

6.3.1.2   Tribes, genera and species

The biochemical properties of the tribes are summarized in *Tables 25* and *26*.

The biochemical properties that distinguish the genera of tribes are shown in *Tables 27–33*.

The presence of members of the family Enterobacteriaceae is used as an indicator microflora in food-microbiological investigations.

190

Table 25

Tribes and genera of the family *Enterobacteriaceae* according to Edwards and Ewing (1972)

TRIBES	GENERA
Escherichiae	*Shigella Escherichia*
	(*Escherichia coli* including Alkalescens–Dispar)
	Edwardsiella
*Salmonellae**	*Salmonella*
	Arizona
	*Citrobacter** (including Bethesda–Ballerup)
Klebsiellae	*Klebsiella*
	Enterobacter (including *Hafnia*)
	Aerobacter
	Serratia
	Pectobacterium
Proteae	*Proteus*
	Providencia

* Old name: *Escherichia freundii.*

Table 26

Classification of the family *Enterobacteriaceae* according to biochemical reactions (Edwards and Ewing, 1972)

Biochemical tests	*Shigella–Escherichia*	*Salmonella–Arizona–Citrobacter*	*Klebsiella–Enterobacter–Serratia*	*Proteus–Providencia*
Indole production	d	–	–	d
Methyl-red reaction	+	+	–	+
Voges–Proskauer reaction	–	–	+	–
Growth in ammonium–citrate medium	–	+	+	d
H_2S production	–	+	– or (+)	d
Decomposition of urea	–	–	– or (+)	d
Growth in Müller's KCN medium	–	d	+	+
Phenylalanine-deaminase test	–	–	–	+

Symbols: + = 90% of the strains *positive* within 1 or 2 days; – = 90% of the strains *negative;* (+) = late positive; d = various biochemical types.

Note: H_2S formation is variable at *S. typhi, S. paratyphi-A* and a few rare species. *Proteus vulgaris* and *P. mirabilis* may give a positive Voges–Proskauer reaction if incubated at 25 °C.

Table 27

Differentiation within *Shigella–Escherichia*

Biochemical tests	Genus Shigella			Genus Escherichia
	A and B	C	D	
	subgroup			
Motility	−	−	−	+(−)
Lysine–decarboxylase	−	−	−	d
Arginine–dihydrolase	−	− or (+)	−	d
Ornithine–decarboxylase	−	−*	+	d
Christensen's citrate	−	−	−	+(−)
Decomposition of mucid acid	−	−	−	+(−)
Fermentation of glucose (gas)	−**	−	−	+(−)
Decomposition of lactose	−	−	(+)	+(−)
Decomposition of salicin	−	−	−	d (60% +)

Symbols: * = some cultures of *Shigella boydii* are positive; ** = certain biotypes of *Shigella flexneri* produce gas; + = positive within 1 or 2 days; − = negative; +(−) = most strains positive, negative variants may occur; (+) = late positive; d = different biochemical types may occur.

Note: Typical *E. coli* strains are easily differentiated from *Shigella*. Non-motile *E. coli* variants, some of which belong to the Alkalescens–Dispar group, require more detailed investigation. The biochemical tests aimed at differentiating *E. coli* from *Shigella* should be considered as a whole; Shigellae are much less active biochemically than *E. coli* strains. The strains decomposing many carbohydrates with acid formation within 24 h (maltose, rhamnose, xylose, sorbitol and dulcitol) definitely do not belong to the genus *Shigella*.

Table 28

Differentiation within *Salmonella–Arizona–Citrobacter*

Biochemical tests	Salmonella	Arizona	Citrobacter
Fermentation of malonate	−	+	d
Lysine–decarboxylase	+*	+	−
Arginine–dihydrolase	(+)**	(+)	(+)
Oenithine–decarboxylase	+**	+	d
Growth in KCN medium	−	−***	+
Liquefaction of gelatin	−	(+)	−
Decomposition of lactose	−	+ or (+)	+ or X
Decomposition of dulcitol	+	−	d

Symbols: * = *S. paratyphi* A: negative; ** = *S. typhi* and *S. gallinarum*, negative; *** = Arizona 021 positive; + = positive in 1 or 2 days; − = negative; (+) = late positive; X = late and irregularly positive; d = different biochemical types.

Note: Most of the *Salmonella* strains ferment dulcitol immediately. *S. typhi*, *S. pullorum*, *S. paratyphi*, *S. cholera-suis* and certain other *Salmonella* species do not ferment dulcitol. None of the members of the genus *Arizona* ferment dulcitol.

192

Table 29

Differentiation within the group *Salmonella–Arizona–Citrobacter* on the basis of decomposition of organic acids

Biochemical tests	Salmonella	Arizona	Citrobacter
Decomposition of mucid acid	+	−(+)	+
Decomposition of D-tartarate	+	−	−
Citrate utilization	+	−	−
Decomposition of malonate	−	+	−

Symbols: + = positive within 20 h; − = negative; −(+) = the majority of the strains are negative; positive strains may occur.

Note: The reactions included in this Table are characteristic of the great majority of the strains belonging to the respective genera. Strains behaving differently may occur.

Table 30

Differentiation within *Klebsiella–Enterobacter–Serratia* by biochemical tests

Biochemical tests	Klebsiella	Enterobacter cloacae subgroup	Enterobacter aerogenes subgroup	Hafnia 37 °C	Hafnia 22 °C	Enterobacter liquefaciens 37 °C	Enterobacter liquefaciens 22 °C	Serratia
Fermentation of glucose with gas	+	+	+	+	+	+	+	d
Fermentation of inositol with gas	+	−	+	−	−	+	+	−
Fermentation in Stern's broth with gas	+	−	+	+	+	d	(+)	−
Fermentation of cellulose with gas	+	+	+	+	+	d	(+)	−
Decomposition of sorbitol	+	+	+	−	−	+	+	+
Decomposition of raffinose	+	+	+	−	−	+	+	−
Lysine–decarboxylase	+	−	+	+	+	d	+	+
Arginine–dihydrolase	−	+	−	−	−	−	−	−
Ornithine–decarboxylase	−	+	+	+	+	+	+	+
Liquefaction of gelatin	−	(+)	X		−		+	+
Motility	−	+	+	d	+	+	+	+

Symbols: + = positive in 1 or 2 days; − = negative; (+) = late positive; X = late and irregularly positive; d = different biochemical types.

Table 31

Differentiation of *Enterobacter liquefaciens* from strains belonging to the genera *Hafnia* and *Serratia*

Biochemical tests	Enterobacter liquefaciens		Hafnia		Serratia
	37 °C	22 °C	37 °C	22 °C	
Fermentation of glucose with gas	+	+	+	+	d
Fermentation in Stern's broth with gas	d	(+)	+	+	−
Fermentation of inositol with gas	+	+	−	−	−
Fermentation of cellulose with gas	d	(+)	+	+	−
Decomposition of sorbitol	+	+	−	−	+
Decomposition of raffinose	+	+	−	−	−
Decomposition of ramnose	−	−	+	+	−
Liquefaction of gelatin		+		−	+
Fermentation of arabinose	+	+	+	+	−

Symbols: + = positive in 1 or 2 days; − = negative; (+) = late positive; d = different biochemical types.

194

Table 32

Differentiation within *Proteus–Providencia*

Biochemical tests	Proteus				Providencia
	P. vulgaris	*P. mirabilis*	*P. morganii*	*P. rettgeri*	
Indole production	+	−	+	+	+
Growth on medium containing ammonium citrate	d	d	−	+	+
H$_2$S production	+	+	−	−	−
Decomposition of urea	+	+	+	+	−
Liquefaction of gelatin	+	+	−	−	−
Lysine–decarboxylase	−	−	−	−	−
Arginine–dihydrolase	−	−	−	−	−
Ornithine–decarboxylase	−	+	+	−	−
Decomposition of mannitol	−	−	−	+	−
Decomposition of maltose	+	−	−	−	−
Decomposition of adonitol	−	−	−	+	d
Decomposition of inositol	−	−	−	+	d

Symbols: + = positive in 1 or 2 days; − = negative; d = different biochemical types.

Table 33

Differentiation between biochemical groups classified as subgenera I, II and III, according to Kauffmann (1961)

Biochemical test	Salmonella subgenus		Arizona–Salmonella subgenus III
	I	II	
Decomposition of lactose	−	−	+ or X
Decomposition of dulcitol	+	+	−
Liquefaction of gelatin	−	+	+
Decomposition of d-tartarate	+	− or X	− or X
Decomposition of l-tartarate	d	−	−
Decomposition of i-tartarate	d	−	−
Citrate utilization	+	$+^2$	$+^2$
Decomposition of mucid acid	+	$+^{1-2}$	d
Decomposition of malonate	−	+	+

Symbols: + = positive; − = negative; $+^2$ = positive on the 2nd day; X = late and irregularly positive; d = different biochemical types.

6.3.1.3   Enumeration of Enterobacteriaceae

The total count of Enterobacteriaceae in food has been recommended as an indicator of the contamination of the food by intestinal material. If in a heat-treated food, the *E.coli* cells have been killed, yet salmonellae and other members of the family Enterobacteriaceae have survived, the determination of the total Enterobacteriaceae count per 1 g or 1 ml of the food will supply vital information to the food inspector.

The food sample to be tested is homogenized in saline (1 part of material in 9 parts of saline containing 0·1% peptone), and serial dilutions are prepared from the homogenate (2.3.4.1).

In the presence or absence of any member of the Enterobacteriaceae (e.g. in 1 g of the sample) is to be assessed ("Present–absent test"), then 1 ml of the homogenate is used to inoculate each of three replicate culture tubes containing a liquid enrichment medium, selective for the family [12]. The tubes are incubated for 24 h at 37 °C and a loopful inoculum is spread on the surface of crystal violet—neutral

196

red—glucose agar [71. v. 102, 86]. After incubation at 37 °C for 24 h five suspected colonies (red colonies with or without red halo) are isolated for further confirmation (oxidase test, glucose fermentation) whether they belong to Enterobacteriaceae or not.

6.3.1.3.A Determination of the Enterobacteriaceae count by the dilution technique

If it is expected that the number of Enterobacteriaceae cells is between 1 and 1000 per 1 g or 1 ml of the sample, then inocula of 1 ml are transferred from the 10^1, 10^2 and 10^3 dilutions into three replicate tubes for each dilution, containing a liquid enrichment medium [12]. The tubes are then incubated at 37 °C for 24 h. The growth in the tubes be confirmed at the same way as mentioned above in section *present–absent test*. The most probable number (MPN) is calculated from the highest dilution showing bacterial growth, using the MPN table.

6.3.1.3.B Colony counting

If the Enterobacteriaceae count is expected to be about 1000 per 1 g or 1 ml of sample, then inocula of 1 ml from the 10^2, 10^3, 10^4 and 10^5 dilutions are transferred into two empty Petri dishes for each dilution. Selective agar [102] is added by pour plating method and incubated at 37 °C for 24 h. The Enterobacteriaceae count is calculated from the number of typical colonies appearing in the plates.

Only those plates can be used for counting on which 10–300 red colonies of about 0·5 mm diameter have developed. Plates in which there is a high density or confluence of colonies must not be used. If less than half of a plate is affected in this way, the non-confluent colonies are counted and the result is calculated for the whole surface of the medium (e.g., if an average of 8 colonies have developed per 1 cm^2 and the surface of the medium is 64 cm^2, the number of colonies on the whole surface is taken as 512.

6.3.1.3.C Calculation of the Enterobacteriaceae count

(1) Calculation of the most probable number (MPN)

Growth of oxidase-negative, glucose-positive organisms in the enrichment media is regarded as a positive indication of Enterobacteriaceae.

The Enterobacteriaceae count per 1 g or 1 ml of food is calculated on the basis of those cultures and dilutions giving positive results (5.2.1).

(2) Calculation of the Enterobacteriaceae colony count

Those plates that have been inoculated with the largest amounts of the sample and show the growth of 10–300 colonies are taken into account.

If more than 75% of the colonies are oxidase-negative and glucose-positive, then all colonies are enumeratad; otherwise the colony count is calculated from the percentage of oxidase-negative and glucose-positive colonies. The means for pairs of plates inoculated with the same dilution are taken.

If the colony number per plate is less than 100, it is rounded up or down to the nearest multiple of 5.

If it is greater than 100 and the final digit is not 5, the number is rounded up or down to the nearest multiple of 10.

If the number is greater than 100 and the final digit is 5, it is rounded up or down to the nearest multiple of 20.

The count per 1 g or 1 ml of food is calculated by multiplying the colony number per plate by the dilution of the respective inoculum.

(a) In the case of present or absent test the result is expressed as follows: members of Enterobacteriaceae family were or were not present in 0·1 g or 0·1 ml of the food.

(b) Enterobacteriaceae count for 1 g food

The colony count is expressed in terms of a number between 1·0 and 9·9 multiplied by a power of 10, e.g. a colony count of 15,000 is presented as $1·5 \times 10^4$ per 1 g food.

If the Enterobacteriaceae count is less than 100, the result is presented as $< 1·0 \times 10^2$ per g.

If the MPN is less than 3 per g, then it is reported that Enterobacteriaceae were not detectable in 0·1 g of food.

6.3.1.4   *Salmonella*

Members of the genus *Salmonella* are Gram-negative, aerobic, non-sporing rod-shaped cells which develop on the commonly used culture media and reduce nitrate to nitrite. Usually they are peritrichous, but cells without flagella as well as flagellated non-motile forms also occur. They do not ferment adonitol, lactose or sucrose, they do not produce indole and they do not hydrolyse urea. They are negative in the Voges–Proskauer reaction and positive in the methyl red test. The majority of the species in Kauffmann's (1960) subgenus II liquefy the gelatin, whereas those of subgenus I do not. Glucose is fermented by all species, in most cases with gas formation. Salicin is not fermented immediately, but delayed fermentation may occur.

According to an international agreement, some species which have salmonella O and H antigens are placed in the genus *Salmonella*, even though this may not be consistent with the biochemical reactions.

198

Since serological tests play a decisive role in the identification of *Salmonella* species, the species of this genus may alternatively be referred to as serotypes. Identification involves the determination of two antigens, namely the heat-resistant O or somatic antigen, and the heat-sensitive H or flagellar antigen. Both the H and O antigens comprise a number of partial antigens.

6.3.1.4.A   Cultivation

6.3.1.4.A.a   Direct inoculation

For direct inoculation, material can be taken from the surface of a sample of food, or after flaming the surface, from the interior of the sample. If the surface layers cannot be separated from the interior of the sample, homogenized (or "average") samples can be used. Homogenization is carried out in saline containing peptone.

Elective and differential culture media can be used, e.g. modified Drigalski's agar [28] or brilliant green–phenol red agar [13]. These strongly inhibitory media must be used in conjunction with media that are weakly inhibitory or non-inhibitory to other microbes, especially if bacteria causing specific paratyphus in animals are to be detected; these strains grow only weakly in the enrichment culture media, if at all. In such cases the use of Drigalski's medium containing one-tenth of the usual amount of crystal violet is recommended.

Cultures are incubated at 37 °C for 16–24 h.

If direct inoculation is omitted, some *Salmonella* strains may escape detection.

6.3.1.4.A.b   Pre-enrichment

The purpose of pre-enrichment is to restore partially damaged cells of *Salmonella* to full viability ("Resuscitation of injured cells"). Liquid, non-selective culture media are used for, *viz.*
   (i) universal basal broth without glucose [32],
   (ii) lactose broth [77],
   (iii) buffered peptone water [111],
   (iv) ten-fold diluted 1/15 M phosphate buffer of pH 7·0 [108].
Cultures are incubated at 37 °C for 16–24 h.

Pre-enrichment cultivation is essential in the examination of foods for export and foods produced by new processes, the hygienic conditions of which are not yet fully known.

Surface samples

If the surface of the food material is separable, 25 samples each of about 1 g are mixed to make a 25-g sample and this is homogenized with pre-enrichment medium. Swabs taken from equipment used in food production lines are moistened first of all with saline containing peptone or broth [77], and then the surface of the equipment, etc. is rubbed with the wet swab. The swab is then transferred to the pre-enrichment medium.

Core samples

25 samples of 1 g are taken from the interior of the food material and are homogenized before culture in pre-enrichment medium.

Average samples

Irrespective of the quality of the food, 25 g are homogenized and cultured in pre-enrichment medium. Since average samples are taken from liquid, semi-solid or greasy foods, or from foods in the form of small tablets or capsules, the samples are powdered or minced as necessary to ensure homogeneity before testing. Otherwise, there is no reason to expect differences between superficial and deep contamination in foods of such description.

Pre-enrichment and/or enrichment (depending on the properties of the food) are necessary because the numbers of *Salmonella* organisms may be very low in foodstuffs and the organisms may be injured by the preservation treatment; furthermore, *Salmonella* may be outnumbered by other members of the family *Enterobacteriaceae*.

6.3.1.4.A.c Selective enrichment

Two different selective media [146, 11] are inoculated either directly from the food material itself, or from pre-enrichment culture. The selective media are incubated at 37 °C or at 42–43 °C.

6.3.1.4.A.d Subculturing from selective enrichment media

Subcultures are prepared on two selective solid media [13, 28] and incubated at 37 °C. Colonies characteristic of *Salmonella* appearing on the diagnostic culture media (pink colonies on medium [13], bright blue colonies on medium [28]) are isolated and subjected to further testing.

6.3.1.4.B   Identification in the health control laboratory*

To check a diagnosis of *Salmonella*, colonies of the organisms are subcultured and their biochemical and serological properties are determined.

6.3.1.4.B.a   Obligatory biochemical tests

Each type of suspect colony must be subcultured on the following selective and differential media:

(1) slope agar (stab culture and surface culture) [166];

(2) semi-liquid glucose agar [52] (simultaneous testing for glucose fermentation and motility);

(3) liquid medium containing urea (to detect urease activity) [20];

(4) tryptophane broth as modified by Ljutov [153], to test for indole production;

(5) Nógrády and Rodler's double-layered polytropic medium for simultaneous detection of H_2S production and decomposition of urea, glucose, lactose and sucrose [107];

(6) ONPG reagent for the beta-galactosidase reaction [108];

(7) glucose broth (for observing cultural properties and preparing culture for the O_1 phage test) [32];

(8) finally, colonies are tested by the Voges–Proskauer reaction.

6.3.1.4.B.b   Slide agglutination with *Salmonella* O polyvalent screening serum

The *Salmonella* O polyvalent screening serum contains all the agglutinins corresponding to the A–E + G + I + L *Salmonella* O groups.** The test should be performed on all cultures showing the characteristic biochemical reactions of *Salmonella (Table 34)*. Two loopfuls of saline (marked O and C) are placed on a degreased slide at a distance of 1–2 cm apart. A small amount of 18–24-h agar culture of the isolate is cautiously suspended in each drop with a loop or needle, and then O polyvalent screening serum is mixed with the O droplet, using an inoculating needle. The slide is waved several times and placed over a dark background. The reaction is positive if a characteristic, finely granulated agglutination is observed in the droplet. No agglutination indicates a negative reaction. Agglutination in the saline (drop C) indicates autoagglutination. If the result is ambiguous, it is

---

* Samples submitted from the field are routinely examined in the health control laboratory, and results are evaluated according to well-defined standards.

** The A–E + G + I + L polyvalent O serum is the most suitable because the most commonly occurring salmonellae belong to these groups.

Table 34

Basic reactions in *Salmonella* diagnostics

Salmonella	
Slope agar	Characteristic non-spreading colonies
Motility	+ except *S. gallinarum-pullorum*
Glucose decomposition	+ positive with gas
Urea hydrolysis	−
Indole production	−
Nógrády–Ródler polytropic medium	alkali/acid with gas H_2S+++
ONPG test	−
Growth in nutrient broth	characteristic growth with even turbidity
Voges–Proskauer	−

recommended that the reaction be carried out with preheated O antigen. Preheating means that growth from a 24-h culture is washed in saline boiled for 1 h, and centrifuged before use as the O antigen. The H antigens being thermolabile, are inactivated by the heating while the thermally stable O antigens remain intact. The agglutination obtained after heating is very fine, and the coarse agglutination often observed with some strains can be avoided in this way. For the screening of meat and meat products it is recommended that the diagnostic polyvalent O serum be used according to the accompanying instructions.

6.3.1.4.B.c Vi agglutination on slide*

Strains containing the Vi (virulence) antigen are not agglutinated by the polyvalent O screening serum or any other O serum. The Vi antigen is thermolabile, i.e. it is inactivated after incubation for 1 h at 60 °C. Heating makes bacterial suspensions agglutinable with O sera.

---

* Vi antigen is demonstrable in 98% of the *S. typhi* strains, but it is absent in the other salmonella strains of practical importance.

The Vi antigen is very important in the identification of *S.typhi,* the causative agent of human typhoid fever. If, therefore, colonies grown on selective and differential culture media are found to be characteristic of *Salmonella,* and the slide agglutination with the polyvalent O screening serum is negative, the following additional tests should be performed: (i) agglutination with Vi serum; (ii) agglutination with polyvalent O screening serum using culture washed with saline and kept at 60 °C for 1 h; (iii) agglutination with 9,12 O serum of *Salmonella* group D. A fine-textured agglutination with this serum is indicative of *S. typhi.* Identification of *S. typhi* is confirmed by a positive H agglutination when living culture is exposed to H-serum.

Salmonella diagnosis is established if the isolate behaves characteristically for *Salmonella* in the reactions of the short biochemical test series, and gives positive slide agglutination with the polyvalent O screening serum.

6.3.1.4.B.d Biochemical confirmation of the salmonella strains in polyvalent O serum

In order to confirm that an agglutinating strain is a *Salmonella*-type organism, the short biochemical test series is first of all put into practice. Sometimes an accurate diagnosis requires differentiation between the genera *Salmonella, Arizona* and *Citrobacter.* In such cases, the following tests are performed:

(1) lysine decarboxylase activity
(2) liquefaction of gelatin
(3) decomposition of dulcitol
(4) growth in a medium containing KCN
(5) decomposition of mucate
(6) decomposition of organic acids
   a) tartarate
   b) citrate
(7) sodium malonate test (see Table 28)
(8) fermentation of lactose
(9) decomposition of sucrose
(10) decomposition of adonitol
(11) decomposition of salicin.

Tests 1–7 are carried out in the reference laboratory in which the isolate is identified. The following procedures must be carried out in the control laboratories before diagnosis of *Salmonella* is established.

(1) pre-enrichment
(2) selective enrichment
(3) spreading on selective and differential media to isolate single colonies
(4) the short series of biochemical tests
(5) slide agglutination with polyvalent O screening serum
(6) O_1 phage test
(7) decomposition of lactose
(8) decomposition of sucrose
(9) decomposition of adonitol
(10) decomposition of salicin.

6.3.1.4.B.e O_1 phage test

Suspect colonies or 24-h stab and smear cultures developed on agar slopes are subcultured in glucose broth [32] (incubation at 37 °C for 16–24 h). From the broth culture, a loopful (loop diameter 4 mm) is transferred to the surface of a dried agar

Table 35

Frequency of positive *Salmonella* reactions

Reaction	Expected reaction	Percentage frequency of positive reaction
Decomposition of glucose with acid formation	+	100·0
Fermentation of glucose with acid and gas formation	+	91·9
Decomposition of lactose	−	99·2
Decomposition of sucrose	−	99·5
Production of H_2S	+	91·6
Hydrolysis of urea	−	100·0
Production of indole	−	99·0
Active motility	+	×
O_1 phage test	+	70·0
Agglutination in the A−E+G+I+L group O screening serum	+	99·8
Lysine–decarboxylase	+	94·6
Beta-galactosidase	−	98·5
Voges–Proskauer test	−	100·0

× except *S. gallinarum-pullorum* and species without flagella.

plate [166]. The plate is inoculated at two points as far apart as possible. When the two inoculation sites have dried, one drop of O_1 phage is added to one of them, the other serving as an untreated control. The plate is incubated at 37 °C for 5 h, then maintained at +4 °C for 24 h. After the initial incubation, and after the second low temperature incubation, the plate is examined under a magnifying lens of ×8 to ×10 magnification.

The test is positive if lysis (transparency) or a number of plaques of various sizes have appeared on the spot where the O_1 phage was added, the bacterial growth having remained undisturbed on the control spot. The test is negative if the growth is undisturbed at both sites. Frequency of positive Salmonella reactions are shown on *Table 35*.

6.3.1.4.C   Examination in the reference laboratory*

a) Compulsory tests:

(1) ability to grow on a culture medium containing ammonium citrate as the sole carbon source
(2) lysine decarboxylase activity
(3) methyl red test
(4) Voges–Proskauer reaction
(5) ONPG tests
(6) decomposition of arabinose
(7) decomposition of mannitol
(8) decomposition of dulcitol
(9) decomposition of inositol

b) Optional tests:

(1) oxidase activity
(2) catalase activity
(3) ability to grow in the presence of KCN,
(4) sodium malonate test
(5) liquefaction of gelatin
(6) hydrolysis of arginine
(7) ornithine decarboxylase activity
(8) phenylalanine deaminase activity
c) The isolate is matched with one of the O groups according to Kauff-mann–White, and the partial O antigens are determined by means of O antisera;

---

* A laboratory adequately equipped for the re-examination by experienced laboratory workers of equivocal diagnoses obtained by the control laboratory.

d) the phase-1 and phase-2 H antigens are determined by means of H antisera;

e) the biochemical tests which characterize the species within the O serogroup are carried out;

f) from the results of the above tests the species of the salmonella isolate is identified.

6.3.1.4.D   Methods of examination

The sample which is to be examined may be a surface sample reflecting the external state of contamination, a core sample containing the microflora of the interior of the food material, or an "average" sample, i.e. a sample of a liquid, semi-solid or oily substance, or an encapsulated, particulate or homogeneously minced substance in which superficial and deep contamination cannot be investigated separately.

6.3.1.4.D.a   Direct inoculation onto selective media [13, 28]

Surface sample. A piece weighing a few grammes is cut off with sterile forceps and scissors, and its surface is touched on the surface of the medium.

Core sample. The outer, contaminated, layer of the sample is removed (by flaming and cutting, or peeling in the case of an integument-covered sample) and a piece weighing a few grammes is cut out under aseptic conditions and used as inoculum with the help of a loop.

Average sample. (Liquid, presumed to be heavily contaminated) 0·1 ml is smeared on the surface of the medium using a sterile glass rod.

(Liquid, presumed to be mildly contaminated.) A dilution series is prepared, and each dilution is filtered through a membrane filter which is then placed on the medium.

(Semi-solid or oily substance.) A loopful (loop diameter 3 mm) is spread on the medium.

(Powdered, tablet-form or coarsly-minced, homogenized materials.) The sample is ground in a mortar with a little sterile saline and 1-ml portions are spread on selective culture media.

6.3.1.4.D.b Pre-enrichment

Surface sample. If the surface layer can easily be removed, a 25-g sample of the latter (25 × 1 g) is homogenized by grinding in a mortar or blender (in case of section 6.3.1.4.A.b (iv) a 50-g sample is processed). Wood and metal surfaces are wiped with a swab moistened with saline or glucose-free nutrient broth, and the swab is then placed in the pre-enrichment medium and incubated.

206

Core sample. A 25-g sample (25 × 1 g) or 50 g (see above) is homogenized in a mortar or blender and portions are pre-enriched (6.3.1.4.A.b).

Average sample. A sample weighing 25 g or 50 g (see above) is homogenized, if necessary, before cultivation in a pre-enrichment medium. Tablets are powdered, and gelatin and gelatin coverings are melted in a water bath at 42–45 °C.

6.3.1.4.D.c Selective enrichment

Surface, core, and average samples

a) Direct inoculation from the sample

Depending on the quantity and quality of the material to be tested, two 25-g samples, each of 25 × 1 g, are homogenized in a mortar or a blender in 9 parts (2 × 225 ml) of selective enrichment medium and incubated at 37 °C and 43 °C.

If the surface area to be sampled is not large enough to provide 50 g of material, as large a sample as possible is taken and suspended in nine times its weight of enrichment medium; replicate cultures are incubated at 37 °C and 43 °C, respectively.

b) Inoculation from pre-enriched material

Selective enrichment medium is inoculated as described in 6.3.1.4.A.c.

Selective enrichment can be combined with different pre-enrichment procedures [see 6.3.1.4.A.b (i), (ii), (iii) and (iv)].

Pre-enrichment in glucose-free nutrient broth [6.3.1.4.A.b (i)] followed by selective enrichment (6.3.1.4.A.c).

This method of enrichment is used in cases of swab samples.

The method is applied to the control of industrial hygiene and the detection of superficial *Salmonella* contamination in large consignments of (mostly imported) meat samples; the method is practicable even if the storage refrigerator is far away from the laboratory.

Swab samples are submitted for testing in sterile, stoppered test-tubes. Aliquots of 10 ml of glucose-free nutrient broth are measured into the tubes which are incubated at 37 °C for 3 h, or preferably for 6 h. Then 1 ml of the culture is transferred from each tube into 9 ml of a selective enrichment medium (Bierbrauer's medium [11], the tetrathionate medium of Müller and Kauffmann [146], or the Stokes and Osborn [129] selenite–brilliant green medium). Replicate cultures are incubated at 37 °C and 43 °C. Subsequently, selective and differential culture media are inoculated by spreading the inoculum on plates. These are incubated at 37 °C for 18–24 h. Brilliant green–phenol red medium [13] is always used as one of the selective and differential media.

If the samples have been taken from heavily contaminated surfaces after pre-enrichment, it is advisable to make second selective enrichment culture by transferring one part of the first selective enrichment culture to 9 parts of medium after 18–24 h incubation. The second enrichment is incubated further 24 h at 37 °C (pre-, first- and second enrichment).

Pre-enrichment in lactose broth [6.3.1.4.A.b (ii)] followed by selective enrichment (6.3.1.4.A.c).

The pH is adjusted to 7·0 with 1,0 M NaOH or 1,0 M HCl; the use of indicator strip is sufficient.

Twenty-five g of the sample are placed aseptically in a 500 ml flask and 225 ml lactose broth are added.

The pH is adjusted after thorough shaking.

Incubation is carried out at 35–37 °C for 24–48 h.

Before incubation, 0·1 ml of the culture is transferred into 10 ml of selenite–cystine broth [130], and 0·1 ml is transferred into 10 ml of a selective enrichment medium containing tetrathionate [146].

Incubation is carried out at 35–37 °C for 18–24 h.

Subcultures are prepared in selective and differential media, one of which should always be brilliant green–phenol red [13].

Incubation is carried out at 35–37 °C for 18–24 h; plates recorded as negative after 24 h are incubated for a further 24 h.

Pre-enrichment in peptone water [6.3.1.4.A.b (iii)] followed by selective enrichment (6.3.1.4.A.c).

A homogenized sample weighing 25 g is placed in a 250 ml flask, and 100 ml buffered peptone water are added. The mixture is thoroughly shaken, and incubated at 37 °C for at least 6 h, and not longer then 24 h. After incubation, 25 g of the mixture are subcultured in 225 ml of tetrathionate enrichment medium [146], and another 25 g are subcultured in 225 ml of selenite–brilliant green (Stokes–Osborne) broth [129].

Incubation is carried out either at 37 °C for 3 days, or at 43 °C for 1 day.

Subcultures are prepared in selective and differential media, one of which must be brilliant green–phenol red [13].

Pre-enrichment in 10-fold-diluted 1/15 M phosphate buffer, pH 7·0 [6.3.1.4.A.b (iv)], followed by selective enrichment (6.3.1.4.A.c).

The material to be tested is homogenized and 50 g are suspended in 2000 ml of sterilized 10-fold-diluted 1/15 M phosphate buffer pH 7·0 (1 part 1/15 M buffer is added to 9 parts water) and the pH is then adjusted.

Incubation is carried out for 18–24 h at 37 °C.

Two 25 ml samples of the pre-enriched material are each transferred into 200 ml of selective enrichment medium containing tetrathionate [146].

Incubation is carried out for 18–24 h, one culture at 37 °C and one at 43 °C.

Subcultures are prepared in selective and differential media, one of which must be brilliant green–phenol red.

Incubation is carried out at 37 °C for 18–24 h.

Selective enrichment according to the ISO standard (Addendum to 6.3.1.4.D.c.a).

200 ml tetrathionate enrichment medium are added to 25 g of the sample, and the suspension is blended for 30 minutes in a sterile blender.

200 ml of selenite–brilliant green (Stokes–Osborne) broth [129] are added to 25 ml of the sample and the suspension is blended for 30 minutes.

The homogenates are transferred to sterile flasks and incubated that in medium [146] at 37 °C for 3 days, or 43 °C for 1 day, and that in medium [129] at 37 °C for 3 days.

In each case subcultures are prepared using brilliant green–phenol red medium [13] and another suitable selective medium. In the author's laboratory, the latter medium is Drigalski's agar [28] where meat or meat products are examined.

Incubation is carried out at 37 °C for 48 h.

6.3.1.5   Isolation of *E. coli I* and coliform bacteria

6.3.1.5.A   Coli count and titre

Escherichia coli comprises strains of Gram-negative non-sporing rod-shaped bacteria belonging to the family Enterobacteriaceae.

The *E. coli* organisms are facultative anaerobes which ferment lactose with gas formation both at 30 °C and 44 °C, and produce indole from tryptophane at 37 °C and 44 °C. The Voges–Proskauer reaction and the oxidase test give negative results, the methyl red reaction is positive, and there is no growth on media containing ammonium citrate.

The coliform bacteria are Gram-negative, non-sporing and rod-shaped; they ferment lactose with acid and gas formation during incubation for 48 h. Coliforms grow aerobically in agar media containing bile or bile salts.

The coli count refers to the total number of *E. coli* and coliform cells per 1 ml or 1 g of sample.

The coli titre is the total titre of *E. coli* and coliform bacteria, i.e. the minimal quantity of the sample in which such bacteria can be detected. The highest dilution found to show the presence of coliforms is regarded as the order of magnitude of the numbers of coliform bacteria present in the sample. In the case of a 10-fold dilution series, the orders of magnitude refers to 1–9, 10–99, 100–999, etc., coliforms per 1 g or 1 ml of sample.

The most probable coli number means the total number of *E. coli* and coliforms per 1 g or 1 ml samples as determined by the MPN method.

E. coli and coliforms are natural inhabitants of the alimentary tract of man and the mammals, and, therefore, their presence is indicative of infection or contamination of faecal origin. In food microbiology, *E. coli* and coliforms are relied upon as an indicator flora of faecal contamination. Since other members of the Enterobacteriaceae family (e.g. salmonellae and shigellae), as well as parasites and viruses may flourish in the intestines, the consumption of food contaminated with coliforms may present a risk of infection by intestinal pathogens.

The presence of *E. coli I* (faecal coli) always indicates faecal contamination, while the presence of coliform indicates that only in general. Since *E. coli I* cells are much less able to survive outside the intestinal environment, than coliforms, coliforms persisting in soil and surfaces do not necessarily indicate a contamination from a faecal source, in the sense of implying immediate contact with faeces.

Coliforms include, *E. coli I,* the species *Enterobacter* and *Klebsiella,* some strains of which are able to multiply or persist in materials of non-faecal origin.

The presence of coliforms, and numbers of coliforms, can serve in food microbiology as indicators of contamination that has occurred after manufacture, as well as contamination arising from inadequate hygiene or shortcomings in processing technology.

Only a small proportion of coliforms are pathogenic to man. The enteropathogenic strains have characteristic antigen structures, and, therefore, in suspect cases agglutination tests are carried out with the corresponding O and OK sera.

If it is suspected that coliforms may be present in an injured form, pre-enrichment is necessary; the sample must be incubated in peptone water at 37 °C for 2 h, and then subcultures are prepared by mixing equal volumes of inoculum and one of the following: double-strength brilliant green–lactose–bile broth [14], MacConkey broth [85], or Kessler and Swenarton gentian violet–lactose–peptone broth [66]. Replicates are incubated at 30 °C and 44 °C, respectively.

6.3.1.5.B   Eijkman's fermentation test

In this test, selective culture media containing lactose are used (*viz.* lauryl–sulphate–tryptose broth [79], brilliant green–lactose–bile broth [14], and/or gentian violet–bile–lactose–peptone broth [66]). The bile salt (or bile) inhibits the propagation of aerobic spore-formers. Obviously, coliforms are also slightly inhibited by bile or bile salts, but in the presence of these substances false positive reactions are practically eliminated.

The appearance of gas bubbles in Durham tubes suggests the presence of coliforms. To minimise errors of identification, subcultures should be prepared on the following media: modified Drigalski's medium containing bile or bile salt, but without crystal violet [28], MacConkey agar [86], eosin–methylene blue agar [35], Klimmer's agar [68], Endo agar [33]. The test is positive if lactose-fermenting colonies develop on these media.

The fermentation test should be carried out at both 30 °C and 44 °C with readings taken after 24-h and 48-h incubation.

Lactose fermentation is the first criterion for differentiating between *E. coli* type I and coliforms. If gas is formed after 24-h or 48-h incubation at 30 °C, i.e. if the fermentation test is positive, subcultures are prepared in another medium which is appropriate for the fermentation test (e.g. MacConkey broth or Kessler and Swenarton's broth). The subcultures are inspected after incubation at 44 °C for 24 h and 48 h. If the fermentation test is positive at 44 °C as well, the presence of *E. coli* type I is suggested; if, on the other hand, the test is positive only at 30 °C, the isolate may be regarded as a coliform provided it meets the criteria of coliforms (*Table 36*).

Table 36

Classification of coliforms according to biochemical reaction

Bacterium	Gas formation at 44 °C in lactose–bile salt medium	Indole	Methyl red	Voges–Proskauer reaction	Growth on ammonium citrate medium
E. coli type I (E. coli) faecal type	+	+	+	−	−
E. coli type II	−	−	+	−	−
Intermediary type I	−	−	+	− *	+
Intermediary type II	−	+	+	− *	+
Enterobacter aerogenes type I	−	−	−	+	+
Enterobacter aerogenes type II	−	+	−	+	+
Enterobacter cloacae	−	−	−	+	+
Irregular type I (faecal type)	−	+	+	−	−
Irregular type II	+	−	+	−	−
Irregular type VI	+	−	−	+	+
Other irregular types	various reactions				

* Occasionally weak reactions.

6.3.1.5.C Differentiation between *E. coli* I and coliforms

If the fermentation test is positive at both 30 °C and 44 °C, subcultures are prepared on MacConkey agar [86], or on modified Drigalski's medium containing 5 g sodium taurocholate and lactose per 1000 ml, but without crystal violet [28]. The colonies which develop in the subcultures are subjected to examination by the so-called IMViC reaction (a test for indole production at 37 °C and 44 °C, the methyl red reaction, the Voges–Proskauer reaction, and testing for growth on medium containing ammonium citrate).

E. coli type I organisms are members of the family Enterobacteriaceae which ferment lactose at both 30 °C and 44 °C, form yellow colonies on modified Drigalski's medium [28], form indole at both 37 °C and 44 °C, give a negative response to the Voges–Proskauer reaction (2.8.16) and a positive response to the methyl-red reaction (2.8.15) and do not grow in the presence of ammonium citrate

(2.8.18). The presence of organisms showing these properties indicates contamination of intestinal origin.

From those culture tubes which are positive for the fermentation test, the most probable number (MPN) of *E. coli* or coliforms is calculated with aid of the Hoskins table. Usually 1 ml of the homogenized substance and 1 ml from each of its five consecutive dilutions are inoculated into 3 tubes of selective media (10-ml quantities). If it is necessary to take 10 ml or 50 ml as inoculum, the selective medium (Kessler–Swenarton, MacConkey, or brilliant green–lactose–bile broth) must be of double strength.

6.3.2   Clostridia

6.3.2.1   *Clostridium botulinum*

6.3.2.1.A   Food poisoning by *C. botulinum*

When *C. botulinum* multiplies in food, it produces a toxin which causes botulism poisoning if the food is consumed. This form of poisoning is characterized by high mortality rate, a non-febrile course, and a nerve-paralysis effect. Visual disturbances, dysarthria, and a disturbance of the swallowing action may be accompanied in severe cases by weakness, respiratory difficulty, and circulatory disturbances. Abdominal pain and meteorism with or without vomiting generally occur, and obstipation is more common than diarrhoea.

Latency usually lasts more than 6 h and the characteristic clinical picture takes 12–24 h to develop. In the most severe cases, especially of botulism type E, death ensues 20–24 h after the consumption of food containing the toxin, but in the more typical fatal case the illness lasts 2 or 3 days, sometimes more than a week.

C. botulinum is an obligate anaerobe which forms very resistant spores. Seven types are known, designated by the letters A–G. Typing involves the use of monovalent antitoxic sera in the antitoxin–neutralization test. Most human cases are caused by types A, B and E, the F toxin and G toxin may rarely present a health risk tool. The C and D toxins present a health risk only to animals, and, therefore, they are not dealt with here.

Spores of *C. botulinum*, like those of other soil-inhabiting spore-formers, are widespread in the soil and in inshore sea-water, and, therefore, vegetables and fish from contaminated areas are exposed to infection by the spores. Types A and B spores are highly, although variably resistant to heat, whereas type-E spores are more sensitive. The variability of the heat resistance determines not only the epidemiology of botulism and the types of food involved, but also the approach to the laboratory diagnostics. Each type is capable of producing toxin in various types of food. Types A and B sometimes produce toxin at relatively low temperatures, e.g. 10 °C. Type E is active at pH values as low as 4.0.

212

Detection and prevention

The isolation of *C. botulinum* strains and their serotyping are based on the fact that experimental animals can be protected* by neutralizing the toxin with antisera containing homologous antibodies. The close resemblance of the cultural properties and colony morphology of *C. botulinum* to those of some non-toxin-producing clostridia, (in particular *C. sporogenes*) makes detection of the toxin a very important part of the diagnosis.

The choice of procedure is influenced by the urgency of the situation. If a food is suspected of having caused botulism, detection of the toxin and the toxin-neutralization test both must be carried out urgently so as to ensure identification and elimination of the source of infection, and minimize delay in serotherapy. The choice of procedure may also be influenced by the amount of sample available. Sometimes only an empty container with traces of the food, or small amounts of food waste are available.

In urgent cases, the toxin should be detected if possible, directly from the sample. If the sample is not large enough for detecting the toxin, the causative organism should be isolated and identified. The bacterium should also be isolated and identified in the following circumstances:

(1) inoculated mice die with typical symptoms after an unusually short time (6–24 hours), but delayed death are occasionally observed (3–4 days) or

(2) inoculated mice are not protected by any of the type-specific antitoxins. In the latter case, a multiple toxin or unidentified toxin of *C. botulinum*, or a toxin of other origin must be present.

If a food contains no factor that is inhibitory to the multiplication of *C. botulinum*, it is essential to control the internal conditions of the food so that production of toxin in the food be prevented. pH values less than 4·5, a low water activity (a_w) and a sufficient concentration of curing salts, etc., may serve for this purpose. Food preservation techniques must aim to destroy every spore, or prevent sporulation, or create conditions that keep bacterial propagation and toxin production in check. In the case of processed foods in which toxin production is not prevented, two essential requirements must be observed, namely

(a) in the heat treatment, the time and temperature necessary for the inactivation of *C. botulinum* must be professionally determined for each kind of food, each size of container, and each processing procedure;

(b) foods in which the occasionally occurring cells of *C. botulinum* have not been inactivated must be stored in the frozen state or at a temperature ($+4\,°C$ for type E) at which bacterial propagation does not occur.

---

*Against the specific effect of the thermolabile toxin.

6.3.2.1.B   Isolation of *C. botulinum* and detection of toxin production

In the isolation of *C. botulinum*, a pure culture is confirmed by testing with type-specific toxin:

(1) An equal volume of absolute ethanol is added to 1–2 ml of the toxin-producing culture in a sterile test-tube.

(2) The mixture is allowed to stand for 1 h at room temperature.

(3) A large loopful of the mixture is spread on a double anaerobic egg-agar plate [147].

(4) The residual culture is set aside.

(5) The plates are incubated under anaerobic conditions for 48 h at 35 °C.

(6) Typical *C. botulinum* colonies are identified. These are white or pale yellowish, smooth, and irregular in shape. They are 1–2 mm in diameter and are surrounded by a precipitation zone 2–4 mm in diameter. The precipitated zone is covered by a pearly layer (reflecting the colours of the spectrum when viewed under oblique illumination). A similar zone may also appear around the colonies of other clostridia, and, therefore, this characteristic cannot be relied upon for identification.

(7) A few typical colonies are subcultured in TPGYT broth [151] if the presence of toxin E has been confirmed, or in freshly boiled liver broth [89] if the toxin was found to be type A or B.

(8) The TPGYT broth culture is incubated for 5 days at 26 °C, and the liver broth cultures for 5 days at 35 °C.

(9) All the cultures are centrifuged from the liver-broth cultures (the supernatant only) at 7000 rpm for 30 minutes.

(10) The supernatants are filtered through "Millipore" or sterile glass filters, and the sterile filtrates are frozen if storage overnight is necessary.

(11) The sterile filtrates are diluted 1 : 10 and 1 : 100 in gel-phosphate diluent [47].

(12) 4 mice are inoculated intraperitoneally with 0·5 ml of the antitoxin (five-fold-diluted in saline) which is homologous to the specific toxin type already established. 4 mice are left uninoculated as controls.

(13) After 30 minutes, 0·5 ml of each filtrate is inoculated into two antitoxin-protected mice and two unprotected mice.

(14) The mice are observed for symptoms of botulism for 72 h, and deaths are recorded (6.3.2.1.D). If the unprotected mice die while the protected ones survive, the originally demonstrated serotype of the toxin is confirmed.

214

6.3.2.1.C   Cultivation of *C. botulinum* types A, B and E, and the demonstration of toxin production by the isolate

If it is suspected that a food product is contaminated by botulinum toxin or by the bacterium itself (food poisoning having already been demonstrated), it is necessary to establish whether toxin is present and to isolate the bacterium. If a food product is under routine inspection and there is no cause for suspicion, the following procedures are followed.

(If toxin has been found in the culture, the procedure for toxin detection must be followed.)

(1) 3–4 g of the material under test are added to liver broth [89] in three test-tubes (the medium is boiled and cooled in the test-tubes before use).

(2) 3–4 g of the material are added to trypticase–peptone–glucose–yeast extract broth [152] in three test-tubes; the latter is prepared with trypsin which is added under sterile conditions (TPGYT [15].

(3) The inoculated liver broth tubes are incubated for 5 days at 35 °C.

(4) The inoculated TPGYT tubes are incubated for 5 days at 26 °C.

(5) After incubation aliquots are taken from the liver broth cultures and from the TPGYT cultures.

(6) Replicate aliquots are pooled, and centrifuged at 7000 rpm for 30 minutes.

(7) The supernatants are filtered through Millipore or sterilized glass filters (not Seitz filters) to obtain sterile filtrates; these are frozen if storage overnight is necessary.

(8) One part (about 5 ml) of the sterile filtrates are heated at 100 °C for 10 minutes.

(9) Dilutions (1 : 5, 1 : 10 and 1 : 100) of the untreated filtrates are prepared in gel-phosphate diluent.

(10) 2 mice are inoculated with each dilution intraperitoneally and 2 mice are inoculated intraperitoneally with 0·5 ml of undiluted heat-treated filtrate.

(11) The mice are observed for symptoms of botulism for 72 h and deaths are recorded.

Sterile filtrate

diluted without heat-treatment   heat-treated undiluted
       inoculated 0·5 ml   inoculated 0·5 ml
1 : 5   2 mice   into 2 mice
1 : 10   2 mice
1 : 100 2 mice

(12) Death of one or both of the mice inoculated with the untreated filtrate, while the mice inoculated with the heat-treated filtrate survive, indicates that toxin was present in the filtrate.

(13) The result should be confirmed by passive protection of mice (see 6.3.2.1.D)

6.3.2.1.D   The detection of botulinum toxin in food

(1) If the food is sealed in a container, an unlabelled end of the container is cleaned with a brush, using water and soap. The container is rinsed and dried.

(2) The cleaned end of the container is disinfected with 2% iodine dissolved in 70% alcohol. After a few minutes the surface is ignited, avoiding overheating and deformation of the container. The iodine solution is wiped away with a sterile wiper. Place hard-swelled ("bombage") containers in the refrigerator before opening.

(3) The container is opened under aseptic conditions.

(4) The state of the container and its contents is recorded.

(5) Any liquid food material, including material washed from the container is examined for the presence of toxin, without any further treatment.

(6) Suspect residual food is treated in the same way as canned solid food.

(7) The suspect food material (the residual contents of the container or other residual food) is ground aseptically in a pestle and mortar with an equal volume of sterile gel-phosphate solution [47] and a little sand, until a homogeneous suspension is obtained. A larger amount of suspension should also be prepared in a homogenizer.

(8) The suspension is centrifuged at 7000 rpm for 30 minutes.

(9) The clear supernatant is decanted and filtered through a sterile Millipore or sterilized glass filter (not a Seitz filter) to obtain a sterile filtrate. If the filtrate is to be stored overnight, it is frozen.

(10) The pH is adjusted to 6·0–6·2 with 1·0 M HCl solution.

(11) A portion of filtrate is trypsinized. 0·2 ml of 10% trypsin solution is added to 1·8 ml of the adjusted filtrate.

(12) The trypsinized filtrate is placed in a water bath for 45 minutes at 37 °C and then cooled.

(13) The trypsinized extract should not be stored overnight.

(14) A portion of the sterile non-trypsinized filtrate is heated at 100 °C for 10 minutes.

(15) The trypsinized and non-trypsinized untreated filtrates are diluted in gel-phosphate to obtain the following dilutions: 0 (undiluted), 1 : 10 and 1 : 100.

Sterile filtrate

Trypsinized diluted		Non-trypsinized diluted		Non-trypsinized heat-treated (control)
non diluted	0·5 ml	2 mice	2 mice	2 mice
1 : 10	0·5 ml	2 mice	2 mice	
1 : 100	0·5 ml	2 mice	2 mice	

216

(16) 2 mice are inoculated with 0·5 ml from each of the diluted filtrates (trypsinezed and non-trypsinized ones).

(17) Another 2 mice are inoculated intraperitoneally with the undiluted heat-treated filtrates.

(18) The mice are observed for symptoms of botulism for 72 h, and deaths are recorded.

(19) The presence of toxin is indicated by the death of mice inoculated with either the trypsinized or the non-trypsinized filtrates, while all the mice inoculated with the heated filtrate survive.

(20) The result is confirmed by inoculation of protected mice:

(a) 1 : 2, 1 : 10 and 1 : 100 dilutions of the trypsinized or non-trypsinized filtrates (whichever showed the higher titre) are prepared in fresh gel-phosphate. (The trypsinized filtrate must not be stored overnight.)

(b) Mice are inoculated intraperitoneally with 0·5 ml of 1 : 5-diluted antitoxin, 6 with A, with B and 6 with E antitoxin. Another group of 6 mice is left untreated.

(c) After 30 minutes, 0·5 ml of each of the filtrate dilutions is inoculated into two mice from each of the groups of 6 treated with antitoxin or untreated (b).

(d) The mice are observed for symptoms of botulism and deaths are recorded.

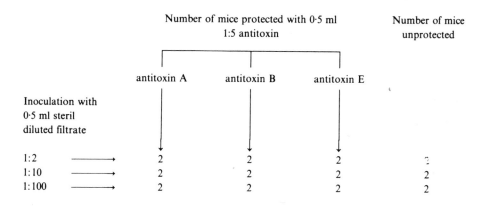

	Number of mice protected with 0·5 ml 1:5 antitoxin			Number of mice unprotected
	antitoxin A	antitoxin B	antitoxin E	
Inoculation with 0·5 ml steril diluted filtrate				
1:2	2	2	2	?
1:10	2	2	2	2
1:100	2	2	2	2

(e) Death of the unprotected mice and survival of those protected by any of the antitoxin treatments A, B or E, indicates the presence of the corresponding toxin, or toxins.

Detection of C. botulinum in smoke-cured fish

(1) The smoke-cured fish (a single fish, or the juice from canned or unpacked fish) is placed in a strong air-tight, plastic bag.

(2) 100 ml of TPGYT broth [151] are added to the sample in each bag so that the sample is covered.

(3) The bag is compressed to expel as much air as possible.

(4) The bag is sealed using a hot iron, or other means.

(5) Bag and contents are incubated for 5 days at 26 °C.

(6) The bag is cut open immediately below the seal and portion (25 ml) of the TPGYT broth is transferred to a centrifuge tube.

(7) Centrifugation is carried out at 7000 rpm for 30 minutes.

(8) The supernatant is filtered through a Millipore filter or sterile glass filter and frozen if overnight storage is necessary.

(9) The sterile filtrate is diluted 1:5, 1:10, 1:100 and 1:1000 with gel-phosphate diluent.

(10) 2 mice are inoculated with each of the dilutions.

(11) The mice are observed for symptoms of botulism for 72 h, and deaths are recorded.

(12) If none of the mice die, the test can be regarded as completed.

(13) If deaths occur, the results are confirmed by using protected mice.

 a) Sterile filtrate is diluted as in (9).

 b) 8 mice are inoculated, each with 0·5 ml of a 1:5 dilution (in saline) of antitoxin E. A further group of 8 mice is left unprotected.

 c) After 30 minutes, 0·5 ml from each dilution of the filtrate is inoculated into 2 protected and 2 unprotected mice.

 d) The mice are observed for symptoms of botulism for 72 h, and deaths are recorded. Death recorded among the unprotected mice and survival of the protected ones indicates the presence of type-E toxin.

 e) If all the mice die the serological neutralization test is repeated with mice protected by antitoxins A and B.

 f) The mice are observed for symptoms of botulism for 72 h, and deaths are recorded. Death recorded among the unprotected mice and survival of those protected with either of the antitoxins A and B indicates the presence of toxin A and B, respectively. If all the "protected" mice have died, further microbiological investigation is necessary.

6.3.2.1.E   Detection of botulinum toxin in blood serum

(1) Sufficient blood is taken from the patient to yield 10 ml serum.

(2) The serum is kept cool until used.

(3) A small sample of the serum is heated at 100 °C for 10 minutes.

(4) Some of the unheated serum is diluted 1:5.

(5) Mice intraperitoneally are inoculated with 0·5 ml of serum, 2 with unheated, undiluted serum, and 2 with unheated diluted serum.

(6) Another pair of mice are inoculated with 0·5 ml heated, undiluted serum.

(7) The mice are observed for symptoms of botulism for 72 h and deaths are recorded.

(8) Death with symptoms of botulism among the mice inoculated with the

unheated serum, and survival of those inoculated with the heated serum is indicative of the presence of toxin.

(9) The result is confirmed using protected mice as follows.

a) Mice are inoculated intraperitoneally, 2 for each antitoxin type, with 1 unit of the antitoxins A, B, C, D, E and F, respectively. Antitoxin G is not available. (Antitoxin diluted in saline to obtain 1 unit per 0·5 ml.)

b) 2 mice are left unprotected.

c) After 30 minutes each of the mice are inoculated with 0·5 ml undiluted serum.

d) The mice are observed for symptoms of botulism for 72 h, and deaths are recorded.

e) Death of unprotected mice and survival of any of the protected mice is indicative of the presence of toxin or toxins corresponding to the antisera given to the surviving mice.

6.3.2.2 *Clostridium perfringens*

C. perfringens can cause food poisoning if consumed in meat or poultry that has not been stored under refrigeration between heat treatment and consumption. During mild heat treatment, oxygen is reduced to a level that allows obligate anaerobic spore-forming clostridia, such as the spores of *C. perfringens*, to survive and multiply. Illness is experienced 8–15 h after consumption, the symptoms being strong intestinal spasms, meteorism, diarrhoea, and less frequently nausea and vomiting. Recovery is usually rapid, but occasionally because of dehydration and circulatory disturbances, severe symptoms may develop.

If possible, all residual food should be submitted for examination, without freezing. Freezing, however, may be unavoidable if the sample must be stored for one or two days before it is tested; the freezing may kill 99% of the vegetative cells of *C. perfringens*, and even cooling without freezing may kill some 90%.

In suspected cases of food poisoning, counts of the numbers of *C. perfringens* cells present should be obtained. In confirmed cases, the *C. perfringens* count is usually high in the contaminated food (10^6 per g or per ml) provided that the test is performed immediately after the onset of illness. However, the number of living *C. perfringens* cells may rapidly decline, owing to

(1) freezing of the food and increasing acidity in the food,

(2) an uneven distribution of the organisms in the food, and

(3) an uneven temperature of heating.

Based on the toxins produced, 5 types of *C. perfringens* have been differentiated, namely types A to E. Food poisoning in man is caused by types A and C. The large number of food poisoning cases found to be caused by different strains of *C. perfringens* of type A, suggest that this type causes the characteristic human disease symptoms if present in the consumed food.

6.3.2.2.A   Isolation procedure

The procedure may be based on

(1) the characteristic haemolysis either of blood agar [159], or of selective horse-blood agar containing Neomycin sulphate, and/or,

(2) the development of black colonies in sulphite agar medium, which can be made selective by an addition of Polymyxin or D-cycloserine [137].

6.3.2.2.B   Verification of the presence of *C. perfringens*

After suspect colonies have been isolated, their identity with *C. perfringens* must be confirmed. If the number of the *C. perfringens* cells is low, the MPN method is employed.

It should be noted that according to recent investigations, enterotoxin is not produced in the contaminated food but is released in the intestines, first of all in the ileum, after spore formation; furthermore, some strains of *C. perfringens* do not produce lecithinase. Confirmation, therefore, is usually based on nitrate reduction, haemolytic activity, catalase test, liquefaction of gelatin and the absence of active motility.

However, about 30% of strains do not reduce nitrate. Even the egg-yolk (lecithinase) reaction may be negative when applied to *C. perfringens*.

A type-A strain of *C. perfringens* may be

(1) a typical toxin-producing strain with heat-resistant spores (D value 30–60 minutes at 100 °C), showing weak or nil haemolytic activity in blood agar;

(2) a beta-haemolytic strain with heat-sensitive spores;

(3) a weakly haemolytic or non-haemolytic strain with heat-resistant spores.

An anaerobic strain can be identified as *C. perfringens* if it blackens semi-liquid agar containing sulphite [136] gives a characteristic lecithinase-positive reaction on lactose–yolk agar [78] (2.8.26), reduces nitrate (2.8.7), brings about beta-haemolysis in blood agar plates (2.8.33), and not produces catalase (2.8.28); the cells are Gram-positive (2.6.4.1), short, rod-shaped and non-motile.

6.3.2.2.C   Enumeration of *C. perfringens*

The *C. perfringens* count refers to the number of *C. perfringens* cells in 1 g or 1 ml food material.

The *C. perfringens* count is most reliably obtained by the most probable number (MPN) method. Six 1-ml portions from each of the 10^0, 10^1, 10^2, 10^3, 10^4, 10^5 and 10^6 dilutions of the homogenized sample are inoculated into semi-solid sulphite-reduction tubes (42 in all) which are brought to the boil and cooled to 45 °C before inoculation [136]; (alternatively, liquid sulphite–polymyxin medium [134]). Three of the inoculated tubes in each group of replicates are maintained at 65 °C for 15 minutes to kill the contaminating flora, while the other 3 remain untreated. Then all 6 tubes are incubated at 46±0·5 °C In the heat-treated media, the degree of sporulation of C. perfringens can be determined.

From the cultures showing blackening (positive sulphite reduction), subcultures are prepared on two further media, namely blood agar plates containing 400 μg D-cycloserine/ml, and sulphite-reduction–egg-yolk agar plates containing 400 μg D-cycloserine/ml. Incubation is carried out at 46±0·5 °C. If beta-haemolytic colonies have developed on the agar plates, and opalescent, Nagler-positive and sulphite-reducing colonies have developed which fit the description of C. perfringens in other respects also (see 6.3.2.2.B), the MPN for the vegetative cells and the MPN for the spores are found by means of the Hoskins table.

The C. perfringens count is reported in the same way as the E. coli I and coliform count.

6.3.2.2.D   Methods of anaerobic cultivation in the diagnosis of C. perfringens

Media suitable for the cultivation of C. perfringens:
    (A) Media of the enrichment type
        (1) Liver broth as modified by Takács [89].
        (2) Holman's medium as modified by Takács [58].
    (B) Selective liquid media
        Sulphite–polymyxin medium [134].
    (C) Semi-liquid differential medium
        Semi-liquid deep glucose agar as described by Takács [52].
    (D) Semi-liquid agar
        Semi-liquid sulphite-reduction agar as described by Takács [136].
    (E) Selective agar media
        (1) Blood agar with 400 μg D-cycloserine/ml [27], (incubation at 46± 0·5 °C).
        (2) Sulphite-reduction–egg-yolk agar with 400 μg D-cycloserine/ml [137], (incubation at 46±0·5 °C).
        (3) Crystal-violet–blood agar [72].
    (F) Differential agar medium
        Blood agar plates [159].
    (G) Culture media used in verification of the diagnosis
        (1) Egg-yolk agar for the demonstration of the lecithinase-positive and lipase-negative characteristics [147].
        (2) Lactose–egg-yolk–milk agar [78].
        (3) Medium for the demonstration of motility and nitrate reduction [97].
        (4) Medium for the demonstration of H_2S production and liquefaction of gelatin [5].

6.3.2.2.E   Biochemical tests necessary for the diagnosis of *C. perfringens*

Fermentation of	Reaction
glucose	acid + gas formation
maltose	acid + gas formation
lactose	acid + gas formation
sucrose	acid + gas formation
D-mannitol	negative
salicin	negative

Production of	
H_2S	positive
indole	negative
Reduction of nitrate	positive
Liquefaction of gelatin	positive
Lecithinase activity (Nagler reaction)	positive
Effect on blood agar	beta-haemolysis
Motility	negative
Gram straining	positive (short, thick rods, single or in pairs, rarely in short chains)
Catalase activity	negative

If the growth occurring in liquid enrichment media, or semi-liquid medium containing glucose, or semi-liquid sulphite-reduction medium, is suspected to be that of *C. perfringens* (short, thick, Gram-positive cells, no aerobic growth), a rapid diagnosis is necessary. It is, therefore, advisable to spread inoculate colonies simultaneously on medium [27] and medium [137] and incubate one plate of each medium under aerobic conditions, and another under anaerobic conditions, in either case at 46 ± 0.5 °C. The growth that develops aerobically reflects the density of the contaminating flora and the aerobe/anaerobe proportions.

6.3.2.2.F   Enumeration of mesophillic clostridia and determination of their titre in sulphite-reduction medium

20-ml aliquots of medium [136] are dispensed into test-tubes. A decimal dilution series from 10^0 to 10^6 is prepared from the sample to be tested. 6 test-tubes are inoculated (1 ml) from each dilution. (To expel the oxygen, the medium in each tube is heated to 100 °C and before inoculation is cooled to 45 °C). Three tubes are maintained at 80 °C for 20 minutes after inoculation. (This is a thermal "shock"

222

treatment to induce sporulation and bring about the destruction of vegetative cells. The cultures are cooled immediately in tap water and allowed to solidify.)

In the thermally untreated cultures, both the spores and the vegetative forms will initiate colonies, whereas in the heated cultures colony development arises only from spore germination, i.e. the number of spores can be determined in the latter. Both series are incubated at 37 °C for 48–72 h.

To confirm the clostridium diagnosis, subcultures are prepared on blood agar plates containing 400 μg D-cycloserine/ml, and on sulphite-reduction–egg-yolk agar medium containing the same concentration of D-cycloserine. Two plates of each medium are inoculated, one for aerobic incubation, and one for anaerobic incubation. The characteristics of the anaerobically incubated cultures are noted. The type of haemolysis occurring and the presence or absence of lecithinase activity (Nagler's reaction) can be observed, and the clostridia can be separated from bacteria which may grow feebly under aerobic conditions, even in the presence of 400 μg D-cycloserine/ml. If opalescence alone is observed (Nagler-positive colonies), all the tests necessary for clostridium diagnosis must be performed (see above, 6.3.2) and C. perfringens must be distinguished from C. bifermentans.

If, in addition to opalescence, a pearly layer appears on the surface of colonies growing on medium [137], the presence of C. botulinum is a possibility, and it is necessary that C. botulinum is distinguished from C. sporogenes.

Mesophilic clostridia can be enumerated using the sulphite-reduction-positive, i.e. "clostridium-confirmed" culture tubes, by using the Hoskins table.

6.3.3   Detection and identification of *Staphylococcus aureus*

6.3.3.1   General characterization of the bacterium

The genus Staphylococcus and the related genus *Micrococcus* are members of the family Micrococcaceae (Baird-Parker, 1975). The bacteria belonging to this family are Gram-positive, and the cells are spherical, showing characteristic grouping arrangements and clustering. The Micrococcaceae are oxidase-negative and catalase-positive, growing aerobically as well as anaerobically; they reduce nitrate to nitrite and decompose various carbohydrates to different degrees.

Basic differentiation between the two genera is based on the decomposition of carbohydrates. The staphylococci are facultative anaerobic bacteria which decompose glucose both by fermentation and oxidative metabolism, while the anaerobic micrococci metabolize glucose solely by the oxidative pathway.

It is very important to select an appropriate pH indicator for demonstrating the decomposition of glucose. Since some micrococci only ferment glucose very weakly, indicators such as bromthymol blue, which indicate a pH change around the level of pH 7·0, may give false positive results. Bromcresol purple, on the other hand, with its pH response around pH 6·0, provides a means of making a distinction between the two genera.

Glucose is fermented by a number of microbes besides staphylococci. In the diagnosis of *S. aureus*, Gram-positive cocci producing catalase in a culture medium containing 1% glucose are tested for the ability to ferment glucose. The test is carried out essentially as recommended by Hugh and Leifson [59] or using a medium rich in nutrients as recommended by Baird-Parker [9]. Deep agar media are melted and immediately cooled to 40–45 °C before use. A heavy inoculum taken from a single isolated colony in a pure culture is transferred into two deep agar tubes, one for anaerobic and the other aerobic cultivation. A long inoculating needle is used and the bottom of the tube should be touched with the tip of the needle. In order to establish anaerobic conditions in the anaerobic tubes the medium is covered with a layer of sterile paraffin wax, melted and cooled to 40–45 °C before use. Alternatively, 2% agar can be used as an overlayer [3]. Both the anaerobic and aerobic cultures are incubated at 37 °C for at least 48 h. A change of the originally lilac colour of the medium to yellow, with greater transparency, indicates acid formation. These changes already become apparent after incubation for 24 h, and there is no reason to incubate the cultures beyond 5 days.

Members of the genus *Staphylococcus* will decompose glucose with acid formation in both tubes, and at all depths in the deep agar. Micrococci, on the other hand, will not give rise to a colour change in the anaerobic culture, except possibly immediately below the overlayer. In the aerobic cultures the colour reaction is usually positive, although some *Micrococcus* strains give rise to such a week and/or protracted production of acid that the colour reaction may not take place.

Table 37

Differentiation between *Staphylococcus* and *Micrococcus*

Test	Micro-organism and reaction			
	S. aureus	*S. epidermidis*	*S. saprophyticus*	*Micrococcus*
Acid formation from glucose				
oxidative	+	+	+	∓
fermentative	+	+	+	∓
Acid formation from mannitol				
oxidative	+	∓	∓	∓
fermentative	+	−	−	−
Phosphatase test	+	±	∓	∓
Coagulase test	+	−	−	−
Heat-resistant endonuclease test	+	−	−	−
Novobiocin	S	S	R	S/R

Symbols: + = positive reaction; − = negative reaction; ± = often positive; ∓ = often negative; S = sensitive; R = resistant.

Three species are distinguished within the genus Staphylococcus (Baird-Parker, 1975), namely *S. aureus. S. epidermidis* and *S. saprophyticus*. Of these, *S. aureus* is important in food bacteriology, and it is, therefore, essential to be able to distinguish with certainty *S. aureus* from the less important species.

The tests most widely used for the differentiation of *S. aureus* from the other two Staphylococcus species and from Micrococcus are listed in *Table 37*, after Baird-Parker (1975).

Of these reactions, the phosphatase test and the aerobically decomposition of mannitol deserve some attention. Both reactions involve the use of differential media. It is clear from *Table 37* that aerobically decomposition of the mannitol is characteristic of *S. aureus*. Combined positivity of both, differentiates *S. aureus* strains from *S. epidermidis* ones. The fermentation of mannitol is very important indication of the presence of *S. aureus*. If it is necessary to differentiate *S. aureus* and *S. saprophyticus*, this may be achieved on account of the resistance of the latter to Novobiocin.

6.3.3.1.A   The coagulase test

Of the reactions listed in *Table 37*, a positive coagulase reaction provides unequivocal evidence that an isolate is *S. aureus*. According to current knowledge, the same result indicates general pathogenicity of a *Staphylococcus* isolate. However, coagulase-negative stains may also be pathogenic.

The coagulase reaction can be performed on slides or on test-tubes. As the slide technique is simpler, it is widely used in routine work. The reaction is in fact a plasma agglutination rather then a coagulation reaction; the so-called clumping factor (the factor responsible for the reaction) is not identical with fixed coagulase. However, a close parallelism has been found between the production of the two factors.

The blood plasma used in the reaction is taken in an anticoagulant. Rabbit plasma or pooled human plasma is used. Four volumes of rabbit plasma are mixed cautiously with one volume of sterile, 3·8% sodium citrate solution. The mixture is centrifuged and the clear plasma is separated and diluted by adding an equal volume, or twice the volume, of sterile saline. Then one volume of a 0·2% thiomersal (merthiolate) solution is added per 19 volumes of the diluted plasma. The plasma thus prepared can be freeze-stored in tubes for a long time. Care should be taken to preserve sterility at each step of the procedure, since microbiologically contaminated plasma cannot be used.

Pooled human plasma, i.e. a mixture of blood plasmas from many donors, is prepared in the same way as the rabbit plasma, except that it is not diluted with saline.

Plasma kept in the refrigerator must be pre-warmed to room temperature before use.

Slide reaction. One drop of plasma and one drop of saline are placed on a clean slide, a few cm apart. Neither the plasma nor the slide might not be at a lower temperature than room temperature. A generous quantity of bacterial culture is taken in a loop from a well-isolated colony and mixed on the centre of the slide using the loop; one part of the bacterial mass is suspended in the plasma, the remainder in the saline. The test is positive if the bacteria and the plasma immediately show aggregation while no aggregation takes place in the saline. The reaction cannot be evaluated if aggregation appears in both drops. The result is taken as negative if both suspensions remain homogeneous, or a faint reaction similar to a serological reaction appears on gentle agitation of the slide. The reaction of cultures that have developed on blood agar may be ambiguous or delayed, thus necessitating repetition of the test with a culture grown on nutrient agar. The cells of some strains are enclosed by a capsule-like substance which covers the receptors responsible for the reaction as carried out on the slide. For this reason, strains whose characteristics otherwise conform to those of *S. aureus* but which give a negative slide reaction should be re-examined for coagulase activity by the test-tube procedure. The tube reaction is not interfered with by the presence of capsules.

Test-tube reaction. Rabbit plasma or pooled human plasma is used. 0·5 ml of 1 : 10 diluted plasma is measured into each of two tubes. The strain to be identified is pre-incubated in brain-heart infusion broth [4] or in nutrient broth [61] at 37 °C for 24 h. A loopful (0·1 ml) of the pre-incubated culture is mixed in the plasma in one of the tubes. The plasma in the other tube remains uninoculated, and both tubes are incubated at 37 °C. A positive reaction can usually be observed after 2 h, but if the result appears to be negative, incubation is continued and the tubes inspected after 4, 6, 16–18 and 24 h. According to the recommendations of the International Standards Organization (ISO), a positive reaction appearing later then 6 h after the commencement of the incubation cannot be accepted. Other authors, Baird-Parker (1966) among them, accept any positive reaction observed during the first 24 h, the only requirement being that an unequivocal coagulation is observed. The reaction cannot be evaluated if coagulation takes place in both tubes as a result of microbial contamination of the plasma.

In the slide and tube procedures, the saline drop, and the uninoculated plasma, respectively, are indispensable controls. The reliability of both tests can be increased by examining known positive and known negative strains at the same time.

For an account of the phosphatase test, see 2.8.31. The decomposition of mannitol can be tested as described above for glucose.

6.3.3.1.B   Detection of heat-stable DNase

Cultures of *S. aureus* contain a deoxyribonuclease (DNase) which decomposes DNA even in pre-heat-treated cultures. The presence of this enzyme provides a reliable criterion for the positive identification of *S. aureus*. In the following, the method published by Lachica et al. (1971) is briefly described.

Colonies suspected of being *S. aureus* are selected from culture on a solid medium. The colonies are inoculated into brain-heart infusion broth [4], and the inoculated media are incubated at 37 °C for 24 h. 3 ml of toluidine–DNA agar [TDA, 148] are pipetted on to the surface of a clean slide and allowed to gel in an even layer. Wells 2 mm in diameter are made in the gelled agar. Cultures of those strains which require to be identified are placed in boiling water for 15 minutes and a drop of heat-treated culture is added to the wells. The slides are incubated at 37 °C for 2 h. If the reaction is negative after 2 h, incubation is continued for another 2 h. A positive reaction is indicated by a bright pink zone around the well.

The method is sufficiently sensitive to detect DNase at a concentration of 0·005 µg ml^{-1}. Thus the reaction can be used as a quantitative assay and this may be of importance in testing heat-treated foodstuffs which have been heavily contaminated by *S. aureus*, but in which the bacterium has been killed by the heat treatment. In these cases, the presence of heat-stable DNase in considerable amounts is indicative of heavy contamination, at some times, by staphylococci. A disadvantage of this approach, however, is that the extraction of DNase from foods, like the extraction of enterotoxins, is a complicated procedure.

S. aureus grows rapidly in brain–heart infusion broth [4] if the inoculum is small and incubation is carried out at 43·5 °C; the diagnosis of *S. aureus* is confirmed by a positive DNase reaction.

6.3.3.1.C   Haemolysis

In culture media containing blood cells, most strains of *S. aureus* cause clearly visible haemolysis of the beta type (complete haemolysis is characterized by clearing of the medium and increasing transparency around the colonies). However, haemolysin production is not an essential criterion for identifying an organism as a strain of *S. aureus*. The evaluation of the reaction may be difficult on account of the fact that a number of different haemolytic substances are produced by some strains and the amounts of these substances are variable. Furthermore, strain differences occur across the haemolytic spectrum. The relationships existing between haemolytic substances and blood cells are indicated in the following table.

Haemolytic substance	*Sensitive erythrocytes*
Alpha-toxin	Rabbit and sheep
Beta-lysin	Ovine and bovine
Delta-lysin	Rabbit, ovine, bovine and human

With respect to the broad spectrum of the delta-lysin, one might expect that any of the listed erythrocytes would be suitable for the test. However, delta-lysin is produced in low quantities, and the use of human and rabbit cells might lead to false negative results in the testing of a strain producing a small amount of delta-lysin that is otherwise detectable with ovine or bovine erythrocytes. The use of sheep erythrocytes is, therefore, more reliable, and if these are not available, bovine cells may be used. If a strain produces only beta-lysin, the phenomenon of "cold–warm" haemolysis may occur; haemolysis begins during incubation at 37 °C, but it cannot be visually confirmed unless the culture is re-incubated for 24 h at 4 °C. As a result, the haemolytic zone has a double contour.

Media containing blood may be inoculated directly, or as a first step an enrichment medium or a plate culture is inoculated by streaking to obtained isolated colonies. Incubation is carried out at 37 °C for 24 h, then at room temperature or in the refrigerator for a further period of 24 h. Haemolysis is easily visible beneath and around individual colonies.

6.3.3.1.D Microscopic characteristics

Microscopic characteristics and the morphology of the colonies provide important information which assists in making an initial distinction between *S. aureus* and either *S. epidermidis* or Micrococcus. The main characteristics are as follows.

S. aureus: spherical Gram-positive cocci, 0·8–1·0 μm in diameter. In smears prepared from cultures grown on a solid medium, the cells show cluster-like groupings; in smears prepared from culture in liquid medium, the groups are smaller and less regular.

S. epidermidis: Gram-positive cocci, similar in size and shape to cells of *S. aureus;* larger cocci may also occur. The cells may be arranged in larger groups, but cluster-like groups are rarely found, even in cultures grown on solid media.

Micrococcus: Gram-positive or Gram-variable cocci, usually larger than cells of *S. aureus;* cells oval also occur. Cocci may be arranged in pairs, in tetrade groups or larger groups. Cluster-like groupings and chains are not formed. Sarcina cells are larger cocci in a packet-like arrangement.

6.3.3.1.E   Morphological characteristics of colonies growing on plates

S. aureus: the colonies are medium-sized, circular, convex, opaque, with an entire margin and a smooth shiny surface. The central part of the whole of the colony is orange (golden yellow) in colour. Some variants produce flat colonies. Pigment production is often not observed during incubation, but appears only after 24 h at room temperature. Assessment of the colour is facilitated by pulling together colonies of similar appearance on the surface of the medium. Pigment formation is not an essential criterion for *S. aureus* diagnosis; sometimes it cannot be observed with certainty.

S. epidermidis: the colonies resemble those of *S. aureus;* they are whitish, or less frequently grayish–white and pale yellow; they are less strongly convex.

Micrococcus: the colonies are variable in shape from flat to convex; they are white, grayish–white, sometimes with a yellowish hue, or pale pink, smooth opaque, compact and with a shiny surface. Confluence of colonies is common. Colonies which show strong pigment production and develop a rough surface after a few days usually belong to the *Sarcina* group.

The morphological characteristics of colonies growing on blood agar medium are as follows:

S. aureus: the colonies resemble those of the same species growing on nutrient agar; they are usually surrounded by a beta-haemolytic zone of variable width. In the cultures of some strains, the haemolytic zone may show only partial clarification, suggestive of warm–cold haemolysis. After keeping the cultures in the refrigerator for 24 h, the haemolysis becomes complete in the majority of cases and shows the double-contour phenomenon.

S. epidermidis: the colonies are similar to those of the same species growing on nutrient agar medium; they are either non-haemolytic or they produce delta-haemolysin.

Micrococcus: the colonies are similar to those of Micrococcus growing on agar plates, except that they grow more abundantly and do not cause haemolysis.

It should be noted that the above characteristics are less pronounced and more difficult to recognize on media which have been directly inoculated from a food sample. It is, therefore, advisable to subculture from the initial cultures before examining the colonies in detail. The colonies developing in subcultures may be more typical in appearance, and the subcultured strain does not usually lose its pathogenicity.

6.3.3.1.E.a Morphological characteristics of *S. aureus* colonies on selective and differential media

Selective and differential media are widely used for the cultivation of *S. aureus* strains, either directly or after enrichment. For general cultural and diagnostic purposes, the blood agar plate [159] without additions of growth-inhibiting factors, is regarded as most suitable. In the presence of miscellaneous contaminating organisms, however, multiplication of the contaminants on the blood agar may seriously interfere with the recognition and isolation of *S. aureus* colonies. If, on the other hand, the medium contains substances inhibitory to the growth of the contaminating flora, it is possible that the growth of *S. aureus* may also be inhibited, so that the characteristic size and number of colonies normally observed on blood agar plates are not observed on the selective media. On selective and differential media, the basic characteristics of the colonies are almost identical with those of colonies growing on agar or blood agar media; however, there is a number of extra characteristics which depends on the composition of the medium.

Raising the NaCl concentration of the medium to 6–10% is a simple method of making the medium more selective, particularly against *S. saprophyticus*. Other salt-tolerant Gram-positive bacteria may grow well, although spore-formers are usually prevented from swarming. The characteristic colour of *S. aureus* colonies is orange, but after being maintained at room temperature for 24 h, strains of weak pigment production may arise.

The use of inhibitory metals (e.g. tellurites, LiCl) increases the selectivity of the medium and facilitates the isolation of *S. aureus*. Metal compounds are reduced by *S. aureus*, causing the surface of colonies and sometimes also the medium around the colonies to become dark, with a metallic sheen. The incorporation of egg yolk in the medium [147] makes visible the lipovitellin-decomposing effect of the lipase–esterase enzyme of *S. aureus*. On medium [147], the colonies are surrounded by a clear zone in which turbid areas may be present. Selective ingredients may be combined in the same medium, e.g. on media containing both metal salt and egg yolk, clear zones may appear around dark colonies.

6.3.3.2 Detection and enumeration of *S. aureus*

In food microbiology, it is necessary to make a clear distinction between less costly, routine methods of investigation, and reference methods. The former are relatively simple, involving less material and labour; the latter are applied in the grading of food where large amounts of food are to be graded, for example in international trading. However, there is little difference between the sandards demanded of the two types of method with respect to the identification of *S. aureus* and the reliability

230

of the diagnosis. A difference in standards between the routine and reference test may be justified if only the presence or absence of the bacterium is to be established (PA test), or the titre value, the MPN, or the colony number is to be determined.

6.3.3.2.A   Routine test for the presence of S. *aureus* (PA test)

For an account of the collection, preparation and processing of samples, and the preparation of dilutions, reference is made to the relevant preceding chapters. It may be added here that when dairy products (especially cheese) are diluted, the use of citrate diluent is the most suitable; coagulated proteins are more soluble in this diluent than in others.

The sodium-citrate diluent is prepared by dissolving 10 g of sodium citrate in 1 litre of distilled water. It is sterilized by steaming for 15–20 minutes on each of three consecutive days. The diluent is warmed to 37 °C before use.

The aim of the PA test is to establish whether or not S. *aureus* is detectable in a given amount of the sample. For example, the foodstuff to be examined must not contain microbes in 0.01 g or 0.01 ml of the sample, i.e. the staphylococcus cell count must be < 100 g^{-1} (or < 100 ml^{-1}). In this instance, 0.01 g or 0.01 ml of the food is examined and evaluated. If cultivation gives a positive result, the sample is unacceptable at the given dilution; if the result is negative, the sample is acceptable. It follows from procedure used in this test that a negative cultivation result does not justify grading the sample as containing no S. *aureus* at all.

Procedure. 1 ml of diluted material containing the required quantity of the sample, is inoculated into 9 ml thiocyanate enrichment medium [118], and the surface of the medium is sealed with a layer of melted paraffin cooled to 40–45 °C, or with a layer of 2% melted agar [3] cooled to 45 °C. Incubation is carried out at 37 °C for 24–48 h.

After incubation, a loopful of the enrichment culture is streaked on a salted blood agar plate [125] or a blood agar plate [159]. The inoculated media are incubated at 37 °C for 24 h, and then at room temperature overnight. The plates are examined for colonies of S. *aureus*. Thus for 1 enrichment medium and 1 or 2 blood agar plates have been required for the basic examination.

Suspect colonies (or one colony) are subjected to the following tests: the Gram reaction, fermentation of mannitol, and the slide coagulation reaction. If the latter test is negative or ambiguous, the coagulase test is carried out by the test-tube method.

If the Gram-positive cocci are found which are both microscopically and macroscopically characteristic of S. *aureus,* positive for the fermentation of mannitol and for coagulase, then it is reported that . . . "coagulase-positive S. *aureus* were detectable in 0.1, 0.01,. . . etc. g or ml of the food under study". If colonies fitting the description of S. *aureus* are not present, or the confirmatory tests give negative results, it is stated that . . . "in the stipulated quantity of the food, S. *aureus* could not be detected".

6.3.3.2.B Reference test for the presence or exclusion of *S. aureus*

Procedure. 1 ml of an appropriate dilution of the material is pipetted into each of 3 tubes of Giolitti and Cantoni's selective and differential tellurite medium [48]. The medium in the tubes is sealed with an overlayer in 6.3.3.2.B and then incubated at 37 °C for 24–48 h. Those cultures in which blackening or precipitation appears are conditionally recorded as positive for *S. aureus*. In the absence of any change, the reaction is regarded as negative.

Some inoculum from each of the positively-indicating cultures is spread on Baird–Parker plates [10]. The spreading is executed carefully and rapidly with a glass rod, and the rod is changed after each spreading. The plates are left at room temperature for 15 minutes and allowed to dry, protected from contamination. They are then incubated in an inverted position at 37 °C for 30–36 h.

The requirements for the basic examination are for each dilution examined, 3 tubes of enrichment medium and one plate for each positive tube.

After incubation, each plate is examined separately for the presence of colonies of *S. aureus,* these being black, shiny, convex, and surrounded by a clear zone. In the clear zone, a fine black precipitate may appear, mainly from the second day on. If characteristic colonies are present, some are selected for the identification procedure; if more than 30 are present, the number taken for examination should be equal to the square root of the number present. At least 5 colonies should be examined if 6–30 characteristic colonies are present, and all the suspect colonies are taken if their number is 5 or less.

The selected colonies are subjected to the following tests: the Gram reaction, the coagulase slide test (the coagulase tube test if the slide reaction is negative), and the DNase reaction.

In the grading of food samples, the result for each enrichment culture should be evaluated separately, and the size of the inoculum is adjusted to contain a known amount of the undiluted food.

Example: inoculation from 1:10 dilution into 3 test-tubes the positive results for two of which are confirmed by other tests—*S. aureus* is positive in 2×0.1 g (or ml) of the foodstuff.

6.3.3.2.C Routine titration of *S. aureus*

The titre refers to the highest positively yielding dilution. If positive results do not allow a regular sequence in the dilution series, the titre is calculated as follows.

Take into account the possible highest positive dilution (e.g. 10^5), and prepare one dilution series from 10^1 to 10^5. 1 ml is pipetted from each dilution into 1 tube containing 10 ml enrichment medium [118]. After incubation some inoculum of the

232

positive-indicating cultures is spread on 1 blood agar [159] or adequate similar media [123, 124, 125] with a glass rod. The plates are left at room temperature to dry and than incubated at 35–37 °C for 24–36 h. The plates are examined for the presence of colonies suspected as *S. aureus*.

This basic testing requires 1 tube of enrichment medium per dilution and at least the same number of plates.

The result is negative if no positively identifiable colonies are found in the subcultures prepared from any of the enrichment cultures, or if the result of the reference tests is negative. The grading is reported accordingly. *S. aureus* could not be detected, e.g. in 0·1 g or ml of the food; alternatively: *S. aureus*, if present, does not exceed the 10^{-1} g^{-1} or 10^{-1} ml^{-1} level.

A dilution should be regarded as positive if characteristic colonies are present and the reference tests are positive.

If the basic dilution and further dilutions are consistently positive, the titre will be equivalent to the highest dilution giving a positive result. The bacterial count is calculated from the dilution. For example, if the 10^3 dilution is negative, it may be reported that *S. aureus* was found to occur at an order of magnitude of 10^3 in 1 g or 1 ml food. If all the subcultures, including that of the highest dilution, are positive, the order of magnitude corresponding to the highest dilution is reported as being the least titre; e.g., if the highest dilution is 10^4, the order of magnitude of the titre may be said to be at least 10^4 g^{-1} or ml^{-1}. If the sequence of positively reading dilutions is interrupted by negatively reading ones, the most concentrated of the latter should be indicated; e.g., if dilutions 10^1 and 10^4 give positive results, but dilutions 10^2 and 10^3 do not, then an order of magnitude of 10^2 should be regarded as the titre.

6.3.3.2.D   Reference test for the MPN if a bacterial count less than 10^3 g^{-1} (ml^{-1}) is expected

In this test, replicates of both the stock dilutions and the higher dilutions are processed. In the following, it may be understood that replicates are examined and evaluated together. Dilution series are prepared up to 10^3.

From each dilution, 3 tubes of Giolitti–Cantoni enrichment medium [48] are inoculated, 1 ml per tube, as described above (i.e. 9 tubes are inoculated for each dilution series). The remaining steps of the reference test are as given above. Where positive results are obtained, each series of dilutions (6 series altogether) is evaluated separately using the Hoskins table (5.2.1) and the results obtained from the table are averaged. If the *S. aureus* count is <100 per g (ml), it should be rounded up or down to the nearest multiple of 5 (e.g. 35 is indicated instead of 37). A count of over 100 is indicated by a decimal number and a power of 10, e.g. 240 g^{-1} = $2·4 \times 10^2$ g^{-1}.

If the result of each cultivation is negative, it is reported that in 1 g or 1 ml of the food the number of *S. aureus* cells does not exceed 3.

6.3.3.2.E Routine method for obtaining the *S. aureus* count if the latter is expected to be more than 10^3 g^{-1} (ml^{-1})

A series of dilutions is prepared up to 10^4 or higher. Starting with the highest dilution, 0·1 ml amounts are spread on salted blood agar [125], or on blood agar [159] and salted agar [123]. The 10^1 basic dilution need not be subcultured; thus, the basic test requires at least 3 plates. The inocula are carefully and rapidly spread using sterile glass rods, and the plates are allowed to dry. The remainder of the procedure and requirements for identification are as described in 6.3.3.2.B.

Each plate is examined separately for the presence and number of *S. aureus* colonies, and the numbers of confirmed colonies are averaged. In establishing the colony numbers pertaining to individual dilutions, account must be taken of the dilution factor and the size of the inoculum (in each case 0·1 ml, giving a multiplication factor of 10 throughout. The product of these factors is the *S. aureus* count g^{-1} (ml^{-1}) of food. This is rounded up or down to the nearest multiple of 20 and expressed as a decimal and a power of 10.

If the result for each subculture is negative and the highest dilution is 10^2, it may be stated that the *S. aureus* count is less than 1×10^3 per 1 g (1 ml) of the sample.

6.3.3.2.F Reference method for obtaining the *S. aureus* count if the latter is expected to be higher than 10^3 g^{-1} (ml^{-1})

From each sample two separate dilution series are prepared, with dilutions up to at least 10^4. The 10^1 dilutions are included in the test. The number of dilutions to be subcultured is $2 \times 4 = 8$.

Two Baird–Parker plates [10] are inoculated from each dilution of each series, 0·1 ml per plate, as described in 6.3.3.2.B. It should be noted that colony numbers up to 300 can be enumerated per plate, and thus with a dilution of 10^4 it is possible to obtain counts as high as 30 million g^{-1} (ml^{-1}). The demonstration of microbial counts higher than this is usually of no practical importance.

After incubation, each plate is examined separately for the presence of characteristic *S. aureus* colonies. Only those plates should be examined on which the number of characteristic colonies is between 30 and 300.

For the two parallel series of dilutions, the numbers of characteristic colonies per plate are averaged, and the identification procedure is carried out as described in 6.3.3.2.B. Colonies are selected for further examination as follows. If the number of characteristic colonies on a plate is more than 30, the number of colonies selected from the plate is equal to the square root of the former number. The square root is rounded up or down to the nearest integral number; e.g. from a total of 40 colonies, 6 should be randomly collected. If the number of characteristic colonies per plate is between 5 and 30, 5 colonies are taken, and all of the characteristic colonies are taken for examination if there are less than 5 per plate.

If more than 75% of the colonies taken for examination are confirmed as being colonies of *S. aureus*, then the characteristic colonies on the plates may be regarded as being colonies of *S. aureus*, and the number of the latter is multiplied by a coefficient representing the proportion of the selected colonies that have been confirmed as *S. aureus*.

The *S. aureus* count is expressed as a decimal and a power of 10 (e.g. $290,000 = 2 \cdot 9 \times 10^5$).

If the final result is below 10^3, it is reported that *S. aureus* could be detected in the food with a bacterial count of $< 1 \times 10^3$.

If all the plates prove to be negative (i.e. no colony is identified as *S. aureus* on any of the plates), it may be stated that no *S. aureus* could be cultivated from $2 \times 0 \cdot 01$ g (ml) of the food.

6.3.3.3 Addendum

Internationally, several other culture media are also used. Of these, the following deserve mention.

In the institutes of the Public Health Laboratory Service (PHLS) of the United Kingdom, a modified Robertson medium [117] containing 10% NaCl is used. From the enrichment medium, a phenolphthalein phosphate agar plate [40] is inoculated, and the phosphatase activity of the colonies is estimated by a colour reaction. (The plate is held so that the culture is exposed to ammonia vapour while the result is read.) The positive plates are then subjected to the coagulase test (phosphatase-negative variants cannot be detected by this test). In the Soviet Union, salted milk agar [124] is used. In this medium, a clarified zone observed around the colonies indicates proteolysis. If it is feasible, mannitol and phenol red indicator are added to the medium. Trypticase–soybean salted broth [152] is a more widely used enrichment medium. Among selective and differential media, the Vogel–Johnson agar plate [162] and the azide–egg-yolk medium [8] are commonly used.

In the last few years, comparative studies have been carried out in numerous institutes including the Central Institute of Food Bacteriology of the PHLS, to find the best method of detecting *S. aureus* in foodstuffs (Gilbert, 1972). In brief, the results of these studies have shown that the different culture media are of approximately the same value, provided that the tests are performed with care by experienced workers. In routine work, media that are simple to prepare are preferably used not only because this saves time, but also because a source of error is reduced by using simple media.

The hygienic conditions of food processing are unacceptable if *S. aureus* can be demonstrated (from the equipments cleaned down) in the course of the processing. The possible spread of this bacterium through the air has been a matter of debate. In any case, it is advisable to check by exposing 5–10 (or more) blood agar plates, depending on the size of the room, for 15 minutes, and incubating the plates for 24 h in an incubator, followed by a further 24 h at room temperature. Any characteristic colonies appearing should be positively identified. The outcome is acceptable if no *S. aureus* colony grows on the plates, or if such colonies occur only accidentally, i.e. 1 colony occurring on just one out of 5–10 plates. The presence of a greater number of characteristic colonies suggests that the cleaning of the factory is inadequate or, less probably, that a subject carrying *S. aureus* is present, or third that there is some serious hygienic or technical shortcoming in the processing.

The presence in foodstuffs of microbes of the genus *Micrococcus* is, according to present knowledge, harmless. Such microbes are often detectable in foodstuffs of good quality. Micrococci are relatively resistant to heat, and very resistant to desiccation, so that they may form the predominant part of the microflora in powdered foods, for example. When certain groups of microbes predominate, or the microbiological content changes irreversibly in a food product, an exact identification of the dominant microbe may be justified. The micrococcal character of microbes can be established on the basis of the above characteristics.

6.3.4 Detection and identification of *Streptococcus faecalis*, and of group D streptococci

6.3.4.1 General characterization of the microbes

Streptococcus is a genus of the family *Streptococcaceae*. Members of the genus *Streptococcus* are Gram-positive cocci arranged in pairs, or more frequently in chains; their cultures are negative for catalase and oxidase, but ferment a considerable number of carbohydrates. The fermentation is of the homo-fermentative type; glucose is largely decomposed to lactic acid without CO_2 production. Streptococci grow well in the presence of bile or bile salts.

In the practical classification of streptococci, Sherman's criteria are still found to be useful (Deibel and Seeley, 1975). The criteria of classification, the designations, and the main characteristics of the groups, are presented in *Table 38*.

Besides the basic physiological–biochemical classification, the serotyping of streptococci is of importance in medicine. Enterococci, i.e. members of serological group D, are highly variable in their properties. The conditions of Sherman's criteria are met positively by *S. faecalis* and its variants *(S. faecalis* var. *zymogenes* and var. *liquefaciens)*, as well as *S. faecium* and *S. durans*, but they do not apply to *S. bovis* or *S. equinus*. It is important to point out this difference because the comprehensive terms 'type D streptococci' and 'enterococci' are often used in food microbiology, although the examination of food does not extended to every characteristic reaction of all the different variants of the group. Usually, the tests which give a positive result with *S. faecalis* (and its variants) and with *S. faecium* are carried out so that the grading in fact refers to these microbes, rather than to the

Table 38

Differentiation of the main groups within the genus *Streptococcus* on the basis of Sherman's criteria

Sherman's criteria	Groups and reactions				
	Enterococcus		Latic-acid coccus	Viridans	Pyogenes
	S. faecalis S. faecium S. durans	S. bovis S. equinus			
Haemolysis	alpha beta gamma	alpha gamma	gamma	alpha gamma	beta
Growth at 10 °C	+	−	+	−	−
Growth at 45 °C	+	+	−	+	−
Growth at pH 9·6	+	−	−	−	−
reduction of 0·1% methylene blue	+	−	+	−	−
Heat tolerance at 60 °C for 30 min	+	−	±	∓	−
NH₃ formation from arginine	+	−	±	−	+

Symbols: + = positive reaction; ± = mostly positive reaction; ∓ = mostly negative reaction; − = negative reaction.

Table 39

Differentiation of group-D streptococci (enterococci) on the basis of physiological and biochemical characteristics

Test	Streptococcus species and reactions				
	faecalis	faecium	durans	bovis	equinus
Growth at 50 °C	−	+	−	−	−
Growth at 45 °C	+	+	+	+	+
Growth in 40% bile	+	+	+	+	+
Growth at pH 9·6	+	+	−	−	−
Growth in the presence of 6·5% NaCl	+	+	−	−	−
Growth in the presence of 0·05% tellurite	+	−	−	−	−
Growth in TTC medium	+	−	−	±	−
Heat tolerance (60 °C for 30 min)	+	+	+	+	−
Decomposition of sorbitol	+	∓	−	∓	−
Decomposition of mannitol	+	+	−	∓	−
Haemolysis	gamma alpha beta	gamma alpha	gamma alpha beta	gamma alpha	gamma alpha

Note: See footnote to Table 38.

whole Enterococcus group. In routine work, only some of Sherman's criteria are examined, and other properties are evaluated by using tests which differentiate between *S. faecalis* and its variants. From the point of view of food hygiene, the latter are the more important microbes. Physiological and biochemical reactions which differentiate between the group-D streptococcal species and variants are shown in *Table 39*.

Table 39 does not include the reactions used to differentiate among the variants of *S. faecalis*. It may, however, be mentioned that beta-haemolysing strains often occur among strains of *S. faecalis* var. *zymogenes,* while *S. faecalis* itself and its variant *S. faecalis* var. *liquefaciens* only very exceptionally show beta-haemolysis. Bovine and ovine erythrocytes, both commonly used in the demonstration of this type of haemolysis, are sometimes unsuitable in this context. Reference tests should always be carried out on media containing horse erythrocytes; after inoculation, these media are incubated for 24—48 h, and if the results are ambiguous, the cultures are kept in a refrigerator for a further 24 h. Strong liquefaction of gelatin is the main characteristic of *S. faecalis* var. *liquefaciens.*

In order to differentiate between *S. faecalis* variants, the decomposition of aesculine is investigated. An inoculum as small as 0·1 µl is used to inoculate the medium containing aesculine [36]. A positive reaction is indicated by blackening in the medium. A disadvantage of the method, however, is that a negative reaction cannot be ruled out until after 7 days' incubation at 37 °C.

6.3.4.2 Procedures followed if the presence of *S. faecalis* is expected

In the course of routine testing for *S. faecalis,* target tests are rarely prescribed. Nevertheless, it is necessary to be on the look out for the presence of microbes such as *S. faecalis* which, because they occur in large numbers, may considerably influence food quality. Characteristic colonies of *S. faecalis* can usually be observed on blood agar plates. It is particularly important to extend the testing to suspected colonies of this organism if they are present in large numbers, or are observed on plates inoculated with high dilutions of samples.

On blood agar plates [159], colonies of *S. faecalis* are 1–2 mm in diameter, grayish–white, circular, slightly convex, with an entire margin. Haemolysis occurs rarely, i.e., gamma-type colonies occur most frequently. However, on careful inspection, a darker zone can be observed around the colonies. The rare occurrence of alpha or beta haemolysis does not exclude the possibility that a microbe is *S. faecalis* if the colonies are otherwise characteristic of this microbe. The presence of Gram-positive cocci, somewhat oval rather than regularly spherical, supports the diagnosis.

Typical colonies are subcultured on selective and differential media (e.g. medium E_{67} [31], sodium azide medium [98], proteose–peptone (PP–tellurite medium

238

[113f]). The media are incubated at 35–37 °C for 24–48 h, depending on the rate of growth and nature of the colonies. *S. faecalis* grows well on the above media, on which the colonies are of similar morphology to those growing on blood agar. On media containing sodium azide or tellurite, the colonies are black, usually surrounded by a black zone; on media containing TTC (such as the sodium azide medium), the colonies are red. Any microbes that cannot withstand the growth inhibitors grow weakly, or not at all, on these media.

Those colonies of characteristic appearance developing on the selective media are usually tested for catalase activity. They are subcultured on PP slant agar [113a], which is incubated at 37 °C for 48 h. Hydrogen peroxide is then dropped on the culture, and any effervescence indicates a positive catalase reaction.

A microbe may be identified as *S. faecalis* if it shows the typical microscopic features of this species, forms characteristic colonies, and grows well on selective media. The diagnosis may be confirmed by the aesculine-fermentation test.

The tests that need to be performed if variants of the group-D streptococci are to be identified are listed in *Table 39*. Growth at 45 °C or 50 °C is tested by preparing subcultures in Hartley's broth [57] and incubating these for 48 h. It should be noted that freshly isolated of *S. faecium* grow well at 50 °C, but soon lose this ability in repeated laboratory culture. Growth tests are carried out in the following media, the cultures being incubated at 37 °C for 48 h: PP-bile broth [113b], PP broth, pH 9·6 [113d], NaCl–PP broth [113c], a medium [113e] containing TTC, and a medium [113f] containing tellurite. Tests for the fermentation of mannitol and sorbitol [42] are carried out under the same conditions.

Growth in liquid broth media is indicated by turbidity of the medium. If a strain forms a sediment, the medium should be shaken before the result is recorded. On the TTC medium colonies of *S. faecalis* are pink, and on the tellurite medium they are shiny black. A change in the colour of the medium from pink to yellow indicates carbohydrate fermentation with acid formation. To demonstrate heat tolerance, subcultures are prepared, both before and after any heat treatment has been applied to the food material. The result is taken as positive if the colony count shows no appreciable decrease attributable to the treatment. The microbes are identified according to the tests in *Table 39*.

All positive results are reported together with the identification, and the highest positive dilution is also given as the order of magnitude at which *S. faecalis* was present in the sample. The sample is then graded according to the stipulated requirements. If the sample had been collected for general inspection and food identical with the sample is still available, further samples are taken for target examination. If a series of samples is examined and found to be unacceptable, the foodstuff represented by the sample must be graded according to this result. If a sample is taken for routine grading and the result of the tests is negative, the presence of *S. faecalis* is not definitively ruled out because the result is not that of a target study.

239

6.3.4.3   Target tests for *S. faecalis* (and group-D streptococci)

Target tests are preferably carried out using enrichment cultures. 1 ml of the food sample is inoculated into each tube of enrichment medium, or an aliquot of the food is placed in the enrichment medium, as in the testing procedure for *Salmonella*.

Littsky–Malmann's medium [81] can be used as an enrichment medium. Subcultures are then prepared on blood agar [159] and on any other of the selective media listed above, although the PP-tellurite [113f] is the most suitable. Subculturing from blood agar makes possible the identification of group-D streptococci; subculturing from tellurite media makes possible the identification of *S. faecalis* and its variants, and the differentiation of group-D organisms. The details of the procedures are as described in 6.3.4.2.

Enrichment carried out for the purpose of identifying *S. faecalis* and/or its variants can be performed successfully on SE medium [127]. They are incubated at 37 °C for 16–24 h, and a positive reaction is indicated by blackening of the medium.

From the blackened enrichment medium, a loopful of culture is spread on SC plates [128] which are incubated at 37 °C for 24 h. A black network surrounding the colonies indicates a positive reaction. In tests which involve enrichment in SE medium [81] or subculturing in SC medium, the development of a characteristic growth and characteristic colonies, respectively, is taken as evidence of the presence of *S. faecalis* and/or its variant(s). On the basis of this, grading can be established. If both of these media are used, false positive results occur exceptionally rarely. If there is any doubt regarding identification, or for any other reason it is considered necessary, the above-mentioned tests are applied.

The PA test, and determination of the bacterial titre and MPN for *S. faecalis*, are carried out according to the same principles as apply to *S. aureus*. If those tests have been carried out which apply only to the identification of *S. faecalis* and its variants, this fact is included in the report of the results. If tests have been carried out for the purpose of demonstrating group-D streptococci in general, the isolate can be designated precisely where a positive result is obtained.

The hygienic state of a sample is unacceptable if *S. faecalis* is detected. Taking into account the relatively high resistance of this microbe to disinfectants, one contaminated sample among a number of samples taken from the same place may be acceptable. Positive results obtained in repeated tests, however, are a cause of concern.

Phage-typing of group-D streptococci, especially in cases of food poisoning, may provide information which assists in tracing the chain of infection (Hoch and Hérmán, 1971; Hérmán and Hoch, 1971; Pusztai et al., 1972).

6.3.5 Detection and identification of B. cereus

6.3.5.1 General characterization of the bacterium

The genus *Bacillus* is a member of the family *Bacillaceae*. The cells of members of the genus are rod-shaped, spore-forming, and the majority are Gram-positive (Gibson and Gordon, 1975). In culture they produce catalase and form spores, even under aerobic conditions. In contrast, members of the genus *Clostridium*, the other genus of the family *Bacillaceae*, do not form spores under aerobic conditions, though some aerotolerant clostridia may produce catalase.

Within the genus *Bacillus*, three main morphological groups are distinguished. *B. cereus* is placed in group I. The members of this group bear oval or cylindrical spores; the sporangium does not visibly swell and the wall of the spore is thin. The characteristics of *B. cereus* and its variants are as follows. They grow neither at 60 °C nor at a pH value of 6·0, but grow well in glucose broth, even under anaerobic conditions; they produce lecithinase; the cells are large (>0·9 μm in width), and some are vacuolated. It should be noted that lecithinase-negative variants may occur, and that an abundant growth under anaerobic condition in glucose broth is characteristic of *B. cereus*, although poor growth in this medium may be observed with some of the other species of group I.

The microscopic appearance of *B. cereus* is very characteristic. The cells measure 1–1·2 μm × 3–5 μm; in fresh culture, the cells are Gram-positive, in older cultures

Table 40

Differentiation of *B. cereus*, *B. subtilis* and *B. megaterium*

Test	Micro-organism and reactions		
	B. cereus	*B. subtilis*	*B. megaterium*
Active motility	±	±	±
Presence of capsule	+	+	+
Haemolysis	+	+	−
Production of lecithinase	[±]	−	−
Liquefaction of gelatin	+	+	+
Anaerobic growth in glucose broth	+	−	−
Pathogenicity in animals	±	−	−
Voges–Proskauer test	+	+	−
Reduction of NO₃⁻	+	+	−
Acid produced from decomposition of mannitol	−	+	+
Acid produced from decomposition of arabinose	−	+	+

Note: See footnote to Table 38. Sign in brackets = rare occurrence.

16 Kiss

they are sometimes Gram-variable. Some of the rodlets appear to be short in relation to their width, and the ends are angled; the cells may be arranged in chains of various lengths. Most of the spores are ellipsoidal, central or subterminal in cell, and do not cause swelling of the cell wall. *B. cereus* grows between the temperature limits of 10 °C and 48 °C, with an optimum between 28 °C and 35 °C.

Of the common culture media, blood agar [159] is the medium on which consistently characteristic colonies are best observed. After growth at 37 °C for 48 h, the colonies are 4–7 mm in diameter, usually showing beta-haemolysis, and less frequently, alpha-haemolysis. The colonies are grayish–green, flat, occasionally granular, and dull. The margins may be irregular, and sometimes rhizoid colonies may occur.*B. cereus* is usually flagellated, and is thus capable of swarming. Being a facultative anaerobe, it is able to grow under anaerobic conditions, producing a growth which may be mistaken for clostridia. In anaerobic culture, the colonies are 2–3 mm in diameter and transparent, resembling colonies of *C. perfringens*, although the margins of colonies of *B. cereus* are irregular.

The reactions most widely used to differentiate between *B. cereus* and other aerobic bacilli commonly occurring in foods, are listed in *Table 40*.

Table 41

Verification of *B. cereus*

Test	Result
Fermentation of glucose	+
Decomposition of sucrose	+
Decomposition of glycerol	+
Decomposition of salicin	+
Peptonization of litmus milk	+
Liquefaction of gelatin	+
Decomposition of soluble starch	(±)
*Acetyl–methyl–carbinol reaction	(±)
Growth in the presence of 10% NaCl	(±)

    * In the event of a negative reaction, it is advisable to repeat the test in peptone water containing glucose; the acetoin production of Gram-positive cocci and bacilli may be inhibited by phosphate (Cowan and Steel's Manual, 1974).

    Note: See footnotes to Table 38 and Table 40.

The tests most commonly used in the identification of *B. cereus* are summarized in *Table 41*. The data are taken from the publications of Mossel et al. (1967) and Thatcher and Clark (1968), and from a guide published by the National Institute of Hygiene, Budapest (1969).

Tables 40 and *41* clearly show that some strains, though they may be identifiable as *B. cereus*, do not definitively show the characteristic reactions of this species.

These strains are regarded as variants of *B. cereus,* provided that they are similar to *B. cereus* in all other respects. From the food-hygiene point of view, the question of the importance of lecithinase-negative strains deserves attention, since according to Nygren's (1962) hypothesis, only lecithinase-positive strains are of importance in cases of *B. cereus* food poisoning. According to the latter hypothesis, lecithinase liberates phosphorylcholine from foods containing lecithin, and the symptoms of food poisoning may, therefore, be attributable to the phosphorylcholine. However, Nygren's hypothesis has not been substantiated, and, therefore, the occurrence of lecithinase-negative strains in foodstuffs cannot be ignored.

Lecithinase production is examined on egg-yolk agar [93]. One colony is spread over one quarter of a plate with radial movements. The inoculated medium is incubated at 32–35 °C for 48 h. The medium clears around the lecithinase-positive colonies, the surface of the medium becomes oily in appearance and Newton's rings appear owing to the presence of free fatty acids released by the lipase–esterase of *B. cereus.* Keeping the plates at room temperature after the period of incubation results in an expansion of the clear zone, and a white precipitate of the calcium salts of the fatty acids is formed.

The decomposition of a particular carbohydrate is examined using nutrient broth containing the carbohydrate and phenol red as indicator [42]. The inoculated media are incubated at 32–35 °C for 24 h. A positive reaction is indicated if the pale pink colour of the medium turns to yellow.

Nitrate reduction is examined by means of nitrate broth [104]. A loopful of culture is inoculated into this medium, which is then incubated at 32–35 °C for 18–24 h. To indicate the presence of nitrate, Griess–Ilosvay reagent [57] is added to the incubated medium. A positive reaction is indicated by a red or pink colour.

The inoculation of litmus milk [75] and the conditions of incubation are as described for the previous test. *B. cereus* shows a rapid peptonization and little, if any, coagulation.

Stab cultures in medium [5] are incubated at 30 °C for 24 h; gelatin is rapidly liquefied by *B. cereus.*

Starch agar [65] is inoculated by streaking in such a way that separated colonies are obtained; the cultures are incubated at 30 °C for 24 h. After incubation, Lugol's solution [55] is poured on the plates, and the decomposition of the starch is indicated by a pale zone around the colonies.

Acetyl–methyl–carbinol production is examined by means of glucose broth culture [53]. The medium is incubated at 32–35 °C for 48 h, and then an equal volume of Voges–Proskauer reagent [163] is added to the culture. The culture-tube is shaken, and a few creatin crystals are introduced in order to accelerate the reaction. The tube is shaken again and allowed to stand uncovered in a slightly tilted position for 4 h. A positive reaction is indicated by a pink discolouration. By the execution of all the above tests it is possible to determine whether or not variants are present, and to establish the frequency of occurrence of certain variants.

In the identification of members of the genus Bacillus, it is occasionally necessary to carry out serological tests. In the identification of strains of *B. cereus,* these investigations are limited to strains which are the cause of food poisoning (Taylor and Gilbert, 1975).

6.3.5.2   Procedures followed if the presence of *B. cereus* is suspected

If preliminary observations suggest that *B. cereus* is present, the investigatory procedures are essentially the same as those described in 6.3.4.

An indication of the presence of *B. cereus* is usually gained during the inspection of blood agar cultures. If large numbers of characteristic colonies are present, or colonies are discovered in subcultures prepared from high dilutions, egg-yolk plates are inoculated; incubation and interpretation of the results are carried out as indicated above. Smears are prepared from characteristic colonies developing on blood agar and egg-yolk media. The demonstration of haemolytic and lecithinase-positive colonies which are typical of *B. cereus* colonies both morphologically and microscopically, makes the presence of *B. cereus* highly probable, and thus grading can be carried out. In dubious cases and in those cases in which the diagnosis for some reason requires confirmation, the cultures should be tested for the ability to ferment glucose. In the course of these examinations, growth under anaerobic conditions should also be demonstrated. If the microbe proves not to be *B. cereus,* the result is not reported.

6.3.5.3   Target testing for *B. cereus*

Media are inoculated with appropriate dilutions of the sample, the dilutions being prepared with peptone–salt diluent [110]. In general, 0·1-ml amounts of inoculum are taken from the 10^1 and 10^4 dilutions. Blood agar [159] and mannitol–yolk (MY) agar [93] may be inoculated in replicate from the individual dilutions. Cultivation and identification are carried out as above.

MY medium is preferable for cultivation, since the selectivity of this medium can be increased by the addition of an appropriate antibiotic, which is necessary for successful detection of *B. cereus*. The plates are incubated at 32–35 °C for 40 h. Dry colonies with an uneven surface on a purple background are typical of those of *B. cereus;* the colonies are surrounded by a zone containing a white precipitate. The presence of *B. cereus* is indicated if microbes which are characteristic with respect to colony morphology and microscopic structure, are observed on MY medium containing polymyxin B sulphate. The microbial count can be obtained from the colony number, taking into account the degree of dilution.

If it seems necessary, colonies developing on the MY medium may be tested for

244

glucose fermentation, gelatin liquefaction and nitrate reduction, and/or the tests listed above may be performed.

In the course of hygiene inspection, the direct detection of *B. cereus* is not a part of the investigation procedure (with the exception of epidemiological investigations connected with cases of food poisoning). The presence of this bacterium on surfaces, equipment, etc. can be tolerated provided that the total viable count meets the hygienic requirements. Furthermore, the presence of *B. cereus* as a part of the total permissible viable count is acceptable in foodstuffs preserved by heat treatment, provided that the spores are in a resting state, i.e. that multiplication from spores in the foodstuff cannot be demonstrated. With respect to the presence or absence test, titre determination, and determination of the colony count, etc., the procedures given for *B. cereus* may be followed.

6.3.6   Detection and identification of *Pseudomonas aeruginosa*

6.3.6.1   General characteristics

P. aeruginosa, a species of the genus *Pseudomonas* is of great importance in medicine (Doudoroff and Palleroni, 1975). The microbes of this genus are related in many respects with those of the groups *Aeromonas* and *Plesiomonas,* and the genus *Vibrio*. They are closely related to the family *Enterobacteriaceae*. All these bacteria are Gram-negative rods. Members of the family *Enterobacteriaceae* are oxidase-negative, whereas the genus *Vibrio* and related groups, including species of *Pseudomonas*, are oxidase-positive. For further differentiation, the procedure of Huge and Leifson is adopted. Members of the genus *Pseudomonas* are obligate aerobes and decompose glucose by oxidative metabolism. Species of the genus *Vibrio* and those of the groups *Aeromonas* and *Pleisimonas* are fermentative microbes.

The identification of *P. aeruginosa* from among the other members of the genus may be based on the following characteristics (Hendrie and Shewan, 1966):

— Gram-negative rods with single polar flagella are observed.

— A majority of colonies shows no colouration. However, a fluorescent pigment is produced in the medium by some species. Strains forming coloured colonies, strains forming white colonies, and variants producing diffusible pigments may occur within the same species.

The production of a diffusible pigment serves as a criterion for identifying some members of the genus *Pseudomonas*.

Some species are easily differentiated by their range of temperature tolerance.

The recognition and identification of characteristic *P. aeruginosa* colonies can be undertaken easily and reliably, even by laboratory staff of little experience. The presence of this microbe is established on the basis of the morphological

Table 42

Main characteristics of *P. aeruginosa* and *P. fluorescens*

Test	Characteristics	
	P. aeruginosa	*P. fluorescens*
Gram reaction	–	–
Shape	rod-shaped	rod-shaped
Motility	(±)	(±)
Oxidase test	+	+
Mode of growth	aerobic	aerobic
Mode of decomposition of glucose	oxidative	oxidative
Production of fluorescein	(±)	(±)
Production of pyocyanin	(±)	–
Aromatic odour	(±)	–

Note: See footnotes to Tables 38 and 40.

characteristics of the colonies, the positive oxidase and Gram reactions, the characteristic motility, the diffusible bluish–green pigment (pyocyanin) colouring the medium around the colonies, and the odour of tributyrin which resembles that of lime-blossom. However, variants which differ with respect to some of these characteristics occur frequently, in particular one which only produces a greenish fluorescent pigment and which confused with the saprophytic *P. fluorescens* strains that produce the same pigment. The main characteristics of *P. aeruginosa* and *P. fluorescens* are presented in *Table 42*. There is a close similarity between the two species, although less similar variants are frequently encountered. Some of the latter differ from the typical *P. aeruginosa* in more than one respect. Strains that do not visibly produce pigment on solid media are subcultured in nutrient broth [57] and incubated at 37 °C for 24 h. It should be noted that the growth of *P. aeruginosa* in a well-defined pellet is commonly observed in liquid media. After incubation, 1 ml of chloroform is mixed with the culture and the two phases are allowed to separate. A bluish colouration in the chloroform layer is characteristic of *P. aeruginosa* and indicates the presence of pyocyanin; the fluorescein produced by *P. fluorescens* is insoluble in chloroform.

If the pyocyanin test is negative, stimulation of pigment production may be tried. King's medium A [67a] and King's medium B [67b] are employed for this purpose. The former stimulates the production of pyocyanin, the latter supports production of fluorescein. A strain which produces blue pigment or both pigments is a strain of *P. aeruginosa;* production of green pigment only indicates the presence of *P. fluorescens*. Both media can be prepared either as agar plates or agar slopes. The colour of the pigment can be judged with the naked eye, or under an analytical UV lamp, the latter method being the more sensitive. Pyocyanin fluoresces giving a bluish–green colour, and fluorescein gives a yellow colour.

246

Table 43

Detailed characteristics of *P. aeruginosa*

Test	Result
Colony morphology	Characteristic, depending on the medium
	+
Nitrate reduction	(Nitrogen gas formed)
Indole production	−
Decomposition of urea	+
Liquefaction of gelatin	+
Growth at 42 °C	+

Note: See footnote to Table 38

If the isolate is a fluorescein-producing variant of *P. aeruginosa* and correct identification is essential, then further identification procedures are necessary as these are summarized in *(Table 43)*. Of the tests listed in *Table 43* very important is the complete reduction of nitrate (a red coloration cannot be evoked by adding either the reagent or powdered zinc), and growth at 42 °C in nutrient broth are of particular importance. Both reactions enable a distinction to be made between *P. aeruginosa* on the one hand, and organisms which are unimportant from the food hygiene point of view on the other.

It should be noted that some strains of *P. aeruginosa* produce a red pigment (pyorubin) instead of pyocyanin, and other strains may form a melanin-like pigment. The fact that a strain may deviate in its characteristics from the general pattern, does not exclude possible pathogenicity of the strain; the identification of strain is particularly important if pathogenicity seems likely. Acetylamylase production and the decomposition of acetamide should be taken into account, for these activities are characteristic of 97% of the strains. A high resistance to antibiotics should also be considered (Mossel, 1975).

The procedures listed in *Tables 42* and *43* are described in Chapter 2.8.

6.3.6.2   Procedures followed if the presence of *P. aeruginosa* is suspected

The procedures in this respect are the same as those outlined in 6.3.4 (in connection with *S. faecalis*).

The presence of *P. aeruginosa* may become evident during the evaluation of plates inoculated in the course of target tests for other micro-organisms. The following description of colonies of *P. aeruginosa* growing on different media is taken from a guide published by the National Institute of Hygiene, Budapest (1969). Of the colonies that appear on nutrient agar or milk agar plates in the course of the determination of the total bacterial count, those surrounded by a greenish zone and showing a pearly surface deserve further investigation. Such colonies may show the following properties: (a) medium to large in size, circular or slightly oval, slightly turbid, smooth surfaced, with a peripheral flattened zone; (b) size and shape as in (a), but slightly convex, smooth, with a shiny surface and entire margin; (c) large, convex, very mucous.

If colonies with these characteristics have been observed and grading is considered necessary, blood agar [159] and brilliant green agar [13] media are inoculated. The media are incubated at 35–37 °C for 24 h and are then maintained at room temperature overnight. If characteristic colonies are observed after the initial incubation, further incubation at room temperature may be omitted.

On blood agar, characteristic colonies are usually surrounded by a beta-haemolytic zone; a greenish or greenish–brown pigment is produced in the medium, and the colonies give a characteristic odour (after incubation for a few days, the colonies may give an odour of ammonia).

The colonies on brilliant green agar generally appear the same as those on blood agar, except that extremely flat and thin colonies may also occur.

Colonies developing on blood agar are examined for their Gram reaction, motility, oxidase activity, and the solubility of the pigment in chloroform. The oxidase test in this case requires no more than half a minute for development.

A microbe can be identified as *P. aeruginosa* if it shows typical colony morphology and microscopic appearance, as well as the characteristic haemolytic properties and odour, the cultures should be positive for oxidase and the pigment produced by the strain should be soluble in chloroform.

In addition to these characteristics of the colonies, other findings may also draw attention to the presence of *P. aeruginosa*. Thus the characteristics of colonies growing on brilliant green or DC medium [13, 26] (inoculated either directly or from selenite enrichment medium) are particularly significant. (The characteristics of colonies growing on brilliant green medium have been described above.) Suspected colonies of *P. aeruginosa* growing on DC medium are usually large, mucous, viscous, convex, strongly pink in colour, with a shiny surface and an odour unlike that of lime-blossom.

Obviously, colonies suggestive of *P. aeruginosa* may be observed on a number of different media, beginning in most cases with blood agar plates. Colonies developing on this medium are subcultured on both blood agar and brilliant green plates, and further investigation may be carried out as described above. If the results of the initial examination are ambiguous, a complete identification procedure should be carried through.

The negative results of examinations carried out to check a suspected presence of *P. aeruginosa* are not reported. Positive results are recorded together with the order of magnitude of the highest dilution giving a positive reading.

6.3.6.3   Target examinations for *P. aeruginosa*

In the processing of the sample and in the preparation of the stock dilution and serial dilutions, the nature of the foodstuff must be taken into account. From each dilution, 0·1-ml aliquots may be used to inoculate plates, or 1·0-ml amounts to inoculate enrichment media. A dilution range of 10^1–10^4 is the most practicable. Where plates are used, 1 blood agar plate and 1 brilliant green plate are inoculated from each dilution, or a plate of modified cetrimide (GMAC) medium [19] is inoculated. If characteristic colonies are found on the blood agar and brilliant green media, the tests outlined above are carried out.

GMAC plates are incubated for 24 h at 42 °C. Colonies growing in abundance, surrounded by a pink zone on account of ammonia production from acetamide, indicate a positive result. The result is negative if a yellow area is seen around colonies. If on GMAC medium, characteristic abundantly growing colonies are produced at 42 °C, the presence of *P. aeruginosa* can be regarded as highly probable. If confirmation is considered necessary, the cultures should be examined for oxidase activity and decomposition of glucose.

Half-strength selenite medium, i.e. medium [130] diluted with an equal amount of saline can be used as an enrichment medium. It is incubated at 37 °C for 48 h after inoculation.

A procedure which is both sensitive and selective is to inoculate NF broth [105] as the enrichment medium and then subculture on GMAC plates. Aliquots of 1 ml of the dilutions which are to be examined are transferred into NF tubes, or if a larger volume of medium is used, the sample itself may be placed in the medium (c.f. similar tests for *Salmonella*). To increase the selectivity, the 24-h incubation is carried out at 42 °C, preferably in a water bath. Bacterial growth is indicated by an increasing turbidity of the medium. No turbidity occurring after incubation for 48 h is regarded as negative. From the positive cultures, GMAC agar slopes are inoculated by streaking the surface of the medium. The tubes are then incubated at 42 °C for 24 h, whereupon positive cultures are subcultured. The GMAC cultures are evaluated, and if necessary, further examination is carried out as described above.

The details of the presence-or-absence test, determination of the titre, calculation of the MPN, and determination colony counts are the same as those described for *S. aureus*.

Demonstration of the presence or absence of the microbe in known dilutions of the sample, and/or determination of the titre may be regarded as the most important of the tests.

The presence of *P. aeruginosa* on desinfected surfaces is unacceptable. Nor can this organism be tolerated in least of all drinking-water; in this context, *P. aeruginosa* may be regarded as an indicator of faecal contamination. The appearance of *P. aeruginosa* in foodstuffs, especially milk and dairy products, may stem from the environment or from cows suffering from mammillitis. Thus if *P. aeruginosa* is detected in foodstuffs the possibility must be considered that contaminated water has been used, or that there are deficiencies in cleaning and disinfection, e.g. disinfectants have been used that are ineffective against *P. aeruginosa*.

6.4   Yeasts

6.4.1   *Debaryomyces* Lodder et Kreger-van Rij

Vegetative reproduction proceeds by multilateral budding. The cells are variable in shape. Rudimentary, occasionally well-developed, pseudomycelium may develop. Ascus formation is preceded by heterogamous conjugation between mother-cell and bud; isogamous conjugation may also occur. The ascospores (2.8.39) are spherical or oval, and the ascospore wall is thickened in a lath-like arrangement. There are 1, 2, or (occasionally) 4 spores per ascus [2, 87]. Fermentation is slow, weak, or absent. Nitrate (2.8.7) is not assimilated [103].

6.4.1.1   *Debaryomyces hanseni* (Zopf) Lodder et Kreger-van Rij
(= *Debaryomyces kloeckeri* Guilliermond et Péju)

After incubation for 2 days at 25 °C in malt extract [90], the cells are spherical or slightly oval, $2-7 \times 2 \cdot 5 - 8 \cdot 5$ µm, occurring singly, in pairs, or in short chains. A sediment, a surface rim and dry surface skim are formed after incubation at 17 °C for one month.

After incubation for 2 days at 25 °C on malt agar [90], the cells are spherical or slightly oval, $2-5 \times 2-7$ µm, occurring singly or in pairs. After incubation at 17 °C for a month, colonies are grayish–white or yellow, soft, bright, smooth, or finely wrinkled.

On potato extract agar [17], the pseudomycelium is poorly developed or absent, occasionally well-developed.

Ascospore formation is preceded by conjugation between mother-cell and bud. One (rarely 2) spherical, warty ascospore (readily visible in the electron microscope) is formed.

Fermentation D + (very weak) or − ; G + (very weak) or − ; S + (very weak) or − ; M + (very weak) or − ; L −.*
Sugar assimilation: D + , G + , S + , M + , L + or − , R + .
Ethanol + .

* D = glucose, G = galactose, S = sucrose, M = maltose, L = lactose, R = raffinose.

250

Splitting of arbutin: positive.

Assimilation of nitrate: negative.

Growth on vitamin-free medium [160]: negative.

Growth at 37 °C: negative.

Note. Torulopsis candida is regarded as the imperfect form of *D. hanseni*.

6.4.2   *Hansenula* H. et P. Sydow

Asexual reproduction proceeds by the formation of comparatively few buds at various sites on the cell surface. The cells are spherical, ellipsoidal, or elongated-cylindrical, sometimes tapering toward both ends, or filamentous. Both pseudo-mycelium and true-mycelium may develop. The asci resemble the vegetative cells in shape, and contain 1–4 spherical, hat-shaped or Saturn-shaped ascospores. The ring of the latter is generally easily visible, but may be very thin. In the light microscope, the ascospore appears smooth-walled. The species of *Hansenula* are haploid or diploid, homothallic or heterothallic. Their cultures are generally long-lived. They assimilate nitrate, but do not assimilate starch. They do or do not ferment sugars. Skin production occurs or absent. They may produce ethers.

6.4.2.1   *Hansenula anomala* (Hansen) H. et P. Sydow

The cells growing on agar are very variable in size, being 2–4·8 × 2·6–5 μm and forming glistening colonies, or as long as 30 μm and forming dull colonies. The colonies of fresh isolates are shiny, smooth, and waxy; those of strains maintained under artificial conditions are white, powdery in appearance, and rough. In malt extract culture [90], the rough colonies form a gray or white surface skin and a surface rim; the smooth colonies form either a thin surface skin, or no skin.

On Dalmau plates [17], the smooth colonies do not form hyphae, whereas the dull colonies form highly branching pseudomycelia; hyphae with septa are formed notably by lime-white colonies.

The diploid cells develop into asci directly; 1–4 hat-shaped ascospores are formed in each ascus. Fermentation: D +, G + (occasionally −), S +, M + (often weakly, sometimes −), L −, R + (1/3).

Sugar assimilation: D +, G + (occasionally −), S +, M +, L −, R +.

Ethanol: +.

Nitrate assimilation: positive.

Growth on vitamin-free medium: positive.

6.4.3 *Kluyveromyces* Van der Walt emend. Van der Walt

Asexual reproduction occurs by budding. The cells are spherical, ellipsoidal, cylindrical or elongate. Pseudomycelium may develop, but true mycelium does not occur. The ascospores are crescent or kidney-shaped, elongate, blunt, smooth-walled, with one or more ascospores per ascus. Metabolism is both oxidative and fermentative. External vitamin sources are required. Nitrate is not assimilated. A red non-carotenoid pigment may be formed.

6.4.3.1 *Kluyveromyces fragilis* (Jörgensen) Van der Walt
(= *Saccharomyces fragilis* Jörgensen)

After incubation in malt extract [90] for 3 days at 28 °C, the cells are spherical or cylindrical 2–3 × 3·5 − 10 μm, arranged singly, in pairs, or in short chains. A powdery sediment and surface ring are formed after incubation for one month at room temperature. After incubation on malt agar [90], for 3 days at 28 °C, the cells are spherical or cylindrical, 2–5·5 × 3·5 − 11 μm, occurring singly, in pairs, or in short branching chains. Usually a pseudomycelium develops. The colony is waxy, cream-coloured, gleaming, flat, smooth, sometimes divided into sectors, and has an entire or wavy margin. The one-month-old colony is grayish–brown.

On Dalmau plates [17], pseudomycelium tends to develop in aerobic conditions; this is generally rudimentary, rarely highly branched. Occasionally blastospores are formed in small groups along the elongated cells.

In general, diploid cells develop into asci, although ascus formation may arise by conjugation as well. 1–4 kidney-shaped ascospores are formed in each ascus.
Fermentation: D +, G +, S +, M −, L + (rapid), R +.
Assimilation of sugars: D +, G +, S +, M −, L +, R +.
Ethanol: +.
Splitting of arbutin: positive.
Nitrate assimilation: negative.
Growth on vitamin-free medium: negative.
Growth at 37 °C: positive.

6.4.3.2 *Kluyveromyces lactis* (Dombrowski) Van der Walt
(= *Saccharomyces lactis* Dombrowski)

In malt extract culture after 3 days at 28 °C [90], the cells occur singly, in pairs, or sometimes in small groups; they are spherical or oval, 2–6·5 × 3–8 μm. A sediment, a surface rim and a thin skin are formed at room temperature after incubation for a month.

252

After incubation for 3 days at 28 °C on malt agar [90], the cells are oval or elongated, 2–5·5 × 3–7 µm, occurring singly, in pairs, or in small groups. The colony is waxy, cream or brownish in colour, shiny, smooth, sometimes with a wavy margin. After incubation for a month, the colony is darker, often with a pink hue.

On Dalmau plates [17] a rudimentary pseudomycelium may be formed, generally under the surface of the medium.

The vegetative cells develop into asci directly or by conjugation of the ascospores. 1–4 spherical ascospores are formed in each ascus.

Fermentation: D +, G +, S + (rarely −), M − (rarely weak +), L +, R + (rarely −).

Assimilation of sugars: D +, G +, S +, M + (or −), L +, R + (rarely −).

Ethanol: positive.

Splitting of arbutin: positive.

Nitrate assimilation: negative.

Growth on vitamin-free medium: negative.

Note. Torulopsis sphaerica is regarded as the imperfect form of *K. lactis*.

6.4.4   *Metschnikowia* Kamienski

Asexual reproduction proceeds by multilateral budding. The cells are spherical, ellipsoidal, pear-shaped, or cylindrical. The pseudomycelium is rudimentary, rarely absent. The ascus is elongated and club-shaped, stalked, containing 1 or 2 ascospores. The ascospores are needle-shaped, tapering at both ends, and without a flagelliform appendix. A few species parasitize invertebrates, others occur in water or in the soil.

6.4.4.1   *Metschnikowia pulcherrima* Pitt et Miller
   (= *Candida pulcherrima* [Linder] Windisch)

In glucose–yeast extract–peptone water [50], the cells bud multilaterally and are spherical or oval, reaching 2·5 − 5 × 4–7 µm in size after 3 days' incubation at 25 °C. The chlamydospores are strongly refractive containing one or more lipid granules, sphaerical 7–11 µm in diameter, taking a month to develop at 25 °C.

After incubation on glucose–yeast extract agar at 25 °C for one month, the colonies are either cream-coloured or reddish–brown (due to the pigment pulcherrimin), shiny, and smooth-surfaced.

On Dalmau plates [17], a rudimentary pseudomycelium is formed beneath the surface of the medium. Fermentation: glucose generally is the only carbohydrate fermented; some strains weakly ferment galactose.

Assimilation of sugar: D +, G +, S +, M +, L −, R −.

Ethanol: +.

Assimilation of nitrate: negative.

Growth in vitamin-free medium: negative (growth is stimulated by biotin).

Glucose tolerance: 45–50%.

NaCl tolerance: 12–14%.

Maximum growth temperature: 33–38 °C.

6.4.5 *Pichia* Hansen

The cells bud multilaterally and are of various shapes; most of the strains develop pseudomycelia, some of them producing a true mycelium of limited growth. The cell develops into an ascus directly, although isogamous or heterogamous conjugation may also occur. The ascospores are spherical, smooth, or warty, hat-shaped or Saturn-shaped, usually containing one oil droplet. 1–4 spores are formed in each ascus. Fermentation may or may not occur.

Nitrate is not assimilated.

6.4.5.1 *Pichia membranaefaciens* Hansen

After incubation in malt extract [90] for 2 days at 25 °C, the cells are observed to occur singly, in pairs, and in chains; the cells may be oval or more elongated, 2–5·5 × 4·5–20 µm. A surface rim, a sediment, and a dull, dry, smooth or rough surface skin are formed.

After incubation on malt agar [90] for 2 days at 25 °C, the cells are observed to occur singly, in pairs, or in chains; they are oval or more elongated, 2–4·5 × 5–20 µm. After incubation for a month at 17 °C, the colonies are yellowish–white or yellowish–brown, soft, dull, smooth, and flat.

On potato agar [17] the pseudomycelium is well-developed, dendroid.

Ascospore formation occurs by conjugation: 1–4 ascospores are formed in each ascus, ascospores are sphaerical, their walls are thickened in a lath-like arrangement. Colonies with many ascospores appear brownish.

Fermentation: D+ (or very weak, sometimes −), G−, S−, M−, L−.

Assimilation of sugars: D+, G−, S−, M−, L−, R−.

Ethanol: +.

Splitting of arbutin: negative.

Nitrate assimilation: negative.

Growth on vitamin-free medium [160]: positive or negative.

Note. Candida valida is regarded as the imperfect form of *P. membranaefaciens*.

254

6.4.6   *Saccharomyces* Meyen emend. Reess

The cells arise by multilateral budding; they are spherical, ellipsoidal, cylindrical, or elongated. Pseudomycelium may be produced, but true mycelium never occurs. On liquid medium [90], a surface growth may develop after prolonged incubation. A surface skin, if formed, is never powdery, neither dry nor creeping. The ascus does not disrupt at maturity. Usually 1–4 spherical or slightly ellipsoidal ascospores are formed in each ascus. Neither lactose nor nitrate is utilized.

6.4.6.1   *Saccharomyces bayanus* Saccardo
(= *Saccharomyces pastorianus* Hansen)

After incubation in malt extract [90] for 3 days at 28 °C, the cells are spherical or elongated, 3·5–7·5 × 4·5–10 (or up to 17·5) μm; they occur singly, in pairs, or in short chains. After incubation for one month, a sediment, often also a surface rim, is produced.

After 3 days on malt agar [90] at 28 °C, the cells are spherical or elongated, 2·5–7·5 × 4–10·5 (up to 16·5) μm, occurring singly, in pairs, or in short chains. The colony is waxy, cream or brownish-cream in colour, smooth or granular, shiny or dull. After one month, the colony is striated with a wavy margin.

On Dalmau plates [17], a rudimentary, sometimes branching pseudomycelium is formed.

The diploid cells develop into asci directly, forming 1–4, rarely more, smooth, spherical ascospores.

Fermentation: D+, G−, S+, M+ (occasionally slow), L−, R+ (1/3).

Assimilation of sugars: D+, G−, S+, M+, L−, R+.

Ethanol: +, occasionally slow.

Splitting of arbutin: negative.

Assimilation of nitrate: negative.

Growth on vitamin-free medium: variable.

6.4.6.2   *Saccharomyces bisporus* (Naganishi) Lodder et Kreger-van Rij
var. *mellis* (Fabian et Quinet) Van der Walt
(= *Saccharomyces mellis* [Fabian et Quinet] Lodder et Kreger-van Rij)

After incubation in malt extract [90] for 3 days at 28 °C, the cells are spherical or ellipsoidal, 2·5–5·5 × 3·5–7·5 (or up to 13) μm, occurring singly, in pairs, or in short chains. After one month, a sediment and surface rim are formed.

After incubation in malt agar [90] for 3 days at 28 °C, the cells are found to occur singly, in pairs, or in short chains. The colony is waxy, white, cream or brownish-cream in colour, shiny (rarely granular), with a wavy margin.

1–4 spherical ascospores are formed by conjugation, or by the direct development of asci.

Fermentation: D+, G−, S− (or +, weak and slow), M−, L−, R−.

Assimilation of sugars: D+, G+ (or −), S− (rarely +, very weak), M−, L−, R−.

Ethanol: + (slowly).

Splitting of arbutin: negative.

Assimilation of nitrate: negative.

Growth on 60% glucose–yeast extract agar: positive.

6.4.6.3 *Saccharomyces cerevisiae* Hansen
(= *Saccharomyces ellipsoideus* Hansen)

After incubation in malt extract [90] for 3 days at 28 °C, 3 size-groups of cells can be observed, *viz.* 4·5–10·5 × 7–21 μm, 2·5–7 × 4·5–11 (or up to 18·5) μm, and 3·5–8 × 5–11·5 (or up to 17·5) μm. After one month, a surface rim and skin are formed.

After incubation on malt agar [90] for 3 days at 28 °C, the distribution of cells by size is similar to that observed in malt extract. The colony is waxy, cream or brownish in colour (after one month, grayish–brown), smooth, striated, divided into sectors, or granular with a wavy or lobular margin.

On Dalmau plates [17], a rudimentary pseudomycelium is formed, especially under anaerobic conditions.

The vegetative cells develop into asci directly, producing 1–4 spherical or slightly oval ascospores.

Fermentation: D+, G+ (occasionally slow), S+, M+ (occasionally slow), L−, R+ (1/3).

Assimilation of sugars: D+, G+ (occasionally slow), S+, M+ (occasionally slow), L−, R+.

Ethanol: + (or −).

Splitting of arbutin: negative.

Assimilation of nitrate: negative.

Growth on vitamin-free medium: variable.

6.4.6.4 *Saccharomyces rouxii* Boutroux

After incubation in malt extract [90] for 3 days at 28 °C, the cells are spherical or elongated, 3–7·5 × 4–9·5 (or up to 15) μm, or 2·5–6 × 4–9 (or up to 15) μm, occurring singly, in pairs, or in short chains. After one month, a sediment and a surface rim, sometimes a dull surface skin, are formed.

256

On malt agar [90], the cells are similar in size and shape to those growing in liquid culture. The colony is waxy, cream or brownish in colour, smooth or granular, sometimes with wavy margin.

On Dalmau plates [17], a rudimentary pseudomycelium may be formed, under anaerobic conditions.

Asci usually arise by conjugation, rarely from diploid cells; 1–4 (most frequently 2) spherical or ellipsoidal ascospores are formed in each cell.

Fermentation: D+, G−, S−, M+ (occasionally slow), L−, R−.

Assimilation of sugars: D+, G+ (rarely –), S− (or +), M+, L−, R−.

Ethanol: + (or −).

Splitting of arbutin: negative.

Assimilation of nitrate: negative.

Growth on vitamin-free medium: negative.

6.4.6.5 *Saccharomyces uvarum* Beijerinck
(= *Saccharomyces carlsbergensis* Hansen)

After incubation in malt extract [90] for 3 days at 28 °C, the cells are spherical, ellipsoidal or elongated, occurring singly, in pairs, or sometimes in short chains. The cells can be divided into 3 groups with respect to cell width, *viz*. 4–10 × 5·5–16 (or up to 25) µm; 2·5–6·5 × 5–11·5 (or up to 22) µm; 3·5–8 × 5–11 (or up to 20) µm.

Very long cells (up to 30 µm) may occur in each group. After incubation for a month at room temperature, a sediment and surface rim are formed.

After incubation on malt agar [90] for 3 days at 28 °C, the cells are similar in size and shape to those growing in liquid culture. The colony is waxy, cream or brownish in colour, smooth or granular, with a wavy margin; no change is observed during further incubation for a month.

On Dalmau plates [17] a very rudimentary, pseudomycelium may be formed.

The vegetative cells develop into asci directly, with 1–4 spherical spores per cell.

Fermentation: D+, G+ (occasionally slow), S+, M+ (occasionally slow), L−, R+ (complete).

Assimilation of sugars: D+, G+, S+, M+, L−, R+.

Ethanol: +.

Splitting of arbutin: negative.

Growth in vitamin-free medium: variable.

6.4.7 *Candida* Berkhout

The cells are spherical, oval, cylindrical or elongated; ogival, tapering and flask-shaped cells do not occur. Asexual reproduction occurs by multilateral budding; bipolar budding does not occur. Pseudomycelium, true mycelium and chlamydospores may occur. Arthrospores, ascospores, teliospores and ballistospores never occur. A visible carotenoid pigment is not formed. Extracellular polysaccharide formation (positive iodine reaction) may occur. Many species give rise to alcoholic fermentation.

6.4.7.1 *Candida kefyr* (Beijerinck) van Uden et Buckley
(= *Cryptococcus kefyr* [Beijerinck] Skinner)

After incubation in glucose–yeast extract–peptone water [50] for 3 days at 25 °C, the cells are oval, 3·5–9 × 6–14 μm.

On glucose–yeast extract–peptone agar [50], after incubation for a month at 25 °C the colony is dull, cream or yellowish in colour.

On Dalmau plates [17], an abundantly branched pseudomycelium appears, with very little blastospore formation.

Fermentation: D+, G+, S+, M−, L+, R+ (or weakly +).

Assimilation of sugars: D+, G+, S+, L+, R+, M−.

Ethanol: + (or weak).

Assimilation of nitrate: negative.

Growth on vitamin-free medium: negative.

NaCl tolerance: 5–7%.

Maximum temperature for growth: 37–42 °C.

6.4.7.2 *Candida krusei* (Cast.) Berkhout

After incubation in glucose–yeast extract–peptone water [50] for 3 days at 25 °C, the cells are oval or cylindrical, 3–5 × 6–20 μm. Powdery, thin growths spreading over the inner surface of the culture vessel are often seen.

After incubation on solid medium [50] for a month at 25 °C, the colony is grayish–yellow, dull, smooth, or slightly wrinkled.

On Dalmau plates [17], a well-developed pseudomycelium appears.

Fermentation: D is the only sugar fermented.

Assimilation of sugars: D+, G−, S−, M−, L−, R−.

Ethanol: +.

Assimilation of nitrate: negative.

Growth on vitamin-free medium: very good.

NaCl tolerance: 5–10%.

Maximum temperature for growth: 43–45 °C.

Note. C. krusei is the imperfect form of *Pichia kudriavzevii*.

6.4.7.3   *Candida lipolytica* (Harrison) Diddens et Lodder

After incubation in glucose–yeast extract–peptone water [50] for 3 days at 25 °C, the cells are oval or elongated, 3–5 × 5–11 µm, up to 20 µm long. A surface skin is formed.

On solid medium [50], the colony is cream-coloured, dull, in some strains slimy or wrinkled.

On Dalmau plates [17], a well-developed pseudomycelium and a true mycelium with septa are formed.

Fermentation: absent.

Assimilation of sugars: D+, G−, S−, M−, L−, R−.

Ethanol: +.

Assimilation of nitrate: negative.

Growth on vitamin-free medium: weak.

NaCl tolerance: 10–14%.

Maximum temperature for growth: 33–37 °C.

6.4.7.4   *Candida utilis* (Henneberg) Lodder et Kreger-van Rij
(= *Cryptococcus utilis* [Henneberg] Anderson et Skinner)

After incubation in glucose–yeast extract–peptone water [50] for 3 days at 25 °C the cells are oval or cylindrical, 3·5–4·5 × 7–13 µm. A thin surface rim may be formed.

After incubation on solid medium [50] for a month, the colony is grayish or cream-coloured, dull, rarely shiny, smooth.

On Dalmau plates [17], a rudimentary pseudomycelium is formed.

Fermentation: D+, G−, S+, M−, L−, R+.

Assimilation of sugars: D+, G−, S+, M+, L−, R+.

Ethanol: + (or weak).

Assimilation of nitrate: positive.

Growth on vitamin-free medium: weak or good.

NaCl tolerance: 6–8%.

Maximum temperature for growth: 39–43 °C.

6.4.7.5   *Candida valida* (Leberle) van Uden et Buckley
(= *Candida mycoderma* [Reess] Lodder et Kreger-van Rij)

After incubation in glucose–yeast extract–peptone water [50] for 3 days at 25 °C, the cells are elongated, curved and sausage-shaped, 2–4 × 4–10 µm, occurring in chains or groups. A dry surface skin may occur.

After incubation on solid medium [50] for a month at 25 °C, the colony is gray, dull, powdery, and smooth or finely wrinkled.

17*

On Dalmau plates [17] a pseudomycelium consisting of long, branched chains of cells appears.

Fermentation: nil, except for weak fermentation of glucose.

Assimilation of sugars: D+, G−, M−, L−, R−.

Ethanol +.

Assimilation of nitrate: negative.

Growth in vitamin-free medium: good or weak.

NaCl tolerance: 2–9%.

Maximum temperature for growth: 33–41 °C.

Note. C. valida is the imperfect form of *Pichia mambranaefaciens.*

6.4.8 *Torulopsis* Berlese

The cells are spherical or oval, rarely elongated. Reproduction proceeds by multilateral budding; pseudomycelium is rudimentary or absent. No ascospores, teliospores, ballistospores, endospores or arthrospores are formed. No visible carotenoid pigment is produced. Extracellular polysaccharide, if produced, does not give a reaction with iodine. Inostilol is not assimilated. Alcoholic fermentation is brought about by many of the species.

6.4.9 *Trichosporon* Behrend

The cells are of various shapes. The pseudomycelium may be well-developed or reduced. True mycelium and arthrospores are always formed. A surface rim and skin commonly occur. Neither ascospores nor teliospores are formed; asexual endospores may be formed. Alcoholic fermentation is weak or nil.

6.4.10 *Brettanomyces* Kufferath et van Laer

Asexual reproduction proceeds by budding. The cells are ellipsoidal, often ogival and elongated. A fully-developed, or sometimes rudimentary pseudomycelium is formed. Ascospores are not formed. Growth is slow and cultures are short-lived on malt agar [90]. Characteristic odours are produced. Under aerobic conditions, acetic acid is rapidly produced from glucose.

6.4.11   *Kloeckera* Janke

The cell is lemon-shaped, oval, sausage-shaped, or irregularly elongated. Asexual reproduction by bipolar budding. Most strains do not form pseudomycelium. Ascospores are not formed. Some sugars are fermented; nitrate is not assimilated. Every species requires inositol and pantothenic acid.

6.4.11.1   *Kloeckera apiculata* [Reess emend. Klöcker] Janke

After incubation in malt extract [90] for 3 days at 25 °C, the cells are lemon-shaped, oval, or elongated, 1·4–5·3 × 2·6–12·2 µm, occurring singly, in pairs, or in groups of 3 or 4; budding is bipolar. A surface rim and sediment are formed after incubation for a month.

  After incubation on malt agar [90] for 4 weeks at room temperature, the colonies of most strains are grayish, sometimes yellowish-cream in colour, smooth, shiny, with irregular margins.

  On Dalmau plates [17], usually no pseudomycelium is formed, or the pseudomycelium is dendroid.
Fermentation: D+, G−, S−, M−, L−, R−.
Assimilation of sugars: D+, G−, S−, M−, L−, R−.
Ethanol: −.
Assimilation of nitrate: negative.
Growth in vitamin-free medium: negative.
Note. K. apiculate is the imperfect form of *Hanseniaspora valbyensis* and *H. uvarum*.

6.4.12   *Rhodotorula* Harrison

The cells are spherical, oval, or elongated, with multilateral budding. Chlamydospore-like cells, and/or pseudomycelium, or true mycelium are formed according to the strain. Neither ascospores, nor ballistospores occur. Red and/or yellow carotenoid pigments are produced. Inositol is not assimilated, starch-like substances are not produced, and fermentation does not occur. The colonies of the capsule-producing strains are slimy; other strains develop waxy, or dry, wrinkled colonies. Cultures on chalk agar produce acid. Gelatin liquefaction usually does not occur.

Table 44

A survey of the most important biochemical characteristics of the yeast species covered in this book

Species	Fermentation of						Assimilation of							Splitting of arbutin	Nitrate assimilation	Growth in vitamin-free medium
	D	G	S	M	L	R	D	G	S	M	L	R	Et			
Debaryomyces hansenii	+, –	+, –	+, –	+, –	–	–	+, –	+	+	+	+, –	+	+	+	–	–
Hansemula anomala	+	+, –	+	+, –	–	+, 1/3	+	+, –	+	+	–	+	+		+	+
Kluyveromyces fragilis	+	+	+	–	+	+	+	+	+	–	+	+	+	+	–	–
Kluyveromyces lactis	+	+	+, –	–, +	+	+, –	+	+	+	+, –	+	+, –	+	+	–	–
Metschnikowia pulcherrima	+	+, –	–	–	–	–	+		+	+	–	–	+	–	–	
Pichia membranaefaciens	+, –	–	–	–	–	–	+	–	–	–	–	–	+	–	–	+, –
Saccharomyces bayanus	+	–	+	+	–	+, 1/3	+	–	+	+	–	+	+	–	–	+, –
S. bisporus var. mellis	+	–, +	–, +	–	–	–	+	+, –	–, +	+	–	–	+, –	–	–	+, –
S. cerevisiae	+	+	+	+	–	+, 1/3	+	+	+	+	–	+	+	–	–	+, –
S. rouxii	+	–	+	+	–	–	+	+, –	–, +	+	–	–	+, –	–	–	+, –
S. uvarum	+	+	+	+	–	+, 1/1	+	+	+	+	–	+	+		–	+, –
Candida kefyr	+	+	+	–	+	+	+	+	+	–	+	+	+	–	–	+
C. krusei	+	–	–	–	–	–	+	+	–	–	–	–	+		–	+
C. lipolytica	–	–	–	–	–	–	+	+	–	–	–	–	+		–	+
C. utilis	+	–	+	+	–	+	+	–	+	+	–	+	+		+	+
C. valida	±	–	–	–	–	–	+	–	–	–	–	–	+		–	+
Kloeckera apiculata	+	–	–	–	–	–	+	–	–	–	–	–	–		–	–
Rhodotorula glutinis							+	+	+	+	+, –	+, –	+, –		+	+

* Et = etanol.

Symbols: D = glucose; G = galactose; S = sucrose; M = maltose; L = lactose; R = raffinose.

6.4.12.1   *Rhodotorula glutinis* (Fresenius) Harrison

After incubation in malt extract [90] for 3 days at 23 °C, the cells are oval, 2·3–5 × 4–10 μm; among populations of some strains, cells 12–16 μm long have been found; these may be as much as 7 μm wide. A conspicuous surface rim and sediment form after a month.

On malt agar [90], the cells are somewhat longer, and after a month, the colonies are coral red, salmon, or orange coloured, shiny and slimy at the surface, or of creamy consistency with irregular margins.

On Dalmau plates [17], pseudomycelium is generally not formed; sometimes both true mycelium and pseudomycelium are observed, both of rudimentary structure.

Assimilation of sugars: D+, G+, S+, M+, L−, R+ (occasionally −).

Ethanol: + (rarely −).

Assimilation of nitrate: positive.

Growth in vitamin-free medium: positive (weak growth can be stimulated with thiamine).

*

The most important of the biochemical characteristics of the above species of yeasts are summarized in *Table 44*.

6.5   Moulds

6.5.1   *Mucor* Micheli

The colony [23, 90] is white or otherwise coloured, standing a few millimetres or several centimetres high. The sporangiophore is simple or branched. The sporangium always has a columella; apophysis does not occur; the sporangial wall breaks down or disintegrates into pieces. The spores are spherical or elongated. Some species form zygospores.

6.5.1.1   *Mucor mucedo* (L.) Fresenius

The colony [23, 90] is silver gray, later yellowish, standing 1 cm high when growing in darkness, and 3–15 cm in the light. The sporangiophore may reach a width of 70 μm. The sporangia developing on longer sporangiophores are 80–200 μm in diameter, those on short ones are smaller. The columella is pear-shaped or cylindrical. The spores are cylindrical or oval 3–7 × 6–13 μm.

6.5.1.2 *Mucor racemosus* Fresenius

The colony [23, 90] is white, yellowish, later brownish, standing 0·2–3 cm high. The sporangiophore branches racemosely or sympodially; strongly refractive gemmae are formed in abundance, even in the hyphae and in the sporangiophore. The average diameter of the sporangia is 70 μm; the columella is spherical or oval. The spores are oval, 3–10 μm in length.

6.5.1.3 *Mucor rouxianus* (Calmette) Wehmer

The colony [23, 90] is white, pale yellow or gray, flat, at most 0·4 cm in height. The sporangiophore is little-branched, sympodial. The sporangium is 20–100 μm, most frequently 50 μm in diameter. The columella is spherical, slightly flattened, 40 μm or less in diameter. Gemmae are formed in abundance. Zygospores are not formed.

6.5.2 *Rhizopus* Ehrenberg

The sporangiophores occur in groups arising from stolons provided with root-like processes (rhizoids) [23, 90]. Less frequently, the sporangiophore develops from the mycelium, or as a side branch of a hypha; initially it is colourless, but later the sporangiophore wall becomes dark brown. Sporangia are generally large, white when young, becoming dark grayish–brown or black with age. The sporangiophore widens like a funnel bellow the sporangium, forming an apophysis. The columella is spherical or hemispherical. Spores are spherical, oval, or irregularly polygonal (in some species striated). Gemmae and zygospores are frequently formed.

6.5.2.1 *Rhizopus oryzae* Went et Prinsen

This species differs from *Rhizopus stolonifer* in that its spores are smaller (7–9 μm), and it develops gemmae abundantly. It grows well at 30–40 °C.

6.5.2.2 *Rhizopus stolonifer* (Ehrenberg ex Fries) Lind.
(= *Rhizopus nigricans* Ehrenberg)

The sporangiophore is initially colourless, later dark brown; 2 to 5 are formed in each group, the groups arising from brown rhizoids. A new group of sporangiophores arises at the point of emergence of stolons growing out from the same site. The young sporangium is white and shiny, later black, in old cultures dull gray,

264

150–350 µm in diameter. The spore is irregularly oval, 10–15 (or up to 20) × 7 µm, with a pattern of striations. Gemmae are not formed. The zygospores are 150–200 µm, black, with sparse warts on the surface.

6.5.3 *Thamnidium* Link

The sporangiophore [23, 90] is tall and well-developed, the tip supporting a sporangium or remaining sterile. The columella resembles that of *Mucor*. On the sporangiophore, short, thinner side branches grow out in whorls. These sporangiophores are dichotomously branched, and at the ends of the side branches, sporangioles without columellae are formed, each containing a few spores. The wall of the sporangium tends to burst open, that of the sporangioles does not open thus. Zygospores may occur.

6.5.3.1 *Thamnidium elegans* Link ex Wallr.

The colony is initially white, later pale gray, sometimes reaching a height of 2 cm. The sporangiophore may be as broad as 32 µm; the sporangium is 40–200 µm with a columella 30–60 × 50–75 µm. The whorled side branches of the sporangiophore are dichotomously branched. The sporangioles are 8–16 µm, each containing 2–6 spores. The spores (those formed in the sporangium and in the sporangioles are identical in shape) are oval, ellipsoidal, smooth-walled, colourless, 4·5–8 × 6·5–15·5 µm. The zygospores are 52–156 µm.

6.5.4 *Byssochlamys* Westling

The asci are spherical, developing without a peridium in groups visible to the naked eye *(Endomycetales)* [23, 90]. The colony is loose, flat, growing optimally at temperatures over 30 °C. Its conidial form is *Paecilomyces*. The phialides are arranged on the conidiophore irregularly or in whorls, sometimes in brushes. In the latter case, the conidiophore resembles that of *Penicillium*.

6.5.4.1 *Byssochlamys fulva* Olliver et Smith (Stat. conid.: *Paecilomyces varioti* Bainier)

The colony grows rapidly in the temperature range 30–37 °C; it is white, dull, later brownish. The underside of the colony is also brown. The asci form spherical groups without a peridium. The ascus is spherical, eight-spored, 11–12 µm in diameter. The ascospore is oval, colourless, smooth-walled, 6–6·5 × 4·3–4·5 µm.

The conidiophore is 2–3 μm broad, otherwise variable in shape and size. The sterigmata (phialides) measure 25×2–3 μm, and are irregularly distributed or arranged in whorls. The conidia are colourless, unicellular, elongated-egg-shaped, 4–9×2.3–2.5 μm, forming very long chains.

Aspergillus repens (Corda) de Bary
The asci are formed in a cleistothecium *(Eurotiales)*. For details see 6.5.14.1.

6.5.5 *Neurospora* Shear et Dodge

The asci are elongated-club-shaped; they are formed inside a dark, true peri-thecium, arranged in a hymenium *(Sphaeriales)*. The ascospore is striated, unicellular, dark, without germ split, mucous coat and appendages. The conidial form is *Monilia,* producing blastoconidia which form branching chains. These fungi are heterothallic, and, therefore, the culture often consists exclusively of the conidial form.

6.5.5.1 *Neurospora sitophila* Shear et Dodge (Stat. conid.: *Monilia sitophila*/Mont./Sacc.)

The colony grows very rapidly at room temperature. The conidial form pre-dominates, and, therefore, the culture is meat-coloured or orange-coloured. It has a large oxygen requirement, and is often found growing on the outer surface of the Petri dish lid. If compatible strains are cultured together, black pear-shaped perithecia are formed. The ascospores are dark, unicellular, lemon-shaped, longitudinally striated, 26×12 μm. The conidia are blastospores developing in simple or branching chains, spherical, oval, or irregularly shaped, 5–14 μm.

6.5.6 *Trichothecium* Link

The conidiophore is erect, simple, sometimes slightly branched, septate, colourless. The conidium is colourless or pale, unequally 2-celled, the upper cell being larger than the lower which terminating in a short beak. The conidia are formed singly, although consecutively produced cells are stuck together with a mucous substance. The mode of formation of the conidia is a complicated process; a small part of the tip of the conidiophore becomes incorporated in the new conidium while the conidiophore continues to grow very slowly. For this reason, *Trichothecium* does not fit into any of the various types of conidium ontogeny. It is, therefore, discussed separately, before other members of the *Deuteromycetes*.

266

6.5.6.1   *Trichothecium roseum* Link

Growth is rapid at room temperature. The colony consists of densely growing conidiophores, the pale or deeper pink colour (sometimes very pale or almost white) depending on the strain. The conidiophore is variable in length, tapering towards the tip, septate. The conidium is pear-shaped, inequally two-celled 12–18 (or up to 29) × 8–10 (or up to 12·5) μm.

6.5.7   *Geotrichum* Link

Conidiophores are not formed; the conidia (arthroconidia) are not formed *de novo,* arising instead by the fragmentation of hyphae. The conidia are cylindrical or rounded, tending to form loose chains.

6.5.7.1   *Geotrichum candidum* Link

The colony is loose, flat, white, mucous, often resembling a yeast colony. The vegetative body is completely fragmented into arthroconidia measuring 8–12 × 4–6 μm.

6.5.8   *Sporotrichum* Link

The colony is white or pale. The conidiophore is not differentiated, arising as a simple side branch of a hypha. The conidium is spherical, oval, or short-club-shaped, articulating on a wide base and not tending to detach itself; it is a typical aleurioconidium, generally with a rough, granular surface.

6.5.9   *Aureobasidium* Viala et Boyer

True, differentiated conidiophores do not occur. Conidia are formed directly by budding from dark basal hyphae, or from denticles on the hyphae. Both outer and inner walls of the conidiogenous cells (blastoconidia) participate in conidium formation. After formation, the conidia bud further like yeast cells, so that the colonies are usually slimy, resembling those of yeasts.

6.5.9.1   *Aureobasidium pullulans* (de Bary) Arnaud
          (= *Dematium pullulans* de Bary)

The colony is dark, especially after ageing, with a mucous mass of colourless blastoconidia. The dark hyphae are septate and slighly constricted, 6–10 μm broad. The blastoconidia are colourless, one-celled, 6–8 × 3–4 μm.

6.5.10   *Cladosporium* Link

The colony is velvety, dark brown or greenish-brown, consisting of a dense mass of conidiophores. The conidiophore is simple or branched. The conidia are blasto-spores, arising in long, often branched, loosely arranged chains. The earliest formed conidia are divided by 1 or 2 septa; those formed later are one-celled.

6.5.10.1   *Cladosporium herbarum* (Persoon) Link

The colony grows satisfactorily at room temperature, not spreading; it is grayish–green or greenish–brown, and generally powdery owing to the mass of conidia. The conidiophore is irregularly branched, septate, 25–225 × 3–6 μm. The conidial chains are branched dichotomously or trichotomously. The conidium has a finely textured surface, the one-celled and the two-celled conidia measuring 4·5–11 (or up to 19) × 4–5 (or up to 7) μm and 9–15 (or up to 20) × 4–7 (or up to 8) μm, respectively. Occasionally, conidia consisting of 3 or 4 cells have been observed.

6.5.11   *Botrytis* Persoon ex Fries

The conidiophore is erect, branching dichotomously, irregularly or it is dendroid. Conidia arising simultaneously over the entire surface of the terminal branches (conidiogenic cells), not forming chains (botryoblastoconidia). Conidia are hyaline or pale. Conidiophore becoming brown with age.

6.5.11.1   *Botrytis cinerea* Persoon ex Fries

The colony grows satisfactorily at room temperature, consisting of a woolly pale gray mycelium, a mouse-gray mass of conidia, and black sclerotia a few millimetres in size. The conidiophore is 11–23 μm broad, becoming brown with age. The conidia are one-celled, 8–12 × 6–9 μm.

268

6.5.12 *Alternaria* Nees ex Wallr.

The mycelium, the conidiophore and the conidium are brown, greenish or greyish–brown, nearly black. The conidium is a typical porospore, which protrudes through a small pore in the external wall of the conidiophore together with the internal wall of the latter. The conidia arise singly or in chains; they are divided by cross and longitudinal septa and have tapering beak-like tips.

6.5.12.1 *Alternaria tenuis* Nees

The hyphae are 3–6 μm wide. The conidiophore measures 5 125 × 3–6 μm. The conidia form long chains and are divided by 0–6 transverse septa and 1–9 longitudinal septa; they measure 7–70 × 6–22·5 μm, with a short beak.

6.5.13 *Fusarium* Link

Microconidia and macroconidia are formed in the colony. The former are one-, two-, or three-celled; the latter are divided by a greater number of transverse septa, are crescent-shaped, tapering at both ends, usually with a characteristic foot cell. The conidia (especially macroconidia) may develop in a sporodochium, in mucous pionnotes. In some species, terminal or intercalary chlamydospores, sometimes sclerotia, are formed.

6.5.13.1 *Fusarium moniliforme* Sheld.

The microconidia are one-celled (rarely two-celled), forming chains; the macroconidia are slightly curved, crescent-shaped, divided by 3–7 transverse septa. The colony is 2·5 cm in diameter after incubation for 4 days, pale peach or violet in colour.

6.5.14 *Aspergillus* Micheli

The young colony [23] is pale, later bright, variable in colour, velvety or woolly. The conidiophore is unbranched, with a smooth, rough, or spiny wall; the tip is swollen, flask-shaped. spherical, or club-shaped, forming a vesicle. The conidiophore is generally thick-walled, originating from foot cell. The vesicle is covered, partly or entirely, by one or two layers of sterigmata (phialides) which may be cylindrical, inverted club-shaped, or guttiform. If the sterigmata are arranged in two layers, those located next to the vesicle are designated primary, while those covering the primary layer are referred to as secondary sterigmata. The conidia are smooth-walled or spiny, developing in chains. Parallel conidial chains form columnar heads, while divergent chains form radial heads. The conidial head consists of the vesicle, the sterigmata, and the conidial mass. Spherical hard sclerotia are often formed. Some species are capable of sexual reproduction by the development of cleistothecia and spherical asci, not regularly arranged. The ascospore is spherical or lentiform, often with an equatorial furrow, and occasionally with a rim. Some species develop so-called "Hülle" cells which have a characteristic shape, although their function is unknown.

6.5.14.1 *Aspergillus repens* (Corda) de Bary

On Czapek's agar [23] growth is slow; the colony is compact, wrinkled, bluish–green, or yellow, with abnormal, usually infertile, cleistothecia. On Czapek's agar containing 20% sucrose, the colony grows rapidly. After 2 weeks, the colony is 5–6 cm in diameter and consists of a greyish–green mat of conidia and yellow–orange areas occupied by cleistothecia. The diameters of the conidial heads range from 125 to 200 µm. The conidiophore is smooth, 500–1000 µm in length. The vesicle is 25–40 µm in diameter. The sterigmata are arranged in a single layer and measure $7–10 \times 3.5–4.5$ µm. The conidium is sphaerical, oval, spiny, 6.5×5 µm. The cleistothecium is yellow, embedded in yellow or orange mycelium, 75–100 µm in diameter. The ascus is 10–12 µm in diameter; the ascospore measures $4.8–5.6 \times 3.8–4.5$ µm and is smooth with a flat equatorial furrow and without a rim.

6.5.14.2 *Aspergillus fumigatus* Fresenius

The colony spreads quickly on Czapek's agar [23], and is velvety-woolly, green, greyish green, or nearly black. The underside of the colony is colourless, rarely yellow or purple. The conidial head is columnar, compact, 400×50 µm; sometimes much smaller secondary heads are also formed. The conidiophore is smooth, not more than 300 µm long, 2.8 µm wide, usually somewhat greenish in colour. The vesicle is flask-shaped, 20–30 µm in diameter, bearing sterigmata on its upper region only. The sterigmata are arranged in a single layer and measure $6–8 \times 2–3$ µm. The conidium is spherical, spiny, 2.5–3 µm in diameter. The fungus grows well at temperatures up to 45 °C.

270

6.5.14.3 *Aspergillus niger* van Tieghem
(= *Sterigmatocystis nigra* van Tieghem)

The colony grows rapidly, producing white or yellowish mycelium. The conidial head is brownish–black, purplish-brown, or black, spherical, radiate, 300–500 (up to 1000) μm in diameter. The conidiophore is generally colourless (sometimes brownish beneath the vesicle), smooth-walled, from 200 μm to several millimetres in length, generally 7–20 μm wide. The vesicle is spherical, 20–50 (up to 100) μm in diameter, colourless, or pale yellowish–brown. The sterigmata are arranged in two (primary and secondary) layers, and measure $20–30 \times 6–8$ μm and $6–10 \times 2–3$ μm, respectively. The mature conidium is rough, spiny, and brown or almost black in colour.

6.5.14.4 *Aspergillus oryzae* (Ahlb.) Cohn

The colony spreads rapidly on Czapek's agar and is initially yellowish–green, later yellowish–brown. The conidiophore is 2 to several mm long, 20–25 μm or less in width, colourless, with rough, granular walls. The vesicle is 50–70 μm or less in diameter. The sterigmata, measuring $15–20 \times 3–5$ μm, are usually arranged in one layer. On the double-layered heads, the primary and secondary sterigmata are 12×5 μm and $10–12 \times 3·5$ μm, respectively. The mature conidium is rough, generally pear-shaped, and of variable size, *viz.* 3×4 μm, 4×5 μm, or 5×6 μm, sometimes even 9–10 μm long. Dark sclerotia may develop.

6.5.14.5 *Aspergillus flavus* Link

On Czapek's agar [23] the colony grows rapidly, initially yellowish–green or green, older colonies brownish–green. The underside of the colony is colourless, later brownish. The conidiophore is usually $400–1000 \times 5–15$ μm, granular, rough-walled, widening uniformly towards the vesicle. The vesicle is dome-shaped or flask-shaped; 10–30 (up to 40) μm in diameter. The sterigmata measure $10–15 \times 3–5$ μm, and are arranged in one layer on the small heads and some of the large heads; on the remaining large heads, the sterigmata are arranged in two layers, the primary and secondary sterigmata measuring $7–10 \times 3–5$ μm and $7–10 \times 2·5–3$ μm, respectively. The conidium is pear-shaped, 3 μm in diameter or $4·4 \times 3$ μm, with granular–spiny walls. A few strains develop sclerotia, which are initially white; the mature sclerotia are dark brown or black.

6.5.14.6   *Aspergillus ochraceus* Wilhelm

On Czapek's agar [23] growth is rapid; the colony is smooth, rarely wrinkled, colourless or yellowish, occasionally with a reddish mycelium. The conidial heads give the colony an ochre colour. The conidiophore is yellowish, variable in size, with rough, granular walls and a large, spherical vesicle. The primary sterigma is 15–30 μm long, and the secondary sterigma measures 7–10 × 1·5–2·5 μm.

6.5.15   *Penicillium* Link

The young colony [23] is pale, later becoming green, bluish–green or greyish–green. The conidiophore is simple (*Monoverticillata* section), or variously branched to produce a brush-like effect. The sterigmata are those terminal branches which give rise to the conidia. The structures bearing the sterigmata are the metulae. True branches, if they occur, are those branches occurring below the metulae. The walls of the cells of the brush are smooth or rough, the cells being in parallel arrangement or converging or diverging, forming asymmetrical or symmetrical structures. Sclerotia similar to those of *Aspergillus,* "Hülle" cells, or cleistothecia may occur in some species. According to colony type, different groups of penicillia can be distinguished, *viz.*

(1) Aerial mycelium weakly developed and flat. Colony mainly consisting of conidiophores growing from the substrate, velvety in appearance (*Velutina* subsection).

(2) Surface of colony loose, with floccose aerial mycelium. Conidiophores mostly originating from the aerial mycelium (*Lanata* subsection).

(3) Structure of colony loose; aerial mycelium forming well-defined bundles thus producing a ropy or funiculose surface (*Funicolosa* subsection).

(4) Some, or all of the conidiophores forming bundles, fascicles (coremia) (*Fasciculata* subsection).

6.5.15.1   *Penicillium notatum* Westling

On Czapek's agar [23], the colony grows to 3·5–4 cm in diameter and becomes radially wrinkled after incubation for 12–12 days at room temperature. The conidium mat is bluish–green, with abundant pale yellow transpiration droplets. The underside of the colony is yellow, becoming light brown later. The conidiophore measures 250–500 × 2·5–3 (or up to 3·5) μm and is smooth-walled. The brush includes one or more side branches, the branches measuring 10–15 × 2·5–3 μm. On each branch 3–5 metula develop, measuring 9–16 × 2·5–3 μm. In each whorl there are 4–6 sterigmata, each measuring 8–10 × 2–3 μm. The conidium is spherical, smooth-walled, 3–3·5 μm in diameter.

272

6.5.15.2   *Penicillium roqueforti* Thom

On Czapek's agar [23], the colony develops rapidly at room temperature. After incubation for 10 to 12 days, the intensely sporulating colony is velvety, smooth-surfaced, with a thin arachnoidal margin. The conidium mat is bluish–green or dull green. The underside of the colony is green, bluish–green or black. The conidio-phore measures 100–150 (or up to 200) × 4–6 μm, and has a granular, rough wall. The metula is rough-walled, 12–15 × 3–5 μm. The sterigmata measure 8–12 × 3–3·5 μm. The conidia are spherical, smooth-walled, dark green as a mass, 3·5–5 μm in diameter, occasionally larger.

6.5.15.3   *Penicillium camemberti* Thom

On Czapek's agar [23] growth is slow. After 10 days the colony is white, 2–3 cm in diameter, with a loose structure. Later the colony is greyish–green, the underside colourless or pale yellow. The conidiophore measures 250–600 × 2·5–3·5 μm, growing out from the substrate, or from the aerial mycelium, the latter conidio-phores being 40–200 μm long. The branches measure 12–18 × 2·2–3·4 μm, the metulae measure 9–14 × 2·2–3·3 μm, and the sterigmata (2–5 per whorl) measure 9–14 × 2·2–2·8 μm. The conidium is initially ellipsoidal, later spherical, smooth-walled, measuring 3·5 × 4·5 μm.

6.5.15.4   *Penicillium expansum* Link emend. Thom
(= *Penicillium glaucum* Link pp.)

On Czapek's agar [23], the colony grows to 4–5 cm in diameter after incubation at room temperature for 8–10 days. It is yellowish–green, later dark–green, the underside being colourless, sometimes yellowish–brown. The conidiophores measure 150–400 (or up to 700) × 3–3·5 μm, and are arranged singly or in coremia; the walls are smooth or finely textured. The brush is large, 75–100 μm long, with one or two branches. The branches measure 15–25 × 2·5–3·5 μm, the metula (3–6 per whorl) measure 10–15 × 2·2–3 μm and the sterigmata (5–9 per whorl) measure 8–12 × 2–2·5 μm. The conidium is smooth-walled, 3–3·5 μm in diameter.

7. EXAMINATION OF ENVIRONMENTAL FACTORS RELEVANT TO THE FOOD INDUSTRY

7.1   Water

The bacteria occurring in water may be divided into three main groups: (i) natural bacterial flora of waters (*Spirillum*, *Vibrio*, *Pseudomonas*, *Achromobacter*, *Chromobacterium*, etc.), (ii) soil bacteria (*Bacillus*, *Streptomyces*, *Aerobacter*, etc.) and (iii) enteric bacteria which are present in water contaminated by faecal matter (*Escherichia coli*, *Streptococcus faecalis*, *Clostridium welchii*, etc.).

The bacterial quality of water can be investigated strictly as a means of checking hygiene and suitability for food-processing purposes; e.g. drinking-water must be shown to be free not only of pathogenic bacteria but also of chemical and/or bacteriological traces of faecal contamination. Water which is hygienically acceptable may be unacceptable from the food-technological point of view, which demands the absence of psychrotrophic, lipolytic and proteolytic bacteria as well, i.e. bacteria responsible for the spoilage of foodstuffs.

In the course of the grading of water, tests are carried out for the following:
the total viable count of bacteria developing at 5–7 °C, 20 °C and/or 37 °C;
the presence of coliforms;
the presence of *S. faecalis* and *C. perfringens* and pathogenic micro-organisms which can be isolated from water.

The technique of sampling is determined by the aim of the examination. An average sample that is representative of the water to be examined is required.

Thus tap water should be allowed to flow for 5–10 minutes to discharge the water that has been stagnating in the pipes (in stagnant water bacteria may increase in number). The tap is flamed and the sample collected with aseptic precautions. To examine chlorinated water, 0·1 ml sterile 2% thiosulphate solution per 100 ml of water is placed in the empty sample jar to remove the free chlorine. If, besides testing the water, the origin of some microbial contaminant is of interest, a sample of the stagnating water may be taken and contamination of the tap may be examined by omitting to flame it. Well-water is sampled after part of the water has been pumped out.

The sample is stored in ice and processed as soon as possible.

274

7.1.1　Determination of total viable count

A tenfold or hundredfold dilution of the water is prepared using sterile water or quarter-strength Ringer solution [115]. The total viable count is determined by the pour-plate method from the undiluted sample and its dilutions.

Most of the bacteria present in water grow well in nutrient agar and nutrient gelatin [51]. The temperature and duration of the incubation depend on which group and type of microbe is being sought and on the composition of the culture medium.

Agar plates are incubated at 37 °C for 2 days. Those bacteria originating in the human or animal gut together with putrefying and saprophytes appear first and grow well. Thus the colony count provides information as to the numbers of the contaminating bacteria, i.e. the degree of contamination.

Nutrient gelatin plates are incubated for 4 days at 22 °C. This temperature favours the multiplication of the natural water flora and soil bacteria. Therefore, the viable count obtained at 22 °C indicates the numbers of saprophytes. In order to kill any gelatin-liquefying saprophytes that are present in the sample colonies of these organisms, are touched daily with a silver nitrate stick.

Water samples may be tested for psychrotrophs separately, e.g. by incubating nutrient agar at 5–7 °C for 7–10 days. *Pseudomonas fluorescens* is easily recognized by the greenish colour and sticky consistency of the colonies. If psychrotrophs are present in very low numbers, an enrichment procedure may be applied, e.g. sterile skim milk is added to 10–100 ml water at a concentration of 0·5–2·5%, and incubated at 5–7 °C for 2–3 days. Pour-plates prepared from the enriched cultures then give a qualitative picture of the psychrotrophic flora.

7.1.2　Determination of micro-organisms indicating faecal contamination

A high viable cell count generally indicates contamination of the water with microbes from sources other than the natural microflora. This, however, does not necessarily mean contamination by faecal matter such as sewage. The majority of the microbes may comprise soil bacteria. Since pathogens of intestinal origin are few in number (or not detectable at all) compared with the non-pathogenic inhabitants of the intestines *(E. coli, S. faecalis, C. perfringens)*, it is easier to examine water samples for the presence of the latter. The presence of non-pathogens is an indication of possible contamination by pathogens. Detection of *E. coli* varieties which commonly occur in warm-blooded animals suggests that faecal contamination has very probably occurred. *S. faecalis*, if present, usually shows a low count and soon perishes in water. Water is commonly tested for the presence of this bacterium, for in contrast to the so-called ubiquitous coliforms, *S. faecalis* does not multiply in the environment. The detection of this organism in addition to coliforms suggests fresh faecal contamination. The anaerobic spore-forming bacterium *C. perfringens* remains viable for a long time, and, therefore, its presence indicates a heavy contamination of long duration.

7.1.2.1 Determination of coli titre and coli count

The coli titre refers to the smallest amount of water in which *E. coli* can be detected by cultivation procedures. The coli count (coli index) refers to the number of *E. coli* cells which can be counted in 100 ml water.

According to the Hungarian Drinking-Water Standards (MSZ 22 907-71), the coli titre is determined by cultivation in lactose–yeast broth containing phenol red as an indicator [142]. The test is based on the fact that the majority of the aerobic and facultatively anaerobic bacteria forming gas from lactose during incubation for 24 or 48 h at 37 °C belong to the coli group. The result of this initial test is verified as follows.

Inoculations:

(a) chlorinated piped water 3×50 and 1×10 ml, each into 25 ml of culture medium;

(b) non-chlorinated piped water 5×10 and 1×1 ml, each into 5 ml of culture medium;

(c) water from bored wells, 5×5 ml, each into 2·5 ml of culture medium;

(d) water from dug wells and spring water, 5×1 ml, each into 5 ml of culture medium.

The 50 ml, 10 ml and 5 ml samples are pipetted into a medium containing 1·5% lactose, and the 1 ml samples are pipetted into the same medium diluted 3-fold (lactose concentration 0·5%). Incubation is carried out for 24 h and 48 h at 37 °C. Those cultures showing acid and gas formation are subcultured on Endo medium [33] or eosin–methylene blue agar [35], and bacteria belonging to the coli group are identified. The coli titre or coli count is obtained from tables drawn up according to the above-mentioned Standards. The sample is hygienically unacceptable if coli bacteria are found in two or more of the cultures.

The cell count is determined by filtering an aliquot from the sample of water under vacuum through a sterile filter membrane of appropriate pore size. The membrane is placed on the surface of a selective medium (e.g. Endo agar, or a paper disc saturated with nutrient medium) and incubated at 37 °C. The colonies are counted directly or after staining, and the colony count per 100 ml is calculated.

In different countries, different media are used in the determination of coli and coliform counts. McConkey broth [84] is widely used in the preliminary test; 50 ml and/or 10 ml water samples (sometimes dilutions of these, also) are mixed with equal volumes of double-strength McConkey broth. Five replicates are prepared for each dilution. The most probable number (MPN) of the coliforms can be

276

calculated, using the Hoskins tables, from the number of cultures showing acid and gas formation after incubation for 2 days at 37 °C. One loopful from each positive culture is then subcultured in McConkey broth (warmed to 46 °C before inoculation), and the subcultures are incubated at 44 °C for 24 h. Under these conditions, only coliforms which have originated from warm-blooded animals (including man) form acid and gas (Eijkman's test). Thus *E. coli,* an indicator of faecal contamination, can be quantified from the number of positive tubes using probability tables.

7.1.2.2 Determination of *Streptococcus faecalis*

This test is based on the observation that, unlike other bacteria, *S. faecalis* grows well in media containing sodium azide at a relatively high concentration. Litsky–Mallmann medium [81], containing glucose and sodium azide, is used in this test. Inoculation is carried out as described in 7.1.2.1. Bacteria developing after incubation at 37 °C for 48 h are identified.

7.1.2.3 Detection of anaerobic bacteria

Wilson–Blair agar [164], a medium developed for the culture of anaerobic sulphite-reducing spore-formers, may be used for the detection of anaerobic bacteria, but the authors prefer to use litmus milk [76], because in Wilson–Blair agar many sulphite-reducing clostridia form black colonies and the use of this medium requires large water samples.

200 ml of the water sample are added to 400 ml of rapidly cooled litmus milk from which the oxygen has been expelled by steaming. The mixture is heated to 80 °C and maintained at this temperature for 10 minutes to kill the vegetative cells present. The medium is cooled to about 37 °C, its surface is topped with a layer of 2–3% agar liquified, and it is then incubated at 37 °C for 5 days. The samples are tested every day for acid formation, clotting, and gas formation. The test is quantitative if the same amount of water is placed in each vessel; the number of anaerobic spore-formers can be calculated by using the probability tables.

*

The degree of microbial contamination of sewage and the effectiveness of the purification treatment are tested by bacteriological methods. Usually the numbers of viable cells multiplying at 20 °C and 37 °C, the coli titre, and the numbers of anaerobic spore-formers are determined. Owing to the large numbers of organisms in sewage, samples must be diluted before tested.

7.2 Air

Air is generally sampled in order to obtain a qualitative and quantitative account of the microbial contamination of the atmosphere. The results are evaluated in terms of general environmental hygiene indices related to dust levels and the state of cleanliness.

7.2.1 Koch's sedimentation procedure

This method is based on the tendency of particles suspended in the air to settle on a horizontal surface, in this case a Petri plate with the lid removed. Thus the number of microbes settling on the medium in a known interval can be determined.

Petri dishes containing sterile medium are exposed in the area of investigation by uncovering them for 3, 10 or 20 minutes, depending on the expected level of contamination of the air. The lids are replaced and the plates are incubated. Colonies developing on the medium are counted and the micro-organisms identified; calculations are based on $10\,cm^2$ of surface area. In an industrial unit, the air is usually examined during industrial processing and after settling has been allowed to take place.

The procedure is simple, easy to carry out, and requires no special equipment. It serves for routine control and supplements more accurate determinations of the microbial count. However, the result is of low accuracy, since it depends, among other factors, on the size of the particles to which the bacteria in the air are bound. During the usual time of exposure, the larger particles are collected to a greater extent than the smaller, which settle more slowly.

7.2.2 Procedure based on impact in a forced air flow

Up-to-date procedures are based on the collection and concentration of particles from the air. The amount of air driven through the apparatus is measured in a gas flow-meter or a rotameter, and the microbial count is calculated per unit volume of air. Of these procedures, the most accurate is described in the following.

The principle is shown in *Figure 59*. Air is driven at high speed through a narrow slit, and particles carried in the air flow, on account of their inertia, will come into collision with any surface placed perpendicularly in the path of the air flow. If the surface is that of a solid culture medium, each of the viable bacterial cells that adhere to the medium will form a colony after an appropriate time of incubation. Thus the numbers of viable cells in the air can be determined. The efficacity of the procedure depends, among other factors, on the width of the slit and the distance between the agar surface and the slit. Recently, a number of different sampling

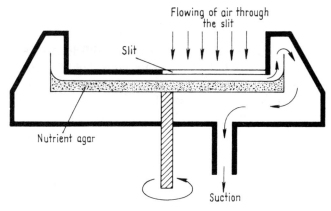

Fig. 59. Diagram of a sampler functioning on the basis of particle impact
(Mórik, 1969)

apparatuses have been developed on this principle. In Hungary, Mórik's slit sampling apparatus is widely used, and this is described in accordance with the author's original publication (1969).

The apparatus consists of two parts, *viz.* (a) a cylindric metal body which contains the driving and speed-regulating equipment and which holds the upper part horizontally, and (b) the upper part which is similar to a Petri dish and is made of plastic. The dish can be covered with a lid which fits with an air-tight seal at the edges. In the lid, there is a radial slit. Nutrient agar [51] or some other solid medium such as malt extract agar [126] is used as the culture medium.

The suction port is connected by a rubber tube to the air pumping equipment. The lid is raised and the Petri dish containing the solid medium is placed in the holder; the lid is closed and sealed tightly at the edges. Adjustments are made to obtain the optimum distance between the slit and the surface of the medium. Rotation of the Petri dish is started at an appropriate rpm, and an air flow rate of 25 dm³ per minute or more is established through the apparatus. Depending on the expected microbial concentration, 0·1–0·2 m³ of air are passed. The minimum volume of air that can be passed is 1 or 2 dm³; in this case the Petri dish should be rotated rapidly because sampling continues for only 5 s. At 6 rpm, the number of revolutions is 0·5 in 5 s, i.e., only one half of the plate surface will be seeded. After an adequate amount of air has been passed, the air-pump is switched off first, followed by the rotation motor. The medium is incubated under conditions appropriate for organisms that are being investigated. Colonies are counted and the results calculated per 1 m³ of air. If necessary, identification of the micro-organisms can proceed by preparing subcultures.

Reliability is one of the advantages of this method, and particles smaller than 1 μm in diameter can be trapped. The only disadvantage is that the apparatus is expensive and its use requires considerable care.

7.2.3   Liquid filtration

In principle, air is bubbled through a liquid from a capillary tube with an aperture of 0·2–0·5 mm diameter. Microbes carried in by the air flow remain in the liquid. The efficiency of the method can be improved by increasing the area of contact between the air and the liquid for the longest possible time. To some degree, the inertial impact of particles plays a role in this technique as well, the degree depending on the construction of the apparatus; if the air arrives in the liquid with a high speed, the bubbles will be finely dispersed in the liquid and the microbes carried in with the air will more readily be retained in the liquid.

Polazaev's impinger *(Fig. 60)* is also widely used, is similar to a gas washing flask. It is provided with a ground glass stopper which forms an air-tight seal. The apparatus and the absorbing liquid are sterilized separately. The liquid is poured into the vessel aseptically and the capillary tube is placed in position. The rate of air flow is 0·1 dm³ per minute, and the total sample may be between 5 dm³ and 50 dm³, depending on the expected degree of contamination of the air. The microbes are enumerated and identified as usual.

A further method of trapping micro-organisms from the air is to pass the air through a solid substance such as fine sand, powdered sugar, sodium phosphate, or a mixture of sodium chloride and magnesium sulphate. After the total sample has been passed through, the solid material is suspended or dissolved in sterile water and the microbe count of the liquid is determined. The efficiency of this approach is poor, since bacteria which do not withstand dessication tend to escape detection. For similar reasons, the filtration of air through cotton wool has not been widely accepted as a routine method.

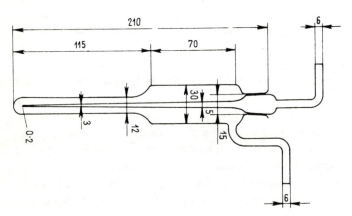

Fig. 60. Polazaev's sampling vessel employing liquid filtration
(Mórik, 1969; dimensions in mm)

7.2.4 Membrane-filtration procedures

The principle of these procedures is the same as that described in 5.2.2.2. The air sample is passed through a filter plate of appropriate pore size, and the plate is then placed on a solid medium on which colonies may grow.

This method is mainly used where pathogenic micro-organisms are involved.

7.3 The microbial contamination of surface

Many techniques are known for examining the hygienic state of surfaces, but they are not altogether satisfactory from the point of view of accuracy.

7.3.1 Contact tests

These tests rely on the microbes adhering to the culture medium when the latter is brought into contact with the contaminated surface.

Of the numerous variants of the method, the Csiszár–Demeter method, the Bacto strip contact method, and ten Cate's agar-sausage method are the most widely known.

7.3.1.1 The method of Csiszár and Demeter

Sterilized, melted medium containing 2·5% agar, or preferably 3·0% agar, is poured into a sterile, rimmed metal trough with a surface area of $10 \, cm^2$, so that the medium fills the trough completely. The solidified medium in troughs can be stored in a box made specially for the purpose, or in Petri dishes.

The test is simply carried out by touching the contaminated surface with the medium and replacing the trough in the box or Petri dish. If large areas are to be tested, samples are taken from more than one site.

The duration and temperature of incubation are determined by the purpose of the test. The degree of bacterial contamination can be assessed after incubation on universal nutrient agar [158] at 30 °C for 2 days, using a fungicidal antibiotic (e.g. Actidion) in the medium. In order to detect moulds and yeasts, cultures are incubated at room temperature for 3–5 days. The colony count is calculated per 1 cm^2 of surface.

A simplified procedure is to sterilize aluminium caps such as those used on milk bottles, fill them with nutrient agar, and store them in sterile Petri dishes until use. The caps, being flexible, are suitable for taking samples from curved surfaces, e.g. from the inner surfaces of milk cans.

7.3.1.2   The Bacto strip contact method

This is an interesting variant of the Bacto strip method. The paper strip containing absorbed culture medium is removed from the bag and pressed against the surface to be tested. After incubation, the colonies that have become red (due to reduction of the TTC in the medium) are counted. The paper strip can be applied to convex surfaces as well.

7.3.1.3   ten Cate's agar-sausage method

Nutrient or plate-count [154] agar is poured into a cellophane tube, sterilized, and the agar filling is than cut into flat discs.

The details of the method are usually as follows. Cellophane dialyser tubes (Kalle AG, Wiesbaden–Biebrich, FRG) measuring 40 mm × 20m or 60 mm × 20 m are used. A length of cellophane tube, tightly tied at one end, is filled with unsterile agar medium to a length of about 20 cm. Then the open (upper) end of the tube is tied and a hole about 5 mm in diameter is cut in the side, above the agar column and below the upper knot. The agar "sausage" is then sterilized. Immediately after it has been removed from the autoclave, the cellophane tube is tied below the hole and is hung up to cool; the cooled agar sausage can be stored in the refrigerator for 1–6 days. The test procedure involves cutting the cellophane open with a flamed hot scalpel or knife at a point where the side of the column is straight (i.e. avoiding the reduced diameter at the end). A 1-cm piece of the agar column is pressed out of the cellophane and the surface to be tested is touched against the cut surface of the agar. The disc projecting from the cellophane is cut off with a flamed scalpel and placed in a Petri dish with the impressed surface facing upward. Agar discs taken directly from the tube serve as controls.

The colonies that develop on the discs are counted. The state of hygiene of the surface is calculated according to a graded scale, the two extreme gradings being a sterile surface (no colonies, and a very heavily contaminated surface (confluent growth).

Grading includes both a qualitative and a quantitative assessment of the contaminating microflora.

The contact tests are simple and labour-saving, requiring only simple apparatus. The impression culture reflects the degree of contamination more reliably than other methods, yet the number of colonies appearing is consistently less than the number of viable cells on the sampled surface. According to Demeter (1967), the ratio of the former number to the latter is approximately 1 : 3, and this factor must be taken into account in the evaluation.

7.3.2   Direct cultivation on the contaminated surface

Transparent glass bottles can be tested by spreading a film of sterile culture medium over the inner surface, and counting the colonies that develop in the film after incubation. The culture medium must contain at least 2%, and preferably 3% agar.

Traces of water are drained from the bottle. The bottle is warmed and depending on its size, 5–20 ml of melted agar at 45 °C are poured into the bottle. The bottle is rotated, if possible, under a spray of water, to develop a uniformly thin film on the inner surface of the bottle (cf. the roll-tube technique). The bottle is closed with a sterile cotton plug or aluminium cap, and incubated for 2–4 days, depending on the temperature. To avoid the movement of condensed water over the film and the spreading of colonies as a result, the bottle is not moved during incubation. Colonies are counted at a minimum of 30 sites on the film, each site consisting of an area of 1 cm^2. Areas affected by condensation water should be avoided. The colony count is expressed per 1 cm^2.

7.3.3   Wash method

A predetermined area is washed with a sterile wet swab made of cotton wool or gauze; the swab is dispersed in sterile water and the viable count of the liquid is determined.

A swab of cotton wool is attached to the end of a small stick, a glass rod, or preferably a stainless steel wire. The swab is placed in a test-tube with the end of the rod or wire fixed to the stopper of the tube, and the unit is sterilized. When required for use, the swab is dipped in 10 ml sterile water or saline in another test-tube, and excess water is expelled by pressing the swab against the side of the tube.

A galvanized iron or plexiglass loop is placed on the surface of the object or foodstuff under examination in order to delimit the area (e.g. 5 × 5 cm) which is to be sampled. The wet swab is moved with heavy pressure back and forth over the sample area so that all parts of the surface are swabbed twice. The swab should be rotated at the same time so that all parts come into contact with the surface. The swab is replaced in the sterile liquid, shaken thoroughly, and all number viable is determined per 1 cm^2 of the swabbed surface.

The reliability of the test is poor. By a single swabbing, and by consecutive swabbings with two cotton swabs, 30–35% and 60% of the viable cells are accounted for, respectively.

7.3.4   Rinse method

This method is mainly used for examining milk bottles and other vessels. The microbes adhering to the inner surface of the bottle are removed by rinsing with a quantity of sterile distilled water. The viable cells thus suspended in the rinsing fluid are enumerated, and the result is calculated in terms of the volume of the bottle. The degree of contamination of the bottle is expressed by the viable count per ml of the contents of the full bottle.

The result greatly depends on the volume of the rinsing fluid and the mode and duration of the shaking, and steps have therefore been taken to standardize these factors. Usually, 2–10% of the bottle's volume is filled with the rinsing fluid, which is sterile water, boiled water, or quarter-strength Ringer's solution. If the bottle had been treated with a disinfectant, the effect of the latter must be neutralized by a suitable means, e.g. disinfectants with an iodine and chlorine content are dealt with by adding 0·05% sodium thiosulphate to the rinsing fluid before sterilization.

The vessel containing the rinsing fluid is sealed and vigorously inverted 10 times. Alternatively, the vessel may be rotated vigorously around its longitudinal axis 25 times, and then around an axis perpendicular to the longitudinal axis, also 25 times. To enhance the efficiency of washing, 0·2% sterile quartz sand can be added to the rinsing fluid. Rinsing can also be carried out in a rotating machine under standard conditions of rotation speed and time of rinsing. The results obtained in this way are more reproducible, although a considerable portion of the microbes are still not accounted for. Thus it is advisable to rinse the bottle with a second portion of sterile fluid.

The rinse method is particularly suitable for the examination of samples consisting of small pieces, such as peas and minced meat. The material is soaked in sterile water, shaken, and the microbial count in the rinse water is determined. The degree of contamination is expressed in terms of 100 g of the sampled material.

7.3.5   Processing of surface scrapings and cut surfaces

These procedures have been widely used, mainly in the surface examination of fatty substances and fatty objects, e.g. bacon, meat, and work-bench tops.

A sterile ring, square or other shape is placed on the surface to delimit the area of examination. Scrapings are taken as evenly as possible from the enclosed area with a sterile scraper, and collected in a vessel. The scrapings are made up with sterile water to 100 g and homogenized in a sterilized Biomix apparatus. The total number of viable cells in the homogenate (or its dilutions) is determined and this number is divided by the area (cm²) of the sampled surface.

Cut surfaces are sampled by stripping away a thin layer from the surface of the material using a sterile scalpel and sterile forceps. The exposed area is measured with a rule, and the surface is washed with sterile saline. The viable cell count per 1 cm² of cut surface is then determined.

8. THE TESTING OF FOOD, FOOD INGREDIENTS AND ADDITIVES

8.1 Milk and dairy products

8.1.1 Sampling and the processing of samples

The prerequisites of microbiological testing include:
  a) sample containment requirements;
  b) rapid refrigerated transport of the samples to the place of examination;
  c) a sampling record book containing all the information required for grading.

The original seal of the sample container (plastic bag, bottle, jar) should be intact. To minimize the risk of microbial contamination in the laboratory before examination takes place, the microbiological tests should always precede the chemical tests.

The sample jars must be clean, sterile, and of the type that can be properly sealed. Filled jars must be closed immediately and kept in ice.

Before a sample is taken from a tap (on a tank, milk conduit, etc.), the nottle of the tap should be flamed thoroughly or sterilized by steaming; some milk is then allowed to run to waste.

Each sample should be transported in an insulated receptacle (a thermos flask, or cooling bag containing ice chips. Care should be taken to prevent samples from freezing. Samples warmer than 8 °C on arrival cannot be accepted for testing, unless, for example, tests are to be carried out for anaerobic spore-formers, or samples preserved with Redofix are to be examined by the resazurin dye-reduction test.

The work record should contain the following information:

The purpose for which the sample was collected (routine sampling, counter-sampling, production phase control, etc.); the exact time, including the hour, of sampling; the place of sampling (e.g. factory, tank, dose dispenser); the serial number of the sample; the exact description of the sample; the quantity and temperature of the sample at the time of sampling; the time of production and issue; the name and location of the manufacturer or issuer; the history of the sample (e.g., whether it has been centrifuged, flash-heated, homogenized, adjusted with cream, stored under given conditions, etc.); any other remarks concerning sampling. The record should be signed by the sampler and a representative of the manufacturer or issuer.

In the laboratory, the sample should be examined immediately. If this is impossible, it may be stored at a temperature of between 0 °C and +6 °C for no more than 12 h.

8.1.1.1   Sampling from different milk products

Milk, flavoured dairy products, sour-milk preparations, starters, sweet or soured cream, buttermilk and whey are carefully homogenized with a pipette before sampling, and the sample should be cooled immediately.

Butter. In the case of large blocks of butter, the surface layer is removed and the sample is taken with a sterile sampler (i.e. a borer specially constructed for this purpose). The sampler is dipped in alcohol and flamed before use. The upper part of the sample is discarded and the remainder placed in a sterile sample jar which can be properly sealed; the jar should be completely filled by the sample. In the case of portioned butter, individual packs are taken from each production batch. If only the surface is to be examined, a surface layer of the desired thickness is collected using a sterile knife.

Cheese. Cheeses of medium or large size, the size or consistency of which does not allow then to be broken into pieces, are sampled with a special sample borer. To prevent the sample from being heavily contaminated by surface bacteria, it is best to wipe the rind of the cheese with alcohol at the sampling site.

The sampler is flamed with alcohol before use. The outermost segment (volume 2 ml) of the sampled cylinder is broken off and replaced in the cheese. The final sample consists of a central segment of the remainder of the cylinder. From small cheeses or block-processed cheese, a slice is taken and cut into pieces which are removed as samples. From boxed, processed soft cheese, one piece is examined from each box. Samples are stored in the cold, in sterile jars.

Canned condensed milk. One or more cans are examined from each batch. In order to reduce the viscosity, cans are placed in a water bath at 45 °C for 15 minutes. Lids are flamed before opening. The sample is removed with a sterile pipette or a flamed spoon.

Milk powder and whey powder. The sample is removed with a flamed spoon or a sterile aluminium tube, and stored in a tightly closed jar.

Ice-cream. Ice-cream is sampled with sterile sample borer, or an unopened packet, carton, etc. is taken.

8.1.1.2   Preparation of samples

Samples are generally processed according to standard procedures, but where special techniques are required, these are described in the following.

Butter. The jar containing the sample is placed in a water bath at 40 °C. In the case

286

of wrapped butter, the wrapping is removed and the sample is placed in a sterile jar and melted in a water bath. The liquified butter is mixed well by shaking, and serial dilutions are prepared; to prevent solidification, pre-warmed pipettes and Petri dishes are used and the diluent is pre-warmed to 40 °C. The emulsion is stabilized by dissolving 0·1% agar in the diluent. Since it has been known that most of the microbes in butter occur in the aqueous phase, some authors take the sample from the separated aqueous fraction.

Cheese. 10 g of the sample are suspended in 90 g diluent. The sample is first of all ground in a mortar with a little of the diluent warmed to 40–45 °C. The pulpy mass thus obtained is processed with further aliquots of the diluent. The tenfold dilution should be milky in appearance, with no visible fragments; otherwise, a new suspension must be prepared. More reliable results can be obtained by using an electric homogenizer (e.g. ATO-MIX or Biomix). The parts of the homogenizer which come into contact with the sample should be kept in 70% alcohol (these parts, therefore, withstand the alcohol treatment). The sample is weighed and transferred together with an adequate quantity of pre-warmed diluent into a sterile wide-mouthed jar (the jar is treated with alcohol and rinsed with sterile water before use). Homogenization is continued until the cheese emulsion apparently becomes uniform.

Stock suspensions are similarly prepared from curd, sweetened condensed milk, milk powder and whey powder.

Ice-cream. The viable count is calculated per unit weight (not volume), because ice-cream varies with respect to its air content, viscosity, and specific weight. The jar containing the sample is placed in a water bath at 45 °C and the melted ice-cream is mixed carefully with a sterile pipette. 10 g of the sample are then added to 90 ml diluent and the mixture is shaken.

Sterile water, sterile saline, or quarter-strength Ringer's solution is used for the dilution. Cheese and milk powder are usually dispersed in a pre-warmed solution of 1% sodium citrate or 1 N sodium carbonate.

8.1.2 Determination of microbial count

8.1.2.1 Determination of total count of milk and dairy products

The total count (viable plus non-viable microbes) for milk and dairy products can be determined by microscopic examination. In the testing of raw milk, Breed's procedure is the most widely used. The same method may be used where a soured milk product, starter culture, etc. is being examined. Pasteurized milk cannot be graded microscopically. In the case of cheese samples, tenfold dilution in the form of a milky suspension is prepared in 1% sodium citrate. On this suspension, 1 ml is mixed with 9 ml of 50% alcohol, and the resulting stock suspension is examined according to Breed. In calculating the microbial count, results obtained from a 1 : 100 dilution of the suspension are also considered.

8.1.2.1.A   Direct enumeration of microbial cells in milk

The sample is mixed carefully and 0·01 ml samples are taken up in a micropipette, a microsyringe, or a circular platinum loop 4 mm in diameter made of wire 0·8 mm thick. The external surface of the pipette is wiped with clean gauze or a filter paper strip, and the milk is blown out on to a clean slide. The sample is smeared quickly and evenly over an area of 1 cm², using a wire bent at 90°. It is helpful to place a ring or square below the slide to indicate the area of 1 cm² in which the cells are to be counted. A circular area (diameter 1·128 cm) is preferable because it is easier to spread the sample evenly over such an area. Alternatively, slides may be used on which areas of 1 cm² are marked out on the slide itself *(Fig. 61)*.

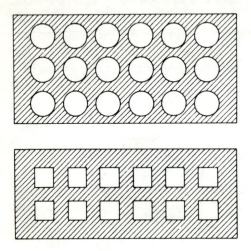

Fig. 61. Breed's slides

The smears are allowed to dry at 40–55 °C in a dust-free place. This treatment should not last longer than 5 minutes, but over-rapid drying should also be avoided. According to the original stipulations, the preparation is defatted, fixed and stained in three steps, but one-step procedures are also widely used. The latter are much simpler, but the outcome is less clear.

Three-step procedure. The slide is immersed in xylol for at least 1 minute, after which the xylol is drained and the slide allowed to dry. The slide is then placed in 96% ethanol, the ethanol is poured off, and the slide allowed to dry. The milk film thus prepared is immersed in methylene blue stain [94], rinsed carefully with water, drained, and allowed to dry in the air.

One-step procedure. Newman's stain [100] is the most widely used. The milk film is dried and immersed in Newman's stain for 15 minutes. The stain is poured off and the slide allowed to dry. Excess stain is removed by careful rinsing and the preparation is allowed to dry.

288

Newman's solution as modified by Charlett [101] contains an addition of basic fuchsin, thereby giving a differential staining effect in which casein forms a pink background against which the cells and microorganisms stain blue.

The smeared, dried milk film is exposed to the Newman–Charlett stain for 12–15 minutes, after which the stain is poured off. No drying is required. The preparation is carefully washed with water, drained and allowed to dry in the air.

To calculate the microbial cell count per ml, it is necessary to know the microscope factor, i.e. the coefficient by which the mean cell count per microscope field must be multiplied to obtain the cell count per ml. The microscope factor can be calculated from the diameter (or area) of the microscope field, which is measured as described in 2.6.5.2. If 0·01 ml of liquid is spread over an area of 1 cm², the microscope factor is

$$\frac{10,000}{r^2 \times 3\cdot142},$$

where r is the radius of the microscopic field measured in millimeters.

Table 45

Coefficients (microscope factors) ($\times 10^3$) for the calculation of microbial counts per ml, for different diameters of microscopic fields (Roeder, 1954)

Objective micrometer unit	Objective micrometer unit (0·1 distance = 0·0001 mm)									
	0	1	2	3	4	5	6	7	8	9
12	884	871	853	844	829	815	802	789	777	765
13	753	742	731	720	709	698	688	678	669	659
14	650	641	632	623	614	606	597	589	582	574
15	566	558	551	543	537	530	523	517	510	503
16	497	491	485	479	473	468	463	457	452	446
17	441	436	431	426	421	416	411	406	402	397
18	393	388	384	380	376	372	368	364	360	356
19	353	349	345	342	338	335	332	328	325	322
20	318	315	312	309	306	303	300	297	294	292
21	289	286	283	280	278	275	273	270	268	265
22	263									

Table 45 gives microscope factors corresponding to different field diameters.

If the tube of the microscope is extended, the magnification can be carried to a small degree. By making adjustments in this way, a coefficient can be obtained in round figures which makes calculation easier. According to *Table 45*, coefficients of 300,000, 400,000, 500,000 and 600,000 are obtained with microscope field diameters of 0·206, 0·178, 0·160 and 0·146 mm, respectively.

Stained films are examined with the same combination of lenses and the same tube length as were used in establishing the microscope factor. Cells are counted in randomly selected fields, using oil immersion. Where the microbial count per field is low and where a greater degree of accuracy is required, a larger number of fields must be evaluated. The USA standards are shown in *Table 46*.

Table 46

The number of microscope fields that needs to be examined as a function of field diameter and cell count

Approximate number of cell clusters/ml	Field diameter (mm)			
	0·206	0·178	0·160	0·146
$3 \times 10^4 - 3 \times 10^5$	30	40	50	60
$3 \times 10^5 - 3 \times 10^6$	20	20	20	30
3×10^6	10	10	10	20

*

Usually clusters and chains of cells are counted, but not individual cells. It is, therefore, more correct to speak of the cell-group count rather than the cell count. To obtain the total cell count per ml in milk (Breed's count), the mean count is multiplied by the microscope factor.

Example. The diameter of the microscope field is found to be 0·202 mm, measured with an objective micrometer.

The microscope factor is thus $\dfrac{10,000}{0 \cdot 101^2 \times 3 \cdot 142} = 312,000$.

The same value is found in *Table 45*.

0·01 ml of a milk sample is spread over an area of 1 cm². The mean cell count per field is found to be 11. Therefore, the total microbe count for the milk is

$$11 \times 312,000 = 3 \cdot 4 \times 10^6 \text{ per ml.}$$

This method has a number of advantages; the sample can be graded within 15–20 minutes and the procedure is quick, simple, and inexpensive in terms of equipment and chemicals. Perfect cleanliness is the only basic requirement, no sterile vessels being needed.

The microbial count can be stabilized at the time of sampling by adding some disinfectant (e.g. 1 drop of 40% formalin solution per 10 ml of milk).

From a microscopic inspection of the microflora, it is possible to approximate the source of contamination; diplo- and streptococci suggest inadequate cooling of the milk; clustered bacteria (mainly cocci) suggest improper cleaning and disinfection

290

of vessels; rod-shaped bacteria suggest contamination with dust or faecal matter. Long-chained cocci together with many leucocytes are characteristic of mastitis-affected milk.

The procedure, however, does not give a high accuracy, the possible sources of error being as follows:

— the sample may be too small to be representative of the batch of milk under test;

— there may be a significant error in measuring out a volume of 0·01 ml;

— the milk film may be unevenly prepared;

— different bacteria have variable staining properties;

— the distribution of the bacteria in the preparation may be uneven;

— the number of fields of view examined may be too small;

— the illumination used in the microscope may be too weak or too strong;

— counting errors can arise from tiredness of the eyes after prolonged use of the microscope.

Errors can be reduced by examining larger amounts of milk, by spreading the sample over an area 2·5 × 2·0 cm. It is not an advantage to use a still larger area, because rapid drying makes it difficult to prepare an even film over such an area.

Sometimes, usually on account of there being impurities on the slide, the milk film becomes detached from the slide during washing.

Errors can be reduced further by increasing the number of fields, for example the cells can be counted in a strip (or strips) terminated by opposite edges of the preparation. In this case, the area that is examined is determined by the width of the preparation.

Example. Microbes are counted in a strip 10 mm long by 0·2 mm wide (area, 2 mm^2), corresponding to 0·0002 ml of undiluted milk. The cell count is multiplied by 5000 to give the count per 1 ml of milk.

The procedure cannot be applied to the examination of pasteurized milk, and the error is relatively large if milk samples of low microbial count (cell counts less than 500,000 per ml) are examined.

The Dreier–Korolev method (Skorodumova, 1949) is a modified Breed method. Briefly, the sample is thoroughly mixed with a yeast suspension of standardized cell density, and the bacterial count is thus calibrated by taking the yeast count as well. It is not, therefore, necessary to spread a precisely measured volume evenly over an exact area, and the milk film may be somewhat thicker. The yeast suspension *(Schizosaccharomyces pombe)* is preserved in phenol solution, which also prevents bacterial multiplication. Thus, the sample does not have to be examined immediately. However, the method has one great disadvantage: if the cell count of the milk is low, then dilution in the yeast suspension may give rise to unreliable counts.

Breed's procedure can be used for the examination of materials other than milk, including other dairy products. To assist the fixation of materials which are poor in protein, and which might otherwise tend to become detached from the slide, sterile horse serum may be added to the sample on the slide. When fibrous substances are examined, a calibrated loop can be used instead of a micropipette.

8.1.2.2   Enumeration of viable cells by cultivation

To determine the viable cell count of milk and other dairy products, the pour-plate method, and less frequently, the limit dilution technique are used. Approximate counts can be obtained by Burri's procedure, the miniature plate method, deep-agar culture, or the Bacto strip method.

Milk and dairy products are examined not only for the total viable count, but also for coliforms, yeasts, and moulds; the presence and numbers of these micro-organisms give an indication of the state of hygiene prevailing during the processing of pasteurized milk and dairy products. Depending on the nature of the product, the numbers of microbes causing spoilage, and the numbers of useful micro-organisms may be assayed.

In the following, a description is given of the culture media and culture conditions most commonly used in the examination of the more important micro-organisms occurring in milk and dairy products.

Total viable count. Polónyi's agar [112] is a frequently used medium which contains milk to support the growth of lactobacilli; owing to the presence of milk protein, the colonies of proteolytic bacteria are seen to be surrounded by a clear zone, and, therefore, these bacteria can be counted directly. One disadvantage is that the presence of milk (in addition to meat extract) causes the medium to be opaque, which makes colony-counting rather more difficult.

As in the meat industry, plate-count agar [154], TGEM agar [155] and lactose broth agar [144] are also widely used in the dairy industry.

If the total viable count is to be determined for raw milk, 30 °C is prescribed as the incubation temperature, and the period of incubation is 48–72 h. However, it has recently been shown that incubation for 72 h results in a significantly higher viable count than that obtained after 48 h. The authors recommend incubation at 22 °C for 1 day, then at 37 °C for another day.

Bacteria forming acid or alkali can be demonstrated in the same medium as that used for the determination of the total viable count [e.g. 144], if the medium contains an acid-base indicator; 5 drops of sterile, cold-saturated aqueous China-blue solution, or 0·2 ml of saturated, aqueous bromthymol blue solution may be added per 100 ml of melted medium. China-blue is the more widely used because the colonies of the acid-formers are dark blue, while those of the alkali-formers are nearly colourless. In the presence os bromthymol blue, the colonies of acid-formers are yellow, those of the alkali-producing bacteria, bluish–green. For clear demonstration of the latter, bromthymol blue is preferable to China-blue.

On culture media containing an indicator, acid-formers give unreliably low colony counts. More reliable results are obtained if calcium carbonate is added instead of the indicator, in which case the presence of acid-formers is shown by a clear zone around their colonies. In medium [16], which contains bromcresol purple and calcium carbonate, the two methods of detection are combined.

Lactobacilli. In raw milk, lactobacilli are negligible in number; in yoghurt and the hard cheeses, on the other hand, these bacteria are of considerable importance. The culture medium should have a pH value which is highly inhibitory to the growth of lactic-acid streptococci.

Gram-negative contaminating flora. The physiological characteristics of the Gram-negative contaminants are very variable. These organisms include the aerobic psychrotrophs *Pseudomonas, Achromobacter, Alkaligenes* and other members of the *Enterobacteriaceae,* the coli-aerogenes bacteria being of special importance; the latter may cause spoilage, and they also serve as indicators of the general state of hygiene of the milk, etc.

The Gram-negative flora can be examined in TTC broth [156] or on Bacto strips. The data given by either of these methods are only approximate.

The TTC test can be used for examination of milk as well as liquid, soured, and solid dairy products. Under well-defined conditions, TTC-reducing microbes are detectable by this test. However, there are a few Gram-negative rod-shaped bacteria which do not grow in TTC broth; some streptococci, on the other hand, produce formazan from TTC, thus giving a misleading result. If the total count is very high and the pH of the sample is low, TTC may be reduced to red formazan even by non-multiplying actively, metabolizing lactobacilli. These possibilities should be taken into account if the TTC test is employed.

In the TTC test, test-tube cultures are inoculated with a quantity of the liquid or dissolved food, under investigation, and the cultures are incubated at 37 °C for 18 h. The positive reaction is indicated by the intense red or brownish-red colour of the formazan.

A positive reaction obtained with the Bacto strip, another test based on the colour change of TTC, also suggests the presence of a Gram-negative contaminating flora. This test is described in detail in 5.2.2.9. After incubation for 8–10 h, the colonies that appear are mainly those of coliforms; after a prolonged incubation, many other Gram-negative bacilli such as *Proteus* and *Pseudomonas,* become detectable; even micrococci may form pink or pale red colonies. The colonies developing on the Bacto strip allow only approximate cell counts to be made; microbes which produce large red spots make counting impossible. For this reason, it is more appropriate to assign a rating to the strips, as follows:

No colonies: −,
a few colonies (e.g. 1–9): +,
10 or more colonies: + +.
inseparable colonies: + + +.

The count of reddish-brown colonies is a measure of the number of coliforms present; if pink and pale red colonies are also taken into account, the result will relate to the entire contaminant flora.

The strips, if dried carefully at 50–60 °C, can be stored for many years if desired for demonstration purposes, etc.

Psychrotrophs. A great many media have been recommended for examining milk for the presence of psychrotrophs. In general, any of the media that are used in the determination of the total viable count are also suitable for the enumeration of psychrotrophs. The media are incubated at 5 °C for 7–10 days. If plates are pre-incubated at 17 °C for 16 h, further incubation for 4–5 days at 5 °C is sufficient. The incubation time may be further shortened if microcolonies are counted (by the miniature plate method or by electrical microcolony-counting).

Coliforms. Coliforms are examined on selective media. The bacteria forming yellow colonies on Klimmer's agar [68], a medium containing trypaflavine as an inhibitor, are usually taken to be members of the coli-aerogenes group.

Other solid media including deoxycholate agar [26], eosin–methylene blue agar [35], and Endo agar [33] have also been widely used. If necessary, confirmatory tests are performed by subculturing those colonies that have developed on agar medium in nutrient broth. A more common practice is to test for coliforms in broth culture (Kessler–Swenarton's broth [66], brilliant green–lactose–bile [14], etc.). Thus a larger amount of sample can be used as an inoculum, smaller numbers of coliforms can be detected, and gas formation can be observed as an indicator of the presence of coliforms.

If coliforms are present in broth cultures, it may happen that tubes inoculated from lower dilutions become turbid without gas formation, while in those inoculated with higher dilutions, gas formation can be observed. The absence of gas formation in the former case may be explained by the large number of concomitant acid-forming bacteria and the inability of coliforms to multiply in the acidified medium. In cultures inoculated with higher dilutions, streptococci are depressed by the selective inhibitor, and the coliforms multiply sufficiently for gas formation to be observed.

Proteolytic micro-organisms. Proteolytic micro-organisms tend to be found most of all in butter, fresh cheese, curd, milk-powder and water samples. Their detection provides a useful indicator to the state of hygiene of milking machines. The use of a gelatin medium for the investigation of proteolytic microbes in water is obligatory in many countries. For other purposes, casein agar [46] is the most widely used medium. These colonies that are surrounded by a clear zone are counted; recognition of the clear zone is facilitated by flooding plates with dilute acetic acid or hydrochloric acid before counting. Frazier–Rupp casein agar [46] can be used for detecting undesirable microbes that may be present in butter. All colonies developing on this medium should be regarded as foreign to good quality butter, irrespective of the presence or absence of proteolytic activity.

294

Lipolytic micro-organisms. Samples that are to be tested for lipolytic micro-organisms are inoculated into media containing fat. Lipolytic colonies give rise to the formation of fatty acids which can be detected with Nile-blue sulphate, or lipolysis may be recognized by the disappearance of fat droplets which otherwise cause the medium to be turbid. Lipolysis is more easily demonstrated in tributyrin agar [149]. Some authors, however, prefer to use triolein or trilinolein instead of tributyrin.

Aerobic spore-formers. These may be expected to occur for, primarily in liquid milk and dairy products. Vegetative cells are killed by heating the sample at 85 °C for 10–15 minutes; then the sample is examined for the presence of aerobic spore-formers by the preparation of semolina agar pour-plates [18], and incubating these at 37 °C for 48 h. The latter medium inhibits the spread of the colonies of spore-formers, thus facilitating the counting of individual colonies.

If, however, the medium is inoculated with undiluted milk (even as little as 0·1 ml) it becomes very turbid and it is, therefore, difficult to recognize and count the colonies, especially those within the medium. In these circumstances the agar surface can be flooded with a 1% aqueous solution of TTC and the plates re-incubated at 35 °C for 2 h; the red-coloured can then be counted without any difficulty.

Anaerobic spore-formers. The detection of heat-resistant anaerobic spore-formers that occur as contaminants of dairy products is based on the formation of gas by these bacteria under anaerobic conditions.

A straightforward approximate method for the detection of anaerobic spore-formers in milk samples is provided by the Weinzirl test.

1 or 2 ml of melted paraffin are pipetted into test tubes which are plugged with cotton wool and autoclaved. Into the test-tubes thus prepared (Weinzirl tubes), are measured 5 ml of the sample, or 0·5 ml of the sample and 4·5 ml of sterile yeast extract–glucose solution [38], or 0·05 ml of the sample and 5 ml of the culture medium, etc. The tubes are placed in a water bath at 80 °C for 15 minutes; this treatment kills the vegetative cells and the melted paraffin floats to the top of the medium to form an airtight layer. 5 replicate cultures are prepared for each dilution and these are incubated at 37 °C for 3 days. In the positive cultures, the plug is expelled by the gas production. The result is evaluated with the aid of the Hoskins table.

Because of the difficulty of cleaning the Weinzirl tubes, many authors prefer the following method.

5 ml of undiluted sample are mixed with 2% 80 °C for 15 minutes. Then 5 ml of yeast extract–glucose medium [38] containing 2% dissolved agar are added to dilutions of the sample (8 ml of each dilution). When the mixtures have solidified, they are covered with 2 ml of 1% agar solution and incubated at 37 °C for 3 days. Disruption of the agar column or the appearance of gas bubbles in the medium indicates a positive result.

Cultivation in glucose–liver broth [88] is more sensitive than any of the above-described methods. Before inoculation, the test-tubes containing the culture medium are heated in boiling water for 20 minutes to expel the oxygen, and then they are quickly cooled. Inocula of 1 ml from the undiluted sample and 1 ml from each of the sample dilutions are transferred into culture tubes. The vegetative cells are killed by heating the cultures in a water bath at 80 °C for 15 minutes; the tubes are quickly cooled and incubated at 37 °C for 5 days. The cultures are inspected daily for gas formation (i.e. foaming). To facilitate the detection of gas formation, the medium can be covered with solid paraffin or agar. Alternatively, 0·12% ferrous sulphate ($FeSO_4 \cdot 7H_2O$) may be added to the liver broth; in this case gas formation will be more rapid and the dirty-grey froth on the side of the test-tube will remain visible for a few days.

The similar use of RCM broth [173] gives more reproducible results.

Propionibacteria. Examination for these microaerophilic bacteria is particularly important in the course of the production of hard cheeses. Deep agar is used as a culture medium. The colonies are brown or red, and are lens-shaped, star-shaped or resembling a kefir fungus in shape.

Yeast and moulds. Their acid tolerance is used as a means of identifying these micro-organisms. Usually, acidified beer agar, malt agar, potato agar, or salted agar is used as the selective medium. The yeast and mould counts are determined in pour-plates. The medium is acidified with lactic acid or citric acid to pH 4·0–3·5 immediately before the plates are poured. If the acidified medium is remelted, it may not solidify again on account of hydrolysis of the agar. The salted agar [126] recommended by the International Federation for Dairying [222] seems to be preferable to the beer agar because it is colourless, and of reproducible composition.

The media and culture conditions required for the detection of the different groups of microbes most commonly tested for are shown in *Table 47*.

Table 47

Microbiological examination of milk and dairy products

Type of microbe detected	Culture medium	Incubation temperature °C	time h
Total viable count	Polónyi's agar [112]	22, then 37	24, 24
	TGE [154]	30	72
	TGEM agar [155]	30	72
	Lactose broth agar [144]	30	48
	Yeastrel milk agar [168]	30	72
Acid- or alkali-forming bacteria	China-blue–lactose broth agar [144]	30	48
	Bromcresol purple–lactose–chalk agar [16]	30	48

(Table 47 cont.)

Microbiological examination of milk and dairy products

Type of microbe detected	Culture medium	Incubation	
		temperature °C	time h
Lactobacilli	Rogosa agar [119]	30	72
	Acetate agar [1]	30	72
Gram-negative flora	TTC broth [156]	37	18
	Bacto strip	37	8 – 10
Psychrotrophs	TGEM agar [155]	5	168
	Yeastrel milk agar [168]	5	168 – 240
	Yeastrel milk agar [168]	17, then 5	16, 96 – 120
Coliforms	Kessler–Swenarton broth [66]	37	48
	Brillant green–bile broth [14]	35	48
	Endo agar [3]	37	24
	Deoxycholate agar [26]	35	18 – 24
	Klimmer's agar [68]	37	24
Aerobic proteolytic bacteria	Casein agar [46]	30	48
Lipolytic bacteria	Tributyrin agar [149]	22, or	180, or
		30	72
Aerobic spore-formers	Semolina agar [18]	35 – 37	48
Anaerobic spore-formers	Milk (Weinzirl test) [38]	37	72
	Liver broth (8)	37	120
	R. C. M. broth [174]	37	120
Propionic acid bacteria	Kurmann's agar [74]	30	288 – 336
Yeasts and moulds	Malt extract agar [126]	27	72 – 96
	Salt agar [122]	27	72 – 96

Note. When a heat-treated milk or dairy product is examined, it is advisable to extend the period of incubation; e.g. to obtain the total count of pasteurized milks requires incubation for 144 h.

8.1.2.3　Determination of viable count and microbial activity by dye-reduction tests

The state of hygiene of large-scale milk production and processing is examined mainly by means of simple reduction tests. These are time-saving and easy to perform. Standard amounts of the milk sample and the dye, usually 1 ml of the dye and 10 ml of the sample, are mixed in sterile test-tubes which are then rubber-stoppered and incubated. Microbial contamination of the milk sample is indicated by a colour change in the dye.

8.1.2.3.A　Methylene blue reduction tests

This is the most widely used of the reduction tests; the redox indicator is methylene blue, a dye which is blue in its oxidized form and colourless in its reduced form. The colour change occurs between $+0.06$ and -0.01 V redox potential.

In Central-European countries the test is performed as follows:

0.1 ml methylene blue solution [94] and 10 ml of the milk sample are pipetted into a sterile or boiled test-tube. The contents of the tube are well mixed while being protected from the light, and incubated in a water bath immediately. If a water bath is not available, the samples can be pre-warmed with water at a temperature of 37 °C before they are placed in the incubator. The time interval between the beginning of the test and decolouration of the methylene blue is recorded. A culture in which at least the lower two-thirds has become colourless is regarded as decolourized. The colour of the uppermost third of the culture may be neglected because the colour of the methylene blue is restored in contact with the air. Apart from any cream that is formed, the bacteria in milk may be regarded as evenly distributed, i.e. de-colourization proceeds at the same rate throughout the milk sample. However, in milk of low microbial count, the degree of decolourization may vary up and down the milk column.

The methylene blue solution contains 5 mg of the dye per 100 ml. It is prepared with boiled distilled water on the day of use, and it is kept protected from the light.

The methylene blue reduction test was modified by Wilson, the modification involving inversion of the samples in the water bath at 30-minute intervals. Since the rate of cream formation in different milk samples is not the same, and the distribution of bacteria between the two phases is also variable, inversion of the samples may be expected to give more reliable results.

A modified test. In the modified test, the decolourization of the milk is more rapid than in the original test; the reduction time is about 75% of the time taken in the original test, and the results are more reproducible and more accurate.

The sample is carefully mixed to ensure that the fat is evenly distributed. Graduated test-tubes are used, and milk is poured in up to the 10 ml mark. Care

298

must be taken to leave at least one inner side of the test-tube unwetted with the milk. 1 ml of methylene blue solution is added without touching the milk in the tube or the wetted side of the tube with the pipette. After three seconds, the dye remaining in the pipette is blown out. The tube is closed with a rubber bung and inverted twice slowly, allowing the air in the rube to rise above the milk level each time. Two control tubes are set up, *viz.,* one containing 10 ml of milk and 1 ml of tap water, and another containing 10 ml of milk and 1 ml of methylene blue solution. Both tubes are placed in boiling water for 3 minutes. The milk samples in the control tubes should be similar to the test samples with respect to fat content and colour. The tubes are inspected at 30-minute intervals over a period of 330 minutes and at each inspection decolourized tubes are removed while partially decolourized tubes are left in the water bath without further inversion. All the remaining tubes are inverted once at each inspection and replaced in the water bath. Anymore vigorous mixing than this would be undesirable because it would promote re-oxidation of the methylene blue. A tube is regarded as decolourized if the milk column is colourless from top to bottom, or only the uppermost 5-mm layer has remained blue. Traces of colour near the bottom may be disregarded if they do not extend more than 5 mm upwards.

Overby's method. This method is also based on Wilson's principle. Methylene blue–calf rennet solution is added to the milk causing the milk to become clotted after 15 minutes. Thus, cream formation and phase separation can be avoided without the need to invert the tubes at intervals.

8.1.2.3.B Resazurin dye reduction tests

Recently, these tests have become more and more widely adopted throughout the world. Resazurin is a redox indicator. Its oxidized form which is pastel blue in colour, is reduced to dihydroresazurin in a redox environment of $+0.2-0.0$ V. The reduction proceeds in two steps while the colour of the indicator changes to lilac, then to pink, finally becoming colourless. It should be noted that besides being a redox indicator, resazurin is a pH indicator as well; it is blue above pH 6·8 and red below pH 5·2. Resofurin is red above pH 6·4 and yellowish–orange below pH 4·8. However, these colour changes are of no importance as regards the microbiological testing of milk.

There are a number of variations of the resazurin reduction test; the time and temperature of incubation may be varied (37 °C for 10, 30, 60 or 180 minutes, or 18 °C for 180 minutes). The authors generally use a combination of 37 °C and 30 minutes.

Test procedure. 10 ml of milk at a temperature of 37 °C are mixed with 1 ml of resazurin solution [114] in a sterile test-tube. The mixture is put in an incubator at 37 °C for 30 minutes, and the milk is graded on the basis of the colour reaction, using a colour table. Mixtures that are pastel blue, intermediate in colour and pale pink are graded as classes I, II and III, respectively. Where a large number of samples is to be handled, the resazurin is added to no more than 10 tubes at a time. The preparation known as Redofix (Szegő, 1969) makes it possible to transport milk samples to a central laboratory within a wide time limit without any special precautions, and without any risk of spoilage. The milk preserved with Redofix (10 ml of milk to 1 ml of Redofix) can be tested at any time from 6 h to 36 h after the Redofix is added; the results are the same as that obtained at the time of adding the Redofix.

8.1.2.3.C   Evaluation of the results of the methylene blue and resazurin reduction tests

The dye reduction tests give an indication of the viable count, the freshness, and the storability of milk. However, the results can be influenced by many factors, including the amount of oxygen dissolved in the milk and the amounts of reducing agents (e.g. ascorbic acid and cysteine) present. Removal of the oxygen or reduction of the oxygen by bacteria, is a prerequisite of reliable dye reduction testing.

With regard to the reducing activity of the bacteria in relation to the viable count of the milk, the following factor should be taken into account.

— reducing activity varies greatly among different bacteria; the activity of some is weak, some show no activity at all, and others, including the typical lactic acid streptococci and the majority of the members of the coli–aerogenes group, are strongly active;

— some bacteria reduce the dye but do not form colonies on agar plates;

— in milk samples containing abundant fat, bacteria may become concentrated in the ascending fat, thus diminishing the reducing activity in the lower region of the milk column;

— the reducing activity of the bacteria is the same regardless of whether the cells are dispersed individually or grouped in clusters, whereas the total count obtained by pour-plating or direct microscopic counting is diminished by clustering;

— a part of the microflora (the psychrotrophs) does not grow at the test temperature within the prescribed incubation period, and thus appears to be inactive in the reduction test;

— in milk samples containing antibiotics or other inhibitors of bacterial growth, the reduction of the dye is slowed down or completely inhibited.

Because of these considerations, the correlation between the true cell count of the milk and the result of the reduction test is not close. Agreement has been found to be closest when average milk samples are examined. Starting from this observation, milk samples have been graded into 4 quality classes, *viz.* class I—cell count less than 500,000 cells per ml; class II—cell count between 500,000 and 4 million cells per ml; class III (poor quality)—cell count 4 million to 20 million cells per ml; class

300

IV (very poor quality)—milk containing more than 20 million cells per ml. The four classes correspond to the following decolourization times in the methylene blue reduction test:

> 330 minutes, 120–330 minutes, 20–120 minutes and < 20 minutes for classes I, II, III and IV, respectively.

The corresponding colours obtained in the resazurin test are pastel blue, pale lilac, pink or pale pink, and colourless.

This classification can now be regarded as outdated. The correlation between the results of cell counts and those of reduction tests is so poor, even when average milk is tested, that grading into four classes cannot be justified. A three-class system has now been introduced in most countries; the system used in Hungary is shown in *Table 48.*

Table 48

The grading of raw milk based on reduction tests

Quality class of the milk	Methylene-blue test: decolourization time (minutes)	Resazurin test: colour of the milk after 30 minutes	Approximate viable count per ml milk
I. Good	> 330	pastel blue	$< 1 \times 10^6$
II. Medium	120 – 330	hues between pastel blue and pink	$1 \times 10^6 -$ 6×10^6
III. Bad	< 120	Lilac-pink to colourless	$> 6 \times 10^6$

The results of the reduction tests decrease in reliability the more the microflora of the milk differs from that of an average source of milk. For example, the reduction tests cannot be reliably applied to samples fresh from the cow, to samples of very low germ count, to milk originating from a single stable or a single cow, or to milk that has gone sour or has been pasteurized. The rate of reduction is influenced by the leucocyte count of the milk, and by climatic factors also.

Pre-incubation of milk samples has been recommended by a number of authors as a means of making the samples more amenable to grading by the reduction test method. Pre-incubation at 12–15 °C, or even at 20 °C, for 18–20 h has been recommended for samples of new milk, weakly-reducing milk, deep-frozen milk, or milk containing psychrotrophs. The correlation between cell count and reduction capacity is improved by pre-incubation, nevertheless the deviations that occur remain considerable.

Because of this poor correlation, it has been recommended that grading which is based on the results of reduction tests be regarded as a separate grading indicative of the level of microbial activity, rather than the microbial count. According to Thomé et al., for example, the methylene blue reduction test is really appropriate only for showing up milk of poor quality. Decolourization with 2 h suggests improper treatment, improper cooling or both.

In spite of these difficulties, the resazurin-reduction test is virtually indispensable for the grading of raw milk. The procedure is simple, inexpensive, rapid, its psychological effect (of controle) is considerable, and the results give an indication of the activity of acid-forming bacteria in the milk; a clue to conditions during milk production and processing is also given by this test.

The reduction tests are not likely to decrease in importance unless the number of microbial cells in milk generally fall below 100,000 cells per ml. There is no reason to examine milk of high quality by the dye reduction method.

Although the two reduction tests show the same degree of microbial activity in high quality milk, the methylene blue test is preferable in this case, since it is very seldom that a sample of high germ count (which cannot be stored for any length of time without changes occurring in it) is graded as class I milk on the basis of this test.

For milk of low quality, on the other hand, the use of the resazurin reduction test seems to be worthwhile. One of the advantages of this test is that the results can be evaluated sooner. Besides, an increased leucocyte count, and the weak reducing activities of microbes in abnormal milk samples are detected with greater sensitivity by this test compared with the methylene blue reduction test.

8.1.2.3.D   The TTC-reduction test

This test involves adding triphenyl-tetrazolium chloride (TTC), a colourless water-soluble compound, to the milk sample. The reduced form of this substance, which is a red formazan derivative, is not sensitive to oxygen. TTC is very rapidly reduced by members of the coli–aerogenes group, and less rapidly by lactobacilli.

Procedure. 0·5 ml of a 1% TTC solution is added to 5 ml of thoroughly mixed milk. The mixture is shaken and examined in diffuse light. The colour reaction first appears where the mixture is exposed to the strongest direct illumination, then later the entire sample becomes red. If the sample is rich in microbes, the TTC will be reduced within a few minutes, and thus the test is a very suitable one for milk of high microbial content. However, the high sensitivity of TTC to light considerably limits its field of use.

8.1.2.3.E   Nitrate-reduction tests

The methylene blue and resazurin-reduction tests provide information essentially about the activity, and approximate numbers of lactic acid streptococci in milk. However, in the grading of raw milk other contaminants, particularly coliforms, need to be accounted for as well.

The nitrate-reduction tests are based on the fact that (a) nitrate is reduced to nitrite by coliforms, (b) the presence of nitrite is easily demonstrable by the

302

Griess–Ilosvay reagent, and (c) nitrate is not reduced by lactobacilli. Since nitrate is reduced by nearly all the Gram-positive and Gram-negative bacteria that are of importance in the dairy industry, it is desirable if possible to eliminate the nitrite-forming non-coliform bacteria (sarcinae, staphylococci, aerobic spore-formers, etc.) if coliforms are to be determined. Attempts to develop effective procedures have, however, failed. For example, it has been found that the activity of Gram-positive nitrate-formers is not totally eliminated when milk is mixed with brilliant green or trypaflavine.

Recently, nitrate-reduction tests have been used to examine milk for the presence of nitrite-producing contaminants, *viz.* coliforms and non-acidformers. Two procedures are described in the following. Kandler's original test is simpler and more rapid than the modified version of Möller and Reimoser. However, according to the latter authors, the modified test is more reliable because the activity of the lactobacilli is decreased.

Nitrate-reduction test according to Kandler (1961)

1 ml of a 1% nitrate solution and 10 ml of milk are measured into sterile test-tube and incubated in a water bath at 37 °C. After 1 h, the contents of the tube are mixed with 2 ml Griess–Ilosvay reagent [56] and shaken thoroughly. A reading is taken after the mixture has been allowed to stand for 5 minutes.

Depending on the colour intensity, milk samples are classified into 3 groups:

no colour changing	class I
pale or moderate pink	class II
intense pink or red	class III

A standard colour has been recommended for making an objective distinction between class II and class III; 1 ml of a mixture made of 1 ml 1 M phosphate buffer, pH 6·3, containing 0.005% phenol-red and 2 ml of 1 M phosphate buffer, pH 6·3 containing 0.005% safranin solution are added to 10 ml of milk. Test samples giving the same, or a more intense colour are assigned to class III.

Those milk samples that are graded as class I samples should be examined once again, 2 h later.

Nitrate-reduction test according to Möller and Reimoser (1969)

5 ml of nitrate–peptone water [103] are added to 1 ml raw milk in a sterile test-tube, and the tube is closed with a metal cap and incubated at 37 °C for 4 h. The sample is removed from the water bath and 2 ml Griess–Ilosvay reagent are added [56]. When the mixture has been allowed to stand for 5 minutes a reading is taken. The milk is of good quality if the test mixture has remained white, acceptable if the

mixture is pale pink or moderately deep pink, and unacceptable if the colour is red or dark red.

The period of incubation may be varied as required.

8.1.3   Tests for keeping quality

The number of microbes causing spoilage, the level of activity of these microbes, and the microbial environment in a given product give some indication of whether the milk or dairy product can be stored successfully. Apart from these indirect indications, direct storage tests are also usually carried out. For the purpose of the latter, the product is kept at the optimum temperature for the spoilage microbes so that the process of spoilage is speeded up; the time taken for signs of spoilage to appear provides a measure of the keeping quality of the product. The principle of this method is objectionable from the microbiological point of view because in most foodstuffs microbes other than those causing spoilage also occur, sometimes in predominant numbers. At elevated temperatures, the numbers of the different microbes as proportions of the total count may change; those causing spoilage may become predominant or they may become depressed in number. It is generally so that spoilage is accelerated in foodstuffs kept at higher temperatures, but it would be incorrect to draw far-reaching conclusions about the storage life at low temperatures from results obtained specifically at high temperatures. A temperature-spoilage rate correlation cannot be assumed, even if the only microbes in the product are spoilage organisms. In such cases (e.g. processed cheese) it must be taken into account that some components of the food inhibit, and others, stimulate microbial growth to a degree that depends on the temperature; for example, NaCl, nisin, etc., show an increased inhibitory effect at elevated temperatures. In spite of this, the results of direct tests are more useful guide to the keeping quality of foodstuffs than indirect indicators such as the cell count and the results of reduction tests. Nevertheless, interpolation from the results of direct tests to find what the storage life might be at lower temperatures, usually requires complex mathematical analysis.

8.1.3.1   Raw milk

The boiling test and the alcohol test have been found to be the most reliable. In both reactions, the protein that has deteriorated because of souring is coagulated.

Boiling test. A test-tube containing 5 ml of milk is placed in boiling water for 5 minutes; it is then set at an oblique angle so that the wall of the tube can be examined for protein precipitation. Precipitation or coagulation means poor quality of the milk-sample examined.

304

Alcohol test. 1 ml of bromcresol purple [15] is added to 1 ml of milk, and shaken. A positive reaction is indicated by the precipitation of protein and a colour change. The alcohol test is more sensitive than the boiling test; a positive reaction can be obtained after incubation at 22 °C for a period of 3 h or less.

Samples should be collected within 2 h of milking, and should be transported to the laboratory in ice. In the laboratory, the samples are placed in a water bath at 22 °C and examined by the boiling test or the alcohol test at 3-hourly intervals. In England and Wales, milk is assigned to class I if souring to a degree indicated by 11–12 SH° in the boiling test takes more than 45 hours at 22 °C; milk is assigned to class IV, on the other hand, if souring takes less than 27 h in the sample test.

The procedure has been simplified for routine tests. When the sample arrives at the dairy plant it is divided into two; in the morning, one part is placed in a water bath at 22 °C, whereas the other part is kept in melting ice until the afternoon, when it is also placed in a water bath at 22 °C. In this way, milk is always graded during day-time working hours.

8.1.3.2   Pasteurized milk

Attempts have been made to obtain information on the keeping quality of milk by incubating samples at 18 °C for 18–24 h and then subjecting them to a reduction test. The correlation between the total count and the storage life is very poor, because the bacteria that develop colonies on plates are not the bacteria that cause spoilage. The number of the latter falls below the threshold of detectability more rapidly than the numbers of other microbes present. The combination of the reduction test with a pre-incubation treatment is based on the fact that an increase in total number is usually accompanied by an increase in the number of latent harmful microbes.

The shelf life of pasteurized milk kept at a temperature higher than 10 °C is reduced first of all by the lactic acid-producing flora (sour coagulation) and by *B. cereus* (sweet coagulation). For the determination of storage life, the boiling test preceded by incubation at an appropriate temperature is the most suitable test, because unlike the reduction tests and determinations of the degree of acidity, this test takes account of any initial sweet coagulation (protein denaturation) as well. For the pre-incubation, temperatures of 15, 20 and 22 °C have been recommended by different authors.

8.1.3.3   Condensed milk

Original cans are kept in an incubator at 37 °C for a week and are then inspected for any sign of swelling. A loopful of the sample is also spread on different agar media and the cultures are incubated at 37 °C.

8.1.3.4   Sterile milk and uperized milk

The sample is incubated at 37 °C for 24 h and an equal amount of 80% alcohol (specific weight 0·846) containing 0·02% neutral red is added to an aliquot of the culture. Precipitation of the protein indicates that storage of the milk for 3 weeks at room temperature would lead to a bitter taste or sweet coagulation of the milk. Precipitation accompanied by souring (red colour) development points to the activity of acid-forming spore-formers *(B. albolactis, B. thermoacidurans)*. However, these bacteria rarely occur in milk. The spoilage of uperized milk after 2–3 days at 22 °C is usually caused by *B. cereus,* which causes fine flocculation in heat-treated milk.

8.1.3.5   Pasteurized sweet cream

1 ml of the sample is added to 10 ml of sterilized skim milk containing 0·01% bromcresol purple. The degree of acidification is indicated by the intensity of colour development after incubation for 16–17 h, and 24–25 h.

8.1.3.6   Butter

Butter is kept at 21 °C for 7 days, or at 15·5 °C for 10–14 days. With melted butter, the test is more sensitive and the time taken to carry out the test can be shortened. After incubation, grading may be carried out on the basis of organoleptic examination or microbiological tests.

8.1.3.7   Processed cheese

From each batch at least 3, and preferably 6 boxes should be examined. Cubes of cheese are removed from each box and incubated at 37 °C for at least 7, and preferably 14 days. The cheese samples are cut up before grading. Expansion, cracking and putrefaction indicate the probable presence of clostridia. It is possible to store cheese for a longer time at room temperature than at 37 °C.

8.1.4   Further tests

8.1.4.1   The efficiency of pasteurization

To determine the efficiency of pasteurization, the sample is examined both before and after pasteurization. The efficiency is expressed as the percentage decrease of the viable count. The samples are taken directly from the milk or cream arriv-

306

ing at, or leaving the pasteurization apparatus. It should be borne in mind that cultures established from pasteurized milk samples must be incubated for a longer time (2—3 days longer) than the usual incubation time.

$$\text{Kill \%} = \frac{a-b}{a} \, 100 \, ,$$

where a is the viable count before pasteurization and b is the viable count after pasteurization.

8.1.4.2   The activity of starter cultures

The activity test is essentially similar to the process of cheese production, carried out under laboratory conditions; the souring of the whey trickling out of the rennet-coagulated milk is measured.

Procedure. 500 ml of milk are transferred to a heat-stable beaker or a wide-mouthed jar, and warmed to 38 °C. 5 ml of the starter culture are added and mixed well. After an interval of 30 minutes 3 ml of rennet are added and mixed well. After 1 h, the coagulum is cut with a clean knife into columns 0.6×0.6 cm at the base. Two hours after cutting, the whey that has oozed out from the coagulum is removed. After another interval of two hours, the whey is poured off again and this time the acidity of the whey is determined. The coagulum is incubated for a further hour, and the whey is poured off and titrated. From the two titres, the activity of the starter culture can be calculated.

In comparative tests, the culture that gives the highest degree of acidification is considered the best. If the difference between whey samples from two different starters is less than 4 degrees of acidity, then the two starters may be regarded as equal in activity.

Cultures that are to be compared should be tested in vessels identical in size and shape, and the coagulum should be cut in an identical manner. Accurate testing requires vessels of standard size and shape. Although the procedure is time-consuming, it is good practice to measure the acidity of all the whey samples poured off. This will enable a souring curve to be drawn for calculating the degree of acidity of the sample, instead of using only two values (Pulay and Zsinkó, 1969).

8.1.4.3   Germination percentage in Roquefort and Camembert cheeses

The germination of conidia from powder-shaped Roquefort and Camembert cultures is tested under standard conditions, and the results are expressed as percentage germination. Beer agar cultures [126] are incubated at 22 °C for 18 h.

Procedure. 1 g of powdered mould preparation is ground in a mortar with 9 ml of saline. Aliquots of 0·1 ml from different dilutions of the sample are pipetted on to the surfaces of Beer agar plates, over the plates spread, and incubated at 22 °C for 18 h. The plates are then examined at 400–600-fold magnification, and those plates on which 15–30 conidia have germinated per microscope field are selected for counting. At least 300 germinating and non-germinating conidia are counted; a conidium is recorded as germinating if its length has enlarged at least 1·5 times.

Conidia that have been dried at a relatively high temperature and those that have been stored for a long time germinate slowly. These conidia are of insufficient activity for cheese-making.

8.2 Examination of meat and meat products

If meat is taken from a carcass without aseptic precautions and is stored without any method of preservation such as cooling, freezing, heat treatment, salting, drying, irradiation, etc., microbes gaining access to the surface of the meat will soon thrive and render the meat unacceptable. The same is true of meat products that have been processed and/or stored under inadequate conditions.

The reasons for rejecting meat or a meat product are usually (a) that it is hazardous to health, or (b) that it has become offensive organoleptically.

A health hazard is caused by the growth of pathogenic micro-organisms in, or on, the meat or meat product; offensive properties are generally caused by bad handling techniques. It must be emphasized that a meat or meat product contaminated by pathogenic bacteria (e.g. staphylococci) may have an entirely acceptable appearance on organoleptic examination, although the health hazard to the consumer is considerable; in other cases (e.g. products contaminated by *Clostridium botulinum*) there may be signs of spoilage.

The purpose of the microbiological examination of meat and meat products is thus to detect both pathogenic microbes and those micro-organisms that cause spoiling.

Some authors distinguish between veterinary–medical and industrial microbiology on this basis, although a clear definition of these cannot be made.

8.2.1 Sampling and the processing of samples

In the course of the grading of meat and meat products, averaged samples or individual samples taken from the surface and from the core are processed.

The site of sampling (surface, core, or bulked sample) is determined by the nature and history of the product. Thus in the case of heat-treated products, the core (thermic or geometric centre) should be examined since this is the least accessible to

308

heat penetration, and consequently the largest numbers of survivors may be expected there. The following procedure has proved satisfactory for taking samples from the core. The surface of the product is flamed and the meat is cut in two with a sterile knife, perpendicular to the longer axis of the piece. Sterile scissors and sterile forceps, or a sterile corkborer, the size of which is determined by the size of the piece of meat, are used for removing the sample. If a corkborer is used, sampling is facilitated by cutting a relatively thick slice in the first instance. Sample should always be taken from the core of cooked sausages and canned food.

In the case of unheated products, such as raw minced meat, bulked samples are taken. When large pieces of raw meat and whole carcasses are being inspected, surface samples are collected first of all.

A *bulked sample* is collected by removing any wrappings present and flaming the surface of the meat. A slice representing the entire cross section of the product is cut and removed. Samples of minced meat are taken with a sterile spoon. A bulked samples are usually taken from non-heat-treated products (raw meat, minced meat, dry sausages, raw-cured products), from items of small size (e.g. Vienna sausages), from sliced vacuum-packed meat products, and from additives.

Preparation of samples. Except in the case of surface samples, 10–250 g should be taken, depending on the amount of material available. Large samples are aseptically minced if necessary, and 10 g of the minced sample are added to 90 ml of sterile diluent containing 0·1% peptone and 0·5–0·9% NaCl. The suspension is homogenized at about 6000–10,000 rpm for a total of 20,000 revolutions, using an Ultra-Turrax or similar apparatus (Barraud et al., 1967). Alternatively, a stomacher can be used. The material is then allowed to stand for 10–15 minutes and 10 ml of the supernatant are added to 90 ml of the diluent. In this way, a tenfold dilution series is prepared.

Many techniques of surface sampling are employed in the meat industry. Sampling from the surface is particularly important for assessing general conditions of hygiene or the hygiene of processing. Thus surface microbial counts are determined for the external and internal surfaces of carcasses after various processing operations, such as scalding and scraping, singeing, evisceration, chilling, freezing, thawing, etc.; the contact surface of processing machines, vessels and utensils, parts of the human body coming into contact with meat, clothes (aprons), etc. are also examined. The following methods of surface sampling are the most frequently used:

(a) *Aseptic excision* of a predetermined area of the surface layer (e.g. 10 cm²), and homogenization without diluent (see above). The microbial count obtained in this way gives the best agreement with the actual count. The disadvantages of the technique are that only small areas can be examined, the procedure is time-consuming and laborious, it damages the material that is being tested, and it cannot be applied of course to the surface examination of equipment.

(b) *Sampling by scraping*. The superficial layer of the meat is scraped off with a sharp scalpel and placed in a diluent medium. If the surface lies horizontally, a metal ring enclosing a known area is pressed on to the surface, some diluent is poured into the ring and the surface layer underneath is scraped off. If the sample can easily be moved, the suspension thus obtained is poured into a jar, otherwise it is sucked up into a pipette.

Of the various methods of surface sampling, swabbing and the contact adhesion technique are the most widely used. Although the recovery rate of micro-organisms is lower than that from excised samples, the two methods (especially the swabbing method) are rapid, they do not involve damaging the raw material and comparatively large areas can be sampled with swabs; consequently, the results are more representative of the general state of contamination of a carcass surface, in spite of being less accurate (Ingram and Roberts, 1976).

(c) *Sampling by swabbing*. A known area (10–100 cm²) is wiped with a sterile cotton swab soaked in sterile diluent; the swab is then homogenized or shaken with the diluent (in the latter case the recovery of cells is lower).

The swab may be made of cotton, or calcium alginate which can be dissolved rather than homogenized in a sodium salt solution. Serial dilutions are prepared from the homogenate (or solution). According to some authors, bacterial growth may be inhibited by alginate, and by sodium haxametaphosphate (Strong et al., 1961). Others, however, have reported good results (Hyatäinen et al., 1975). The swabbing method has a number of advantages, *viz.*, a relatively large total surface area can be sampled in a short time, the meat or meat product is not in any way impaired, and the method can be applied to any type of surface. However, the recovery rate is only about 10% of the total number of microbes (Catsaras et al., 1974).

d) *Sampling by the contact adhesion technique*. According to early descriptions of this method, a sterile slide was pressed on the surface that was to be tested, and the adhering microbes were fixed, stained, and examined microscopically. This method provides an imprecise result, and, therefore, it is used only exceptionally. Its up-to-date equivalent is a combination of this type of sampling with cultivation. The so-called agar-sausage method (ten Cate, 1964) is the most widely used version of the technique (see 7.3.1.3); it is rapid, it can be applied to any kind of flat surface, and it does not impair the material to which it is applied. However, serial dilutions cannot be examined in this way. Furthermore, the recovery rate is low, although not as low as the rate obtained by swabbing (Hess and Lott, 1970).

Those versions of the resazurin and TTC test in which impregnated filter paper strips are pressed against the contaminated surface and then incubated, are, in fact, based on the adhesion principle. These methods will be described later.

Regardless of which sampling method is used, it is of great importance to carry out the test always in the same way. For example, the size of the area to be swabbed should always be the same; on the surface of a carcass, areas of 10 cm² should be swabbed at 5 different sites, the swabs must be identical in size and quality, and the duration of the swabbing action must be constant. The need for standard procedures extends also to the storage and transport of samples. With respect to the cultivation procedure, a standard medium of optimal composition for the microbes to be cultivated is a further important requirement.

Other methods of sampling. Very rarely, liquid samples are also taken in the meat industry, e.g. from curing brines, scalding water, or cooling water involved in the production of semi-preserved foods. Samples can be collected in a sterile jar or by means of a sterile pipette. Samples are also taken by pipette directly from the core of large brawns, after heat treatment. Such samples, however, give an indication of the level of contamination of the liquid jelly itself, rather than of the meat particles in it. One great advantage in taking liquid samples is that early sampling is possible. (Otherwise sampling is usually preceded by storage in the refrigerator overnight.) To have the results of preliminary tests before products are sold is a considerable advantage.

8.2.2 The microbiological examination of the sample

Except for samples taken by the contact adhesion technique, bacterial suspensions or serial dilutions are available as part of the processing of samples for microbiological analysis. The test method may involve direct counting in order to determine the total cell count, or cultivation in order to determine the viable count. Less frequently, the total viable count is estimated on the basis of microbial enzyme activity.

8.2.2.1 Microscopic counting

In the meat industry, direct counting is rarely employed. Microscopic counting is carried out with stained or native preparations, using normal and phase-contrast illumination, respectively. Other procedures are of little use because of the interfering effect of debris and fat droplets. (The microscopic and other assays are described in detail in Chapter 5.)

8.2.2.2 Rapid tests based on enzyme activity

Such tests are only used occasionally, and the results obtained are approximate. The resazurin-reduction test is the most widely accepted test, a modified form of which is described in the following.

Tube test (Losonczy, 1970). 20 g of meat or meat product are minced in 180 ml of a physiological phosphate buffer, pH 7, for 10 s at 10,000 rpm in a knife homogenizer. 9 ml of the supernatant are pipetted into a sterile test-tube and 1 ml resazurin solution* is added and warmed in a water bath at 37 °C for 10 minutes before being placed in an incubator at 37 °C. The time required for the development of a pink colour is recorded. If curing brine is to be examined, 1 ml of this is mixed with 8 ml phosphate buffer, pH 7, and 1 ml resazurin solution. The mixture is allowed to stand in a water bath for 10 minutes and is then incubated at the same temperature until a pink colour develops. The most probable number for the sample is estimated on the basis of the data in *Table 49*.

A further simplification of the method is the resazurin-reduction test on paper strips. A paper strip of area 10 cm², soaked in resazurin solution, is placed on the material to be tested and covered with polyethylene film. Strip and sample are incubated until the pink colour appears. The reaction time recorded at the given temperature is correlated with the viable count. This method is less accurate than the tube test, yet it is useful for checking the various steps of meat processing, e.g. deboning, the chilling of carcasses and cuts, the handling of raw meats, etc. The test cannot be applied during the first two days following slaughter when, owing to the large reducing activity of the fresh muscle, the reduction time is found to be very short even if the viable count is low. Another meat product that cannot be examined by this method is the emulsion-type sausage prepared by chopping and curing with salt.

Table 49

Grading of meat or curing brine, using the resazurin dye-reduction (tube) test

Quality class of the sample	Time necessary for development of pink colour (h)			Approximate viable count per 1 g or 1 ml of material
	pork	beef	brine	
Very good	>7	>8	>12	$<10^5$
Good	6−7	6·5−8	8−12	$10^5−10^6$
Acceptable	4−6	4·5−6·5	3·5−8	$10^6−10^7$
Unacceptable	<2	<2·5	3·5	$>10^8$

8.2.2.3 Cultivation procedures

Besides the total viable count, the presence or absence of bacteria such as *Salmonella* can be determined by cultivation. In food-industrial microbiology, pour-plating, surface cultivation, and determination of the most probable number (MPN) by the dilution technique are the most commonly used methods. Although the latter method is less accurate than the other two, in some cases it is the only method applicable, for example where the viable count is expected to be very low,

*Resazurin solution: 0·005 g resazurin in 100 ml sterile destilled water (use only freshly).

where anaerobic sulphite-reducing bacteria are among those to be counted, and where the total count of aerobic spore-formers is to be determined; these determinations cannot be carried out by other methods in the presence of mixed populations. The principles and techniques of cultivation procedures are discussed in detail in Chapter 5.

The cultivation methods lead to the determination of viable counts, including the total viable count and the viable counts of indicator organisms, or of selected species.

8.2.2.3.A   Total viable count

The total viable count provides information on the general level of contamination of the product, raw material, working surface of equipment, etc. including contamination with pathogenic bacteria and/or bacteria that may cause technical failure.

The informative value of the total viable count has often been questioned because it gives no indication of the probable presence or frequency of occurrence of pathogenic bacteria. Nevertheless, in many situations a high level of general contamination is accompanied by an increased occurrence of pathogens, and thus the total viable count provides a very useful measure of hygienic conditions in the abattoir, and the prevalence of pathogens there; in some situations, however, e.g. in the case of dry sausages, there is no such correlation. For the reasons mentioned above, and on account of the simplicity of the test, the determination of total count is the most generally, sometimes exclusively, applied method.

The purpose of routine tests, i.e. tests of other than those aimed at revealing some specific processing error, is the detection of pathogenic bacteria that may be present in the product. In this respect, the total viable count gives a very rough estimation.

8.2.2.3.B   Examination of the indicator flora

Pathogenic bacteria may be present in meat products in very small numbers which do not reach the threshold of detectability. However, there may be present a characteristic and abundant concomitant flora which thus serves as an indicator flora for pathogens. Bacteria indicative of faecal contamination, and, therefore, of the presence of *Salmonella*, enteroviruses, etc. are usually selected as indicators. However, some bacteria that are occasionally used as indicator micro-organisms may not suit this role, either because they include species of nonfaecal origin (e.g. coliforms, *Enterobacteriaceae* species), or because their faecal origin must be regarded as indirect (enterococci) owing to their high tolerance of external environmental conditions and their common occurrence in food-industrial environments.

8.2.2.3.C   Detection and counting of various species in meats and meat products

If meat is to be examined for the presence of a given species, selective culture media are employed with the aim of obtaining the particular species in pure culture, or at least of distinguishing it easily from other micro-organisms so that it can be counted.

Pathogenic bacteria may be counted, or their titre may be determined, the latter correlating closely with their number. The titre may be related to the quantity present in the sample, or sometimes it is related to values obtained after enrichment procedures have been carried out (in the case of salmonellae, and occasionally staphylococci).

In the following, the occurrence and method of detection of individual species (or groups of species) are detailed. It should be noted that with respect to cultivation in enrichment media, only the lowest dilution is used as an inoculum (surface sampling) or a piece of the product is used, whereas in the case of direct examination, selective culture media are inoculated from serial dilutions.

8.2.2.3.C.a   Pathogenic bacteria

Salmonella

Raw meat, raw meat products, equipment, utensils coming into contact wich these, as well as products that have been inadequately heat-treated may be primarily infected with *Salmonella;* other products may be contaminated with *Salmonella* by secondary infection (post-infection). Taking into account the fact that *Salmonella* is one of the most common food-borne pathogens causing food infection in man, the isolation and detection of these bacteria are essential in all cases except those involving some technical failure. The surfaces of slaughtered animals, raw meat, and non-heat-treated products such as raw sausage and smoked sausage (a product smoked at 40–50 °C, but not heat-treated), require particular attention in this respect. For detailed descriptions of the procedures, see 6.3.1.4.

Staphylococcus

In addition to *Salmonella, Staphylococcus* is an other very common causative agent of human food-borne illness (food poisoning). Staphylococci usually enter foodstuffs from the human body and produce toxin in the food. The main danger arises from the fact that the food causing the poisoning sometimes has quite acceptable organoleptic properties, and from the fact that the endotoxin is highly resistant to heat. It may easily happen that toxin is present in a heat-treated product in which no bacteria can be detected. The testing of meat or meat products, especially raw sausage and brawn, for *Staphylococcus* is generally compulsory. The

314

importance of *Staphylococcus* in the food industry is enhanced by its high tolerance to extreme environmental conditions including heat, cold, freezing, low water activity, presence of nitrite, etc. For methods of detecting the bacterium and the endotoxin, see 6.3.3.

Clostridium perfringens

Sources of *C. perfringens* are soil, water, or human or animal faeces. Since it may cause food poisoning, the role of this organism has attracted increasing interest. Primary isolation, as in the case of other clostridia, is carried out in sulphite agar [133]. For details see 6.3.2.2.

Clostridium botulinum

Food poisoning caused by this species occurs rarely, but a proportion of cases are fatal. Deriving from the soil, *C. botulinum* entering food grows, and may produce toxin in the food. Industrial food products give rise to botulism much less frequently than home-made products. The difference is due to proper hygienic processing and the use of curing additives in the food industry. Cooked, cured products generally represent no risk of botulism if they are refrigerator-stored, owing to the presence of sodium nitrite, a salt known for its germination–inhibiting effect. Examination for *C. botulinum* is a matter of expedience, especially in the case of products not containing nitrite (or products cured for a long time, containing nitrate). For methods of detecting the bacterium and its toxin, see 6.3.2.1.

Bacillus cereus

Like other aerobic bacteria, this is an ubiquitous micro-organism. Studies in Hungary and elsewhere suggest that it may play a role in food-borne disease, especially if it reaches a high cell count in the food (10^5–$10^6 g^{-1}$ or higher). The spores of *B. cereus* may survive the normally applied heat treatment of food, and the bacterium can grow in products that have not been stored under proper conditions. A method for the detection of *B. cereus* is described in 6.3.5.

Enterococci

The possible role of these bacteria is a matter of controversy. Food poisoning is usually attributed to enterococci if a pure culture of enterococci has been isolated from the suspected food. However, enterococci, because of their high resistance to external environmental conditions including acidity, alkalinity, heat, low water activity, high levels of salt and nitrite, etc., may grow and outnumber other bacteria. Occasionally, enterococci are used as starter culture. For their role as indicator organisms see above; a method for their detection is described in 6.3.4.

Enteropathogenic coli and *E. coli* I

In the food industry, enteropathogenic strains occur very rarely and only in unheated products. The method used for detecting these is the same as the method for detecting *E. coli,* and the final identification requiring an immunological assay. This assay and the detection of *E. coli* I (an indicator of faecal contamination) are described in detail in 6.3.1.5.

8.2.2.3.C.b Examination of meat for micro-organisms causing technical failure

Proteolytic bacteria

Proteolysis in meat and meat products is accompanied by a characteristically foul smell. Many different bacteria can cause proteolysis. The pour-plate medium used for the detection for these organisms contains gelatin, meat protein, or milk. Colony counting is carried out on poured plate cultures containing $HgCl_2$ solution or some other protein-precipitating agent; colonies of proteolytic bacteria are surrounded by a clear zone. Under aerobic conditions the main organisms responsible are aerobic spore-formers, while under anaerobic conditions (vacuum-packed meat and meat products), clostridia play the main role. The former are detected and counted by the MPN method (5.2.1) using nutrient broth [138]. The presence of aerobic spore-formers is indicated by the development of a skin on the surface of the broth. The skin does not disintegrate completely on shaking, but tears into small fragments instead. On the basis of the above criterion, aerobic spores can be counted after incubation in inoculated broth, pretreated at 80 °C for 20 minutes. The cultivation and counting of clostridia are carried out on sulphite medium [133], making use of the sulphite-reducing ability of these bacteria; for details see 6.3.2.2.F.

Psychrotrophs

For the detection of psychrotrophs, the culture media used are the same as those used in the determination of viable counts. The temperature of incubation should be such as to have a selective effect, e.g. 5 °C for 7–10 days, although the incubation period can be shortened by introducing microscopic colony counting. The *Pseudomonas* and *Acinetobacter (Achromobacter)* strains which are responsible for superficial spoiling (slime) in raw meat during cold storage, are typical psychrotrophs. Because of their low heat tolerance and the sensitivity of *Pseudomonas* to salt, psychrotrophs play a secondary role in the post-infection spoilage of heat-treated meat products.

Bacteria causing greening

Greenish discolouration occurs almost exclusively in cured, cooked meat products (sausages, hams), since the natural catalase activity of the raw meat, which interrupts the greening process, ceases in the heat treatment. The greening process is based on peroxide production by bacteria. The greenish discolouration sometimes observable on the surface of raw meats is not of chemical origin, but is due instead to the colour of bacteria (e.g. *Pseudomonas*) forming a layer on the meat surface. The bacteria incriminated in the greening of meat products (*Lactobacillus viridescens*, enterococci, pediococci) have the common property of producing hydrogen peroxide under aerobic conditions; in the absence of catalase, the peroxide reacts with the meat pigment, causing greening. The detection of greening bacteria is based on the demonstration of peroxidase activity. Briefly, a bacterium that is suspected of being a greening bacterium is spread on a culture medium containing suspended black manganese oxide (MnO_2) [medium **169**]. In the presence of peroxid the MnO_2 is reduced and its colour disappears; thus if the medium clears around the colonies, the bacterium is regarded as a greening stain. A positive catalase reaction (intense sparkling observed when a loopful of the bacterium is mixed with H_2O_2) is evidence that the bacterium is not responsible for greening.

Bacteria causing surface slime

Under some conditions, slime-producing bacteria may grow and develop a continuous coating on the surface of meats and meat products. The development of surface slime is inhibited by the traditional prolonged smoking, by treatment of the surface with liquid smoke and by the maintenance of low relative humidity. Among slime-producing micro-organisms, micrococci, lactobacilli, *Leuconostoc*, *Microbacterium thermosphactum*, vibrios, *Proteus* species and yeasts deserve mention; on raw meat *Pseudomonas* may occasionally produce slime. In pre-packed products, lactobacilli are the predominant slime-producers. *Pseudomonas* species, micrococci and *Leuconostoc* can be cultivated in nutrient broth or nutrient agar [**138**], yeasts grow on a special medium as described in 6.4, lactobacilli grow on Rogosa's medium [**119**], and *Microbacterium thermosphactum* on Gardner's medium [**170**] (Gardner, 1966). Vibrios and *Proteus* species are isolated on medium [**172**] as described in 6.3.1. For details of the criteria used for identification, see the appropriate tables for the identification of bacteria and yeasts.

Other kinds of spoilage

Cured, cooked and sliced vacuum-packed meat products sometimes have a sourish smell which is generally caused by *Streptococcus faecalis* or *S. faecium*, or occasionally by *Microbacterium thermosphactum*.

It has been reported that enterococci may attack the fibres of muscular tissues, thus loosening the consistency of cured meats and making the meat soft.

A slightly foul cheesy smell is characteristic of a type of spoilage of dry sausages, which occurs if ripening (fermentation) temperature is not appropriate so that at a time when the high water activity of the product, together with the elevated temperature, makes conditions favourable for growth of Gram-negative non-spore-forming proteolytic bacteria, and clostridia. It should be noted that even *Salmonella* can grow under these circumstances. This kind of spoilage can be prevented by ripening at a low temperature, by using carbohydrate-decomposing starter cultures, or by a combination of both of these. A high total viable count obtained from dry sausages does not necessarily mean that there is a microbiological problem, if handling procedures have been followed correctly and the organoleptic properties of the product are acceptable. Viable counts as high as $10^8 \, g^{-1}$ may be accounted for by lactobacilli, or in a wider sense, members of the *Lactobacillus* family, these being necessary to develop the characteristic aroma of the product.

Mould cover on the surface is another phenomena occurring on dry sausages. This may be desirable or undesirable, depending on tradition. Growth of moulds is generally preceded by the growth of yeasts. For the detection of yeasts and moulds see 4.2 and 4.3.

8.2.2.3.D   The microbiological examination of curing brines

Nowadays only a few products such as bacon and country ham are cured in the traditional way with nitrate and NaCl in order to develop the desired organoleptic properties (colour, flavour, tenderness, etc.) This method of curing requires the injection of brine or the use of covering brine, while up-to-date methods use nitrite ($NaNO_2$) instead of nitrate (KNO_3), and tumbling is involved. In modern curing techniques, curing usually does not take longer than 24 h, and since the brine is not used again, there is no need to subject the brine to microbiological tests. However, the use of curing ingredients (water, salt, phosphate, nitrite) of high bacteriological quality is important.

If, on the other hand, curing takes several weeks, some bacteria may be capable of growing in the brine. Thus the useful species may exhibit their nitrate-reducing and flavour-developing properties while harmful ones are spoiling the product.

In nitrate-containing brines, nitrate-reducing and flavour-developing bacteria such as lactobacilli represent the useful species. For the development of flavour, starter cultures may be added to produce the characteristic aroma. Micrococci and vibrios play the main role in nitrate reduction; these organisms are active even at salt concentrations as high as that of curing brines (10–20%). The source of nitrate-reducing strains is usually not the raw meat (Kitchell, 1957), but is more frequently the wall of the vat, or an old brine. It is of interest that most of the bacteria that cause superficial spoilage of cured products *viz.* odour effects and slime, are micrococci or vibrios; these are less salt-tolerant than their useful counterparts, and often do not reduce nitrate. Proteolytic *Proteus* strains may also cause spoilage in curing brines.

Micrococci can be isolated on nutrient agar. Their isolation from brines is facilitated by raising the NaCl concentration of the medium to 5%. Gardner's method [172] is recommended for the selective detection of vibrios. For details of the identification see Gardner (1973).

Proteus strains can be isolated and identified on the selective media used for isolating *Enterobacteriaceae*.

Finally, the determination of total cell counts and total viable counts is also part of the microbiological examination of curing brines. For obtaining the total viable count, Gardner's medium containing 4% NaCl according to the original prescription [71] is used (Gardner, 1968). The salt concentration may be varied between 1% and 20%, depending on the purpose of the examination and the salt concentration in the brine. The detection of salt-tolerant bacteria requires, of course, a salt concentration in the medium higher than that to which the spoilage organisms are exposed. Thus the salt concentration in the brine must be taken into account. Ingram (1957) obtained lower viable counts when brines of high salt concentration were examined on media of low salt concentration.

The incubation temperature is a further important factor; taking the low temperature of curing into account, 20 °C seems to be the most suitable temperature.

If some special form of contamination is suspected in the curing brine, particularly where the latter is of low salt concentration, the testing should include examination for *Pseudomonas*, yeast, *Salmonella*, Enterococcus, *Bacillus*, *Staphylococcus*, etc. (Details of the methods of examination are given in the relevant chapters.) It should be noted that pathogenic species generally do not grow in curing brine, but they may remain viable in brine, and then grow in those products of lower salt content.

8.2.2.3.E   The microbiological examination of additives

As well as salt, spices have been used as meat additives since ancient times. Spices are generally heavily contaminated, and, therefore, they have to be microbiologically examined. The factors usually determined in this respect are the total viable count, the coliform count, the aerobic and anaerobic spore counts, the yeast and mould counts, and occasionally in the context of canned food production, the thermophilic spore count. The testing may be extended further if any processing problem or hygienic defect makes the product suspect of bacterial contamination.

When spice extracts are used, no microbiological problems arise. NaCl, KNO_3, $NaNO_2$, and phosphate additives are usually of acceptable purity. Occasionally, spore counting may be necessary.

Nowadays, protein-based additives (caseinates, soy protein concentrates and isolates, texturized vegetable protein) are extensively used in the food industry. The preparation and processing of these additives precludes the survival of any of the original microflora. Furthermore, the rigorous treatment and conditions of packing guaranteed by the manufacturer usually ensures that the microbiological quality is always high. Thus, the microbiological analysis of such products includes only the viable count and the coliform count. These products usually cause neither hygienic nor technical problems.

The same is true of those additives of the polysaccharide type, such as carragenates, agar-agar, etc., but does not hold for spray-dried blood plasma or gelatin. In dried plasma especially *Salmonella* may occur as a result of primary or secondary contamination. Thus besides the determination of the viable count and coliform count, it is advisable to test for *Salmonella* contamination, and in certain cases, for sulphite-reducing clostridia, too.

8.2.2.3.F   The microbiological examination of lard

Rendered lard may be regarded as a bacteriologically stable product on account of its very low water content. Occasionally an odour suggestive of proteolysis is given off, and this is usually caused by deteriorating impurities derived from the rendering, pressing and dosing equipment, i.e. fibrous material rich in protein may find its way into the fresh lard during the rendering. On the other hand, proteins within the fat or connective tissue may be subjected to proteolysis. This can occur if the fat is rendered at a relatively low temperature, at which neither thermal inactivation nor dehydration due to heating occurs.

In addition to the aspects of spoiling described above, discolouration with a characteristic spotted pattern may develop on large blocks of lard. These changes are caused by yeasts and moulds (for isolation and identification of these, see 6.4 and 6.5), and the process is limited to the surface layer. This type of spoilage only occurs if water droplets in the lard exceed the colloidal size by several orders of magnitude, and if evaporation is inhibited by a water vapour-impermeable wrapping such as polyethylene.

Rancidity of fat generally develops independently of microbiological activity, the more so as it has been observed that the peroxide number of rancid lard may be decreased as a result of microbiological activity (Alford et al., 1971).

Owing to the presence of fat in certain meat products (dry sausages), aroma substances are formed when microbial lipolysis takes place (Cantoni et al., 1967a, b).

8.2.2.3.G Demonstration of the presence of inhibitory substances (antibiotics)

Antibiotics ingested by animals from feed or from drugs are detectable on agar plates on account of the inhibitory effect of these substances on bacterial growth (Katona and Pusztai, 1975).

The following strains can be used as test bacteria:

B. subtilis ATCC 6633: sensitive to streptomycin, chlortetracycline and erythromycin;

S. aureus ATCC 6538: sensitive to penicillin, streptomycin, chlortetracycline, erythromycin and oxytetracycline;

Sarcina lutea ATCC 9431: sensitive to penicillin, chloramphenicol and chlortetracycline;

Escherichia coli ATCC 9637: sensitive to streptomycin and neomycin.

Procedure. Nutrient agar is inoculated from a culture, or with some spore suspension of *B. subtilis*. The pour-plate method is used, the inoculum being calculated to achieve a final cell concentration in the medium of about 10^5 ml^{-1}. An aseptically excised piece of a muscle tissue or fragment of an organ is placed on the agar and the culture is incubated at 30–37 °C. The pH is adjusted to 7·2–7·4 for the detection of streptomycin, and 6·5–6·6 for other antibiotics.

An inhibitory zone indicated by the absence of bacterial growth around the tissue fragment suggests the presence of an antibiotic. The test is semi-quantitative; the diameter of the zone gives some indication of the quantity of the antibiotic substance present.

8.3 Examination of poultry, eggs and egg products

8.3.1 The microbiological examination of poultry

Although poultry is frequently stored in the frozen state, consumers generally prefer to buy refrigerated poultry, and, therefore, refrigerated storage is the most common form of storage for poultry. Microbiological examination of poultry may be necessary for a variety of reasons. The cell count on the surface of the skin gives an indication of processing hygiene and the keeping quality of the poultry. An estimation of disease risk to the consumer may be based on a determination of the numbers of pathogenic bacteria.

The methods of cell counting and detection of the various types of micro-organism are generally the same as those described above for meat and meat products. One essential difference is that swabbing cannot be used as a sampling method because bacteria are not found so much on the body surface as deep in the

21 Kiss

skin tissues and in the feather follicles, where they are not accessible to swabbing. Instead of swabbing, samples are excised from the skin aseptically at different sites, and total counts and total viable counts are obtained. To determine the level of contamination of the musculature, the skin samples are taken from over the breast muscle and leg muscle. Microbial counts are determined in the usual way. The procedures for detecting pathogenic microbes are the same as those described for meat and meat products, except in the case of suspected *Salmonella* contamination. Under these circumstances the entire bird is placed in a plastic bag, and is washed with 300 ml of glucose broth [50]; the washing fluid is incubated at 37 °C for 4 h to enrich any *Salmonella* that might be present. From the incubated medium, 20 ml are added to *Salmonella* enrichment medium (6.3.1.4.A), and the latter is incubated at 37 °C. (For details of the methods see the section dealing with the detection of *Salmonella*.)

8.3.2   The microbiological examination of eggs and egg products

The spoilage of eggs arises from tissue deterioration or from bacterial enzyme activity. The developing egg is generally sterile, but it can become infected while passing along the cloaca, and also in the external environment when microbes adhering to the egg may penetrate it and cause colour changes and offensive odours. Of the microbes causing spoilage in eggs, *Pseudomonas, Proteus,* coliforms, cocci, *Penicillium, Serratia, Cladosporium,* as well as species of the genera *Mucor* and *Aspergillus* deserve mention. The relative importance of these depends very much on storage conditions. *Proteus* is not a characteristic member of the flora or the infected egg. Of the types of spoilage involving discolouration, greening is caused by *Pseudomonas fluorescens,* a reddish-pink effect is caused by *Pseudomonas* species which produce a non-fluorescent pigment, and blackening is caused by *Proteus* species. *Pseudomonas* species and fungi may be present as a concomitant flora in the blackening egg.

Among those bacteria that present a health hazard, *Salmonella* is of primary importance, especially on account of its common occurrence in eggs. In the hen's egg *S. gallinarum* is common, and in the eggs of geese and ducks. *S. typhimurium* occurs with considerable frequency. Possible contamination of eggs with *Salmonella,* and the multiplication of these bacteria that would occur if the eggs were pre-incubated, makes human consumption of such eggs impossible.

Also, the causative agent of fowl tuberculosis *(Mycobacterium avium)* may find its way into eggs. Although this bacterium rarely causes human disease, proven cases have been registered. Therefore, eggs originating from stocks infected by *M. avium* must be rejected.

The allowable level of microbiological contamination of egg products (liquid egg and egg powder) is the same as that which applies to shelled; these products are also

322

Table 50

Microbiological examination of raw materials, ingredients and additives used in the meat industry

Type of products	Test to be performed	Recommended tests	Remarks
Raw meat incl. poultry, plucks, fat	Salmonella Staphylococcus aureus E. coli	total viable count clostridium count, (sulphite reduction) Pseudomonas Acinetobacter aerobic spore-former count coliform count Microbacterium thermosphactum	
Sausages, cold meat	Staphylococcus sulphite-reducing clostridia coliforms total viable count Enterococcus	aerobic spore-former count Lactobacillus count M. thermosphactum Pseudomonas Acinetobacter Micrococcus greening bacteria	in vacuum-packed products in presence of surface slime in presence of greening
Dry sausage	Salmonella Staphylococcus E. coli coliforms	sulphite-reducing clostridia Lactobacillus yeasts moulds	on the surface
Sausage made of plucks; pastes	Salmonella Staphylococcus sulphite-reducing clostridia Enterococcus coliforms, total viable count	Micrococcus Proteus } Acinetobacter	in presence of surface slime
Raw products (sausage cured unripened)	Salmonella Staphylococcus E. coli coliforms	total viable count	
Cured raw (or smoked) meat products (ham, chuck, shoulder, spare rib, bacon)	Salmonella Staphylococcus sulphite-reducing clostridia	Micrococcus Vibrio Total viable count	in cases of organoleptic rejection

(Table 50 cont.)

Type of products	Test to be performed	Recommended tests	Remarks
Cured cooked meat cuts	*Salmonella* *Staphylococcus* (both from the surface)	total viable count *M. thermosphactum* *Lactobacillus* sulphite-reducing clostridia Enterococcus	in vacuum-packed products in cases of softening
Semi-preserved food	*Staphylococcus* sulphite-reducing clostridia, total viable count	aerobic spore count Enterococcus greening bacteria *Salmonella*	in the presence of greening
canned food	sulphite-reducing clostridia count, aerobic spore-former count	*Salmonella* *Staphylococcus* total viable count	in cases of postcontamination due to injury or faulty sealing
Brine	total viable count *Micrococcus* *Vibrio*	yeasts *Pseudomonas* *Acinetobacter* *Proteus*	in cases of spoilage of brine
Ingredients and additives	*Salmonella* *Staphylococcus* sulphite-reducing clostridia, moulds		
Eggs and egg products	*Salmonella* *Staphylococcus* coliforms, total viable count Enterococcus	*Pseudomonas* *Proteus* moulds	in cases of spoilage

sometimes contaminated with *Staphylococcus, Clostridium* and/or saprophytes. Spray-drying is accompanied by heating, but this is insufficient to kill the entire bacterial flora; even salmonellae may survive, and, therefore, the examination of egg products is essential.

The culture media and microbiological tests used in connection with egg and products are the same as those described for meat and poultry. Processing of the material to be tested is carried out as follows.

The surface of the egg is cleaned by brushing with dilute detergent solution and wiped with cotton wool soaked in disinfectant (70% ethanol). The egg is broken and the white is separated from the yolk if necessary. 10 ml of the sample are made up to 100 ml with diluent [110], and further dilutions are prepared if necessary. Frozen eggs should be allowed to thaw before they are examined.

Egg powder is homogenized in a homogenizer or by shaking with sterile glass beads, so as to prevent coagulation.

*

The microbiological tests recommended for the examination of different foodstuffs are shown in *Table 50*.

8.4 Examination of canned products

8.4.1 Storage test for canned foods

In the microbiological laboratories of the food industry, canned food is usually tested for:

— shelf-life (keeping time) and sterility in unspoiled samples;
— the presence of spoiled samples and the possible origin of the spoilage.

Table 51

Number of cans (drawn at random) required to detect various levels of spoilage with indicated degree of probability

Spoilage in lot	Probability*	
	0·95	0·99
0·5	597	919
1	298	458
2	148	228
3	99	151
4	74	113
5	58	90
6	48	74
7	41	64
8	36	55
9	32	49
10	28	44
11	26	40
12	24	36
13	22	33
14	20	31
15	18	28
16	17	26
17	16	25
18	15	23
19	14	22
20	13	21
25	10	16

* Probability that at least one can in the incubated sample will show spoilage.

Note: National Canners Association, Research Laboratories, 1968.

Table 52

Probability* of detecting various levels of spoilage with indicated sample sizes**
(National Canners Association, Research Laboratories, 1968)

Spoilage in lot	Number of cans incubated							
	12	24	48	72	96	120	144	200
0·5	0·06	0·12	0·20	0·31	0·38	0·46	0·47	0·63
1	0·11	0·21	0·38	0·51	0·62	0·65	0·77	0·87
2	0·22	0·38	0·62	0·76	0·85	0·92	0·95	
3	0·31	0·52	0·77	0·88	0·94	0·97	0·98	
4	0·39	0·62	0·86	0·94	0·98			
5	0·46	0·71	0·91	0·97	0·99			
6	0·52	0·77	0·95	0·99	0·997			
7	0·58	0·82	0·97	0·994	0·999			
8	0·63	0·86	0·98	0·997				
9	0·68	0·90	0·99	0·999				
10	0·72	0·92	0·994					
11	0·75	0·94	0·996					
12	0·78	0·95	0·998					
13	0·81	0·96	0·999					
14	0·84	0·97						
15	0·86	0·98						
16	0·88	0·98						
17	0·89	0·989						
18	0·91	0·991						
19	0·92	0·994						
20	0·93	0·995						
25	0·97	0·999						

* Probability that at least one can in the incubated sample will show spoilage.
** Sample consisting of cans drawn at random.

Storage tests, including the incubation of cans, are carried out to determine the risk of spoilage that is present in a canned food.

The temperature and duration of incubation are chosen on the basis of the planned marketing and storage of the product. In assessing the hazards of spoilage due to mesophilic and facultative thermophilic micro-organisms, incubation at 37 °C for a week is thought to be sufficient in the majority of cases, to cover any likely spoilage of non-sterile products during marketing and storage. In the case of products containing material that may be contaminated with obligate thermophiles, cans should be incubated at 50–55 °C to establish the proportion of spoiled cans that may be expected from the presence of these bacteria during marketing and storage under tropical conditions. A period of 10–14 days is thought to be an appropriate incubation time.

The accuracy with which the spoilage level is determined, depends on the number of samples examined from each batch. *Tables 51* and *52* show, according to calculations carried out by the Department of Statistics of the National Canners Association (NCA, USA), how many cans must be incubated per batch in order to determine different spoilage levels with a given degree of probability; also shown are the probabilities of detecting various spoilage levels as the sample size is increased.

The storage test consists of (i) an incubator test aimed at detecting any viable pathogenic, toxin-producing, or spoilage (saprophytic) micro-organisms that may be present, (ii) a microbiological examination to enumerate the microbes causing spoilage, and (iii) a search for signs of spoilage by organoleptic examination.

The commercial, special and tropical storage tests differ with respect to the duration of the incubation and the temperature of incubation.

8.4.2   Investigation of the causes of spoilage in canned foods

8.4.2.1   The main causes for spoilage

The spoilage of canned food may be the result of inadequate heat treatment, or post-processing contamination caused by faulty sealing of the can or contaminated cooling water in the autoclave.

The mechanism of post-processing contamination, the sucking-in of contaminants, the important aspects of the construction of the can in this respect, and the possibilities of preventing post-processing contamination, have been discussed in detail by Put et al., (1972).

Inadequate heat treatment may arise not only from failure to observe treatment schedules, but also from spore numbers higher than those which the sterilization procedure is designed to deal with. High spore counts may result from contamination of the raw material with soil, inadequate washing and sterilization of the processing equipment (especially the filling equipment and containers), contamination of the brine by thermophiles (by the use of microbiologically contaminated sugar or from keeping the brine at a temperature below 90 °C), etc.

8.4.2.1.A   Preparatory procedures in the investigation of spoilage

Before an investigation into the cause of spoilage is begun, a check should be made of the degree of spoilage and number of spoiled cans expected, as well as of the duration and temperature of the sterilization procedure applied. Any deviation from the prescribed procedures should be noted. The external appearance and possible mechanical injury, including rustiness, holes, dents, and any other anomalies should be registered together with the production number.

The cause of spoilage and the origin of viable counts higher than the permitted viable count can be discovered by following the procedure described below (National Canners Association, 1968).

The can is washed thoroughly with soap and water, and any grease around the point of opening is removed with a fat-free cotton swab soaked in 96% alcohol. If the can is intact, the area to be opened is flamed in a Bunsen flame, or the alcohol remaining on the surface after swabbing is ignited. Swollen cans are treated with 60% alcohol, 5% phenol solution, and chlorine water containing 200 mg per litre active chlorine (Stumbo, 1973), or they can be treated with diluted (1 : 1000) corrosive sublimate solution (A.P.H.A., 1966). Flaming should be omitted, since a swollen can may explode under the increased pressure.

The can is opened by piercing its lid with a pointed thin steel lance fixed in an inverted glass funnel by means of cotton. Before use, this device is wrapped in paper and sterilized in a dry oven or autoclave. The can which is now ready for opening is placed in a dish containing disinfectant (Frazier et al., 1968). The area that is to be perforated is covered with the inverted funnel which protects the working-surface and the investigator from any spurting of material from the can, and prevents the content food from being contaminated from the outside *(Fig. 62)*.

It is advisable to cover the can with a sterile Petri dish lid during the sampling operation.

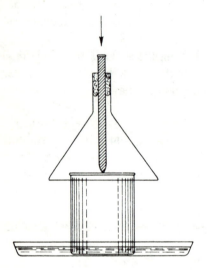

Fig. 62. Punching of can for microbiological examination

The sample can be removed from the opened container using a wide-mouthed pipette or a simple sterile glass tube; solid materials are sampled with a sterile cork borer, forceps, or a spoon. The sample is transferred aseptically to a sterile test-tube or flask.

A smear is prepared from the remainder of the material for microscopic examinaton; the pH of the sample is measured and the external appearance, consistency, and odour of the sample are recorded.

It should be borne in mind that spoilage may be limited to a small part of the volume of the can if the contents are dense (e.g. tomato pulp), or contain a large amount of solid fragments (e.g. hamburgers).

Where a product is dispensed at a high temperature after pre-sterilization and is not re-sterilized after sealing, microbiological contamination may originate from the lid of the can. In such cases the contamination may be limited to the outer layers in contact with the lid.

8.4.2.1.B   Microscopic investigation of spoilage

Microscopic examination is nearly always necessary since some or all of the microbes causing the spoilage may have died out; grading must, therefore, be based on direct microscopic observation (Mossel and Zwart, 1959). Sometimes concomitant bacteria which are inactive in the canned food are isolated instead of those that actually cause the spoilage, and this may be highly misleading without the use of microscopy.

The sample is either suspended and homogenized for microscopic examination, or an impression preparation, i.e. a contact plate, is examined (see 7.3). The latter method has the advantage that the original distribution pattern of microbes, microcolonies, and aggregates of microbes is also observable (Hadlok, 1960). If imperfect sealing, and/or sucking-in is suspected, it is advisable to take the samples from the critical sites of the canned mass, such as those areas in contact with the side or end seams (Kelch, 1960; Kelch and Hadlok 1960 a, b; Stumbo, 1973). If, however, an inadequate heat treatment seems probable, it is preferable to take samples from the thermic centre. The quantitative microscopic method recommended by Mossel and Zwart (1959) is based on homogenization, suspension in tap water and sedimentation, followed by viable counting according to Breed's procedure (see 8.1.2.1.A).

The following stains have been recommended for the staining of microscopic preparations: gentian violet (A.P.H.A., 1966), crystal violet (Stumbo, 1973), and spore staining with malachite green–safranin, or with brilliant green/coriphosphin stain (Lerche, 1957).

8.4.2.1.C   Examination of spoilage by cultivation

The results of these methods are unequivocal, since only those microbes are regarded as having caused spoilage which, when inoculated into an unspoiled preparation, induce the same changes as observed in the spoiled product.

Procedure. The sample is transferred aseptically into a sterile test-tube or flask. Two tubes of bromcresol purple–glucose–tryptone medium [54], and 4 tubes of liver broth [88] freshly boiled and cooled to 50 °C are inoculated. The liver broth cultures are overlain with sterile liquid paraffin, melted paraffin wax, or 2% agar solution. Two of the liver broth cultures are placed in a water bath at 82 °C for 13 minutes to kill the vegetative cells.

One of the glucose–tryptone tubes is incubated at 30–35 °C, and one each of the pasteurized and unheated liver broth cultures is incubated at 50–57 °C. Before tubes are placed in the incubator at the higher temperature, they are pre-warmed in a water bath to the temperature of the incubator. If the pH of the canned food is below 4, incubation at 50–57 °C may be omitted.

Incubation at 50–57 °C is continued for 2 days followed by incubation at 30–35 °C for a week. The cultures are checked daily, and gas formation, acid formation, sediment and film formation, and any other such phenomena indicating microbial growth are recorded. The odour of the anaerobic culture may assist recognition of the type of spoilage.

Those cultures showing intense microbial growth are kept cold to keep them alive; thermophilic bacteria especially tend to die out after rapid multiplication at a favourable temperature.

Microscope preparations made from the incubated cultures are stained with crystal violet. If the food under study is of low pH, it is advisable to examine an unstained preparation as well.

The presence of non-sporing bacteria is suggested by the following observations: microbial growth with gas formation occurs in liver broth; growth occurs in the glucose–tryptone medium without the formation of a surface skin; rod-shaped micro-organisms are observed under the microscope. This result, however, needs to be confirmed by spread-inoculating solid medium, since facultative anaerobic spore-formers tend not to sporulate in liver broth and media containing sugar. Swelling caused by microbiological activity in meat products preserved by a relatively mild heat treatment is most frequently caused by *C. perfringens* (Leistner, 1970) (see 6.4.2.2), while swelling occurring in products canned by a botulinum-cook is caused by the much more heat-resistant *C. sporogenes*. The appearance of growth with skin formation in glucose–tryptone medium is a very strong indication of the presence of *Bacillus* spp. These are all spore-formers. Most of the non-sporing food-spoilage bacteria that grow well in glucose–tryptone medium belong either to the genus *Lactobacillus* (rodlets), or the genus *Leuconostoc* (coccoids or very short rods). The lactic acid bacteria, being for the most part micro-aerophiles, grow very sparsely on the surface of solid media.

Species of *Leuconostoc* grow poorly on culture media containing no sucrose, but in the presence of sucrose show abundant growth accompanied by mucus formation.

It is often useful to check the spoiling effect of micro-organisms isolated in the above manner by inoculating the strain into a sterile sample of the foodstuff under examination. This kind of control is of particular importance in the case of products with a pH range from 4·0 to 5·6.

330

The flat-souring organism *Bacillus coagulans* (*B. thermoacidurans*) grows very poorly in the usual culture media, and, therefore, isolation from flat-sour products should be carried out on a selective medium, *viz.* thermoacidurans agar [145].

8.4.2.1.D Gas analysis and examination for corrosion

If spoilage cannot be demonstrated in a swollen can of food, it may be deemed necessary to carry out gas analysis in cans from the same batch. The simplest method of gas analysis is to hold a glowing match near the point where the swollen can is pierced. A flaring of the glow just after piercing indicates the presence of H_2 gas, whereas glowing will be extinguished if CO_2 is present. If the swelling of the can is not of microbial origin, but is due instead to corrosion (so-called sterile, or hydrogen bombage), the H_2 concentration in the gas space is generally 50% or more at the first appearance of the swelling. A considerable amount of CO_2 may also be present in cases of hydrogen bombage. Furthermore, hydrogen may be produced by bacteria, and, therefore, gas analysis alone is not a means of establishing the cause of spoilage.* The contents, and if present, the laquering of the can should be carefully examined. In the presence of swelling of chemical origin, the inner surface of the can is usually corroded and the liquid phase of the product is clear. Signs of corrosion should be sought first of all on the metal surfaces in contact with the gaseous phase. Particular attention should be given to the search for defects in sealing and rawness of the product, and there may be a brownish discolouration which appears when the product comes into contact with the air. These phenomena suggest that the product has not been properly heat-treated.

8.4.2.2 Key to the cause of spoilage

Using the results of the above examinations, the key given in *Table 53* enables a probable identification of the cause of spoilage to be made.

---

* A gas-chromatographic method has been developed for analyzing the gaseous phase of canned foods (Vosti et al. 1961); the use of gas chromatography has also been suggested for the detection of some of the characteristic metabolites of micro-organisms, e.g. carboxylic acids and carbonyl compounds, thus providing a means of establishing the presence of, and identifying micro-organisms (Henis et al. 1966; Moss and Lewis, 1967).

Table 53
Key to Probable Cause of Spoilage in Canned Foods

Group 1. Low-Acid Foods–pH Range 5.0 to 8.0

Condition of Cans	Characteristics of Material in Cans						Diagnosis
	Odor	Appearance	Gas (CO_2 & H_2)	pH	Smear	Cultures	
Swells	Normal to "metallic"	Normal to frothy. (Cans usually etched or corroded)	More than 20% H_2	Normal	Negative to occasional organisms	Negative	Hydrogen swells
	Sour	Frothy; possibly ropy brine	Mostly CO_2	Below normal	Pure or mixed cultures of rods, coccoids, cocci, yeasts or mold	Growth, aerobically and/or anaerobically at 30 °C, and possibly at 50 °C	Leakage
	Sour	Frothy; possibly ropy brine; food particles firm with uncooked appearance	Mostly CO_2	Below normal	Pure or mixed cultures of rods, coccoids, cocci and yeasts	Growth, aerobically and/or anaerobically at 30 °C and possibly at 50 °C. (If product received high exhaust, only spore formers may be recovered)	No process given
	Normal to sour –cheesy	Frothy	H_2 and CO_2	Slightly to definitely below normal	Rods, med. short to med. long, usually granular; spores seldom seen	Gas, anaerobically at 50°C, and possibly slowly at 30 °C	Underprocessing–thermophilic anaerobes
	Cheesy to putrid	Usually frothy with disintegration of solid particles	Mostly CO_2; possibly some H_2	Slightly to definitely below normal	Rods; usually spores present	Gas anaerobically at 30°C	Underprocessing–mesophilic anaerobes (possibility of *C. botulinum*)
	Slightly off– possibly ammonical	Normal to frothy		Slightly to definitely below normal	Rods; spores occasionally seen	Growth, aerobically and/or anaerobically with gas at 30 °C, and possibly at 50 °C. Pellicle in aerobic broth tubes. Spores formed on agar and in pellicle	Underprocessing–*B. subtilis* type
No vacuum and/or cans buckled	Normal	Normal	No H_2	Normal to slightly below normal	Negative to moderate number of organisms	Negative	Insufficient vacuum, caused by 1. Incipient spoilage 2. Insufficient exhaust 3. Insufficient blanch 4. Improper retort cooling procedures 5. Over fill
Flat cans (0 to normal vacuum)	Normal to sour	Normal to cloudy brine		Slightly to definitely below normal	Rods, generally granular in appearance; spores seldom seen	Growth without gas at 50 °C. Spore formation on nutrient agar	Underprocessing–thermophilic flat sours
	Normal to sour	Normal to cloudy brine; possibly moldy		Slightly to definitely below normal	Pure or mixed cultures of rods, coccoids, cocci or mold	Growth,, aerobically and/or anaerobically at 30 °C, and possibly at 50 °C	Leakage

Group 2. Semi-Acid Foods—pH Range 4.6 to 5.0

	Normal to "metallic"	Normal to frothy. (Cans usually etched or corroded)	More than 20% H$_2$	Normal	Negative to occasional organisms	Negative	Hydrogen swells
Swells	Sour	Frothy; possibly ropy brine	Mostly CO$_2$	Below normal	Pure or mixed cultures of rods, coccoids, cocci, yeasts or mold	Growth, aerobically and/or anaerobically at 30 °C, and possibly at 50 °C	Leakage
	Sour	Frothy; possibly ropy brine; food particles firm with uncooked appearance	Mostly CO$_2$	Below normal	Pure or mixed cultures of rods, coccoids, cocci and yeasts	Growth, aerobically and/or anaerobically at 30 °C, and possibly growth at 50 °C. (If product received high exhaust, only spore formers may be recovered)	No process given
	Normal to sour–cheesy	Frothy	H$_2$ and CO$_2$	Slightly to definitely below normal	Rods–med. short to med. long, usually granular; spores seldom seen	Gas, anaerobically at 50 °C, and possibly slowly at 30 °C	Underprocessing–thermophilic anaerobes
(Note: Cans are sometimes flat) do	Normal to cheesy to putrid	Normal to frothy with disintegration of solid particles	Mostly CO$_2$; possibly some H$_2$	Normal to slightly below normal	Rods; possibly spores present	Gas anaerobically at 30 °C. Putrid odor	Underprocessing–mesophilic anaerobes (possibility of C. botulinum)
	Slightly off–possibly ammoniacal	Normal to frothy		Slightly to definitely below normal	Rods; occasionally spores observed	Growth, aerobically and/or anaerobically with gas at 30 °C, and possibly at 50 °C. Pellicle in aerobic broth tubes. Spores formed on agar and in pellicle	Underprocessing–B. subtilis type
	Butyric acid	Frothy, large volume gas	H$_2$ and CO$_2$	Definitely below normal	Rods–bipolar staining; possibly spores	Gas anaerobically at 30 °C. Butyric acid odor	Underprocessing–butyric acid anaerobe
No vacuum and/or cans buckled	Normal	Normal	No H$_2$	Normal to slightly below normal	Negative to moderate number of organisms	Negative	Insufficient vacuum caused by; 1. Incipient spoilage 2. Insufficient exhaust 3. Insufficient blanch 4. Improper retort cooling procedures 5. Over fill
Flat cans (0 to normal vacuum)	Sour to "medicinal"	Normal to cloudy brine		Slightly to definitely below normal	Rods–possibly granular in appearance	Growth without gas at 50 °C and possibly at 30 °C. Growth on thermoacidurans agar	Underprocessing–B. coagulans
	Normal to sour	Normal to cloudy brine; possibly moldy		Slightly to definitely below normal	Pure or mixed cultures of rods, coccoids, cocci or mold	Growth, aerobically and/or anaerobically at 30 °C, and possibly at 50 °C	Leakage

333

Table 53 (cont.)

Group 3. Acid Foods–pH Range 4.0 to 4.6 (Except Concentrated Foods)

Condition of Cans	Characteristics of Material in Cans				Smear	Cultures	Diagnosis
	Odor	Appearance	Gas (CO₂ & H₂)	pH			
Swells	Normal to "metallic"	Normal to frothy. (Cans usually etched or corroded)	More than 20% H_2	Normal	Negative to occasional organisms	Negative	Hydrogen swells
	Sour	Frothy; possibly ropy brine	Mostly CO_2	Below normal	Pure or mixed cultures of rods, coccoids, cocci or yeasts	Growth, aerobically and/or anaerobically at 30 °C and possibly at 50 °C	Leakage or gross underprocessing
	Sour	Frothy; possibly ropy brine; food particles firm	Mostly CO_2	Below normal	Pure or mixed cultures of rods, coccoids, cocci or yeasts	Growth, aerobically and/or anaerobically at 30 °c and possibly at 50 °C. (If product received high exhaust only spore formers may be recovered)	No process given
	Normal to sour–cheesy	Frothy	H_2 and CO_2	Normal to slightly below normal	Rods–med. short to med. long, usually granular; spores seldom seen	Gas anaerobically at 50 °C and possibly slowly at 30 °C	Underprocessing–thermophilic anaerobes
	Butyric acid	Frothy; large volume gas	H_2 and CO_2	Below normal	Rods–bipolar staining; possibly spores	Gas anaerobically at 30 °C. Butyric acid odor	Underprocessing–butyric acid anaerobes
	Sour	Frothy	Mostly CO_2	Below normal	Short to long rods	Gas anaerobically; acid and possibly gas aerobically in broth tubes at 30 °C. Possible growth at 50 °C	Gross underprocessing–lactobacilli
No vacuum and/or cans buckled	Normal	Normal	No H_2	Normal to slightly below normal	Negative to moderate number of organisms	Negative	Insufficient vacuum caused by: 1. Incipient spoilage 2. Insufficient exhaust 3. Insufficient blanch 4. Improper retort cooling procedure 5. Over fill
Flat cans (0 to normal vacuum)	Sour to "medicinal"	Normal		Slightly to definitely below normal	Rods–possibly granular in appearance	Growth without gas at 50 °C and growth on thermoacidurans agar	Underprocessing–*B. coagulans* (Spoilage of this type usually limited to tomato juice)
	Normal to sour	Normal to cloudy brine; possibly moldy		Slightly to definitely below normal	Pure or mixed cultures of rods, coccoids, cocci or mold	Growth, aerobically and/or anaerobically at 30 °C and possibly at 50 °C	Leakage, or no process given

334

Group 4. Concentrated Acid Foods—pH below 4.6
High Acid Foods—pH below 4.0

Swells	Normal to "metallic"	Normal to frothy. (Cans usually etched or corroded)	More than 20% H_2	Normal	Negative to occasional organisms	Negative	Hydrogen swells
	Normal	Normal to frothy; scorched	All CO_2	Normal	Negative to occasional organisms	Negative	"Frothy fermentation" (This spoilage is limited to concentrated products)
	Normal to sour to yeasty	Normal to frothy; possibly surface growth	Mostly CO_2	Normal to below normal	(1) Short to long rods. (2) Pure or mixed cultures of short to long rods, cocci, coccoids or yeasts	(1) and (2): Growth aerobically and/or anaerobically with gas production at 30 °C and possibly at 50 °C	(1) Underprocessing or leakage (2) Gross underprocessing or leakage
	Sour	Frothy; possibly ropy brine; food particles firm with uncooked appearance	Mostly CO_2	Below normal	Pure or mixed cultures of rods, coccoids, cocci or yeasts	Growth, aerobically and/or anaerobically at 30 °C and possibly at 50 °C	No process given
No vacuum and/or cans buckled	Normal	Normal	No H_2	Normal to slightly below normal	Negative to moderate number of organisms	Negative	Insufficient vacuum caused by: (1) Incipient spoilage (2) Insufficient exhaust (3) Insufficient blanch (4) Improper retort cooling procedure (5) Over fill
Flat cans (0 to normal vacuum)	Normal to sour	Normal to cloudy brine; possibly moldy		Normal to definitely below normal	Pure or mixed cultures of rods, coccoids, cocci or mold	Growth, aerobically and/or anaerobically at 30 °C and possibly at 50 °C	Leakage or underprocessing (If mold present, leakage)

335

8.4.2.3   Investigation of moulds in tomato products according to Howard

In tomato products of known dry matter content, the presence of hyphae, and hyphal counts (hyphae per 1·5 mm² or per 0·15 mm³) are established by microscopy. The Howard count refers to the percentage of microscope fields in which moulds are observed.

Howard or Buerker chambers are used. The former in its usual form *(Fig. 63)* has an elevated platform, which may be oblong, 15 × 20 mm, or circular and 19 mm in diameter. The suspension which is to be examined is transferred to this platform. At either side, there is a glass lath which is exactly 0·1 mm above the level of the platform. The coverslip, which is 0·5 or 1·0 mm in thickness, is laid across the lathes and thus forms a chamber 0·1 mm deep. In one of the glass lathes there are two parallel lines etched 1·382 mm apart; these facilitate calibration of the microscope field. Alternatively, a line interrupted at intervals of 1·382 mm, or a circle of this diameter may serve as a basis for calibration.

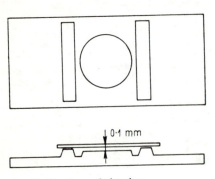

Fig. 63. The Howard chamber

A microscope, monocular or binocular, equipped with a cross-stage and a diaphragm should be used. The most useful magnifying power is given by an ocular of × 10 and an objective of × 10. (Objective: 16 mm; numerical aperture 0·25.) The microscope should also have available objectives of higher magnifying power, i.e. × 20 (8 mm) and ≥ × 40 magnification. A microscope with an inbuilt source of light is preferable. Where an external source of light is used collimation is necessary, placing the collimator 25 cm from the mirror of the microscope; a plane mirror should be used. The light intensity on the mirror surface should be about 1300 lux.

A micrometer scale with a network consisting of 36 squares is placed in the ocular lens. The side length of each square is 1/6 of the aperture of the ocular.

The optics of the microscope are adjusted to obtain a field diameter of 1·382 mm (ocular lens × 10, objective × 10). With this diameter, the field of vision covers an area of 1·5 mm² and a sample volume of 0·15 mm³. The adjustments are carried out with the aid of the markings on the glass lath of the Howard chamber. The tube of the microscope is extended to make the adjustment, or if this is not possible, a narrowing ring of suitable size is placed in the ocular.

When using a Buerker chamber, the diameter of the field of vision can be adjusted in two different ways. Either etched calibration markings on the chamber can be used to adjust the field of vision to a diameter of 1·38 mm, or an objective micrometer can be used to obtain the desired diameter of the field (*Fig. 64*).

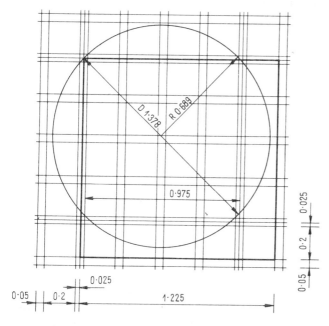

Fig. 64. Adjustment of the Howard chamber to the graduation of the Buerker chamber

Sampling. The size of the sample should be large enough to be representative of the product. For the study of moulds in tomato puree of 28–30 refr.%, at least one can of 0·2 kg should be taken.

Preparation of the material to be examined.

The dry matter content of the tomato concentrate is adjusted as follows. The dry matter content of the original material is determined by refractometry. Starting with 10–15 g of the concentrate, about 1 ml of the juice is pressed out by hand using a refractometer cloth. The juice is collected, except for the first few drops, in a dry beaker of 10 ml capacity. The juice is mixed and examined in the refractometer. *Table 54* gives (on the basis of the degree of refraction) the amount of water with which the concentrate must be diluted to give a refractive index between $n_{20}^D = 1·3445$ and $n_{20}^D = 1·3460$ (7·9–8·8 refr.%).

22 Kiss

Table 54

Relationship between the degree of refraction of tomato paste and the amount of water necessary for adjustment to 8·4 refraction degree

Soluble dry matter		Water (g) necessary to adjust 100 g of the paste to 8·4 refr. % $(n_{20}^D = 1·3454)$
refr. %	refractive index n_{20}^D	
of the tomato paste		
25·0	1·3723	197·6
25·5	1·3733	203·6
26·0	1·3741	209·5
26·5	1·3749	215·5
27·0	1·3759	221·4
27·5	1·3767	227·4
28·0	1·3776	233·3
28·5	1·3784	239·3
29·0	1·3793	245·2
29·5	1·3802	251·2
30·0	1·3811	257·1
30·5	1·3820	263·1
31·0	1·3828	269·0
31·5	1·3839	275·0
32·0	1·3848	281·0

From a concentrate of known dry matter content, 100 g are weighed out into 250-ml beaker. The amount of water required for the dilution is measured into a dry 500-ml beaker. Of this, 100 ml are added to the concentrate in the 250-ml beaker, the contents of which are then well stirred with a spoon and poured into the beaker containing the remainder of the water. After careful mixing of the fully diluted juice, the 250-ml beaker is rinsed with the juice by pouring it from one beaker into the other and back again several times. The result of the dilution is checked by direct refractometry, without using the pressing cloth. If the refractive index of the diluted juice falls outside the given range, the dilution must be corrected. Aliquots of 80–90 ml are poured from the well-stirred solution into 100-ml sample jars and transferred from these to the counting chamber.

Tomato juice and ketchup should be mixed well and examined without dilution.

Procedure. First of all the chamber is checked for cleanliness and accuracy; if Newton's rings are not visible, the chamber must be cleaned and the coverslip renewed.

The contents of the small sample jar are stirred and a droplet is taken on a scalpel and placed on the central platform of the Howard chamber. The droplet is spread

338

evenly, and carefully covered with the coverslip. Bubbles must not be allowed to form. The droplet should not be larger than that required to fill the volume between the central platform and the coverslip. Preparations should not be used which show any obvious unevenness or which do not result in the appearance of Newton's rings.

In the Buerker chamber, a droplet of appropriate size may be placed on both platforms.

The preparation is examined immediately. When the Howard chamber is used, 5 microscope fields are examined in each row of the grid; the selected fields should be distributed to represent every part of the preparation. In the Buerker chamber, the 25 fields should be evenly distributed between the two platforms, and neighbouring fields should be about 1·4 mm apart.

Mould hyphae are searched for in the 25 fields. To this end, the fine focus adjustment is moved continuously back and forth and the illumination is adjusted for best results using the diaphragm. An increase in the magnification may help in recognizing the hyphae.

A microscope field is recorded as positive if the aggregate length of not more than three hyphae exceeds 1/6 of the diameter of the field. The aggregate length is checked with the ocular micrometer. If the filaments in a field are so short that the lengths of 4 or more hyphal fragments must be summed to obtain 1/6 of the field diameter, then that field is recorded as negative, even if disintegrated hyphae are seen in it.

The relationships between the numbers of fields counted, the level of probability, and the confidence interval with respect to determination of the Howard count, are shown in *Table 55* (Pratt, 1950).

At least four Howard preparations are made from each sample, so that at least 100 microscope fields are examined per sample.

Table 55

Relationships between the number of microscope fields examined, the probability level, and the confidence interval in the determination of the Howard count (Pratt, 1950)

Howard count	Range in which the average Howard count is expected to fall if					
	50		100		250	
	fields are examined					
	P=0·05	P=0·01	P=0·05	P=0·01	P=0·05	P=0·01
0	0–7	0–10	0–7	0–5	0–1	0–2
10	3–22	2–26	5–18	4–19	7–14	6–16
20	10–34	8–38	13–29	11–32	15–26	14–27
30	12–44	15–49	21–40	19–43	24–36	23–38
40	27–55	23–59	30–50	28–53	34–46	32–48
50	36–67	31–69	40–60	37–63	44–56	42–58

P = probability of error.

8.4.2.4   A rapid method of estimating the Howard count

A procedure for the rapid estimation of the so-called "multiple mould count" has been developed by the U.S. Department of Agriculture, Consumer and Marketing Service. In this procedure, instead of attempting to obtain an exact Howard count, an assessment is made of whether the Howard count for the sample falls outside the limit values of a predetermined range.

The preparation of samples and the microscopic examination of these are carried out as described previously. Ten microscope fields are examined, and on the basis of this examination, one of the following conclusions are drawn using the Howard multiple count (Table 56).

— the Howard count for the sample is probably below the specified maximum value;

     — the Howard count for the sample is larger than the specified maximum value;

·   — examination of a further 10 fields is necessary.

Thus in the first two cases, because of the mould count of the sample is well below or well above the Critical value, examination of a relatively small number of fields may be satisfactory.

Table 56

Howard mould count, multiple count table (National Canners Association, Research Laboratories, 1968)

n_c***	Howard count															
	10%		20%		30%		40%		50%		60%		70%		80%	
	c*	r**	c	r	c	r	c	r	c	r	c	r	c	r	c	r
10	X	4	x	6	0	7	0	8	1	9	2	10	4	10	5	XX
20	X	5	0	8	2	10	4	13	6	15	8	17	10	19	13	20
30	1	6	3	10	5	13	7	17	11	19	14	23	17	26	20	28
40	1	7	5	12	8	16	11	21	16	25	20	29	24	32	28	36
50	2	8	7	14	11	19	15	25	21	30	26	35	31	40	36	44
60	3	9	9	16	14	22	20	29	26	34	32	41	38	48	44	52
70	4	10	11	18	17	25	24	33	31	40	38	46	45	57	52	60
80	5	11	13	20	20	28	28	36	36	45	44	52	52	62	61	68
90	6	11	15	22	23	31	32	40	42	49	50	58	60	68	70	76
100	8	12	17	23	26	35	37	43	47	57	57	64	66	75	78	83
110	10	13	20	23	30	37	42	47	53	58	63	69	74	80	86	90
120	12	13	24	25	36	37	48	49	60	61	72	73	84	85	96	97

   *c   = maximal cumulative number of positive fields for acceptance.

   **r   = minimal cumulative number of positive fields for rejection.

***n_c = cumulative number of fields examined.

Notes:

  x Too few fields for acceptance.

  xx Too few fields for rejection.

According to the principles on which the test is based, one out of four consecutive samples (or 2 out of 9, or 3 out of 13) may exceed the maximum limit value, if they are acceptable in terms of the next level of Howard count in *Table 56,* and provided that the product generally seems to be prepared from raw materials of satisfactory quality.

8.4.2.4.A Guide to the recognition of hyphae

The most difficult part of the determination of the Howard count is the recognition of hyphae, and successful differentiation between these and pieces of tomato tissue of similar appearance. The observer needs to be very familiar with the microscopic picture of ground tomato tissue and the morphological features of fungal hyphae. The following considerations must be borne in mind during the examination.

Hyphae are generally tubular with parallel walls, *Oospora* and *Mucor*-type fungi often develop tapering hyphae. The growing hypha is completely filled by the protoplasm or it is vacuolated. The protoplasm may seem granular, and the granular appearance may not be affected by the heat treatment although sometimes the protoplasm coagulates forming a non-granular hyaline mass, or plugs interspaced by empty sectors along the filament. The granular effect is most pronounced in *Mucor* and *Rhizopus,* and is hardly observable in other fungi.

The hyphae of most fungi occurring in foodstuffs are septate. In otherwise doubtful cases, the presence or absence of septa may facilitate identification. However, septa are often absent in species of *Mucor*. Most mycelia are abundantly branched and observation of the details of branching may help in identification, although disintegrated hyphae are usually too short to show the branching pattern.

Filaments may be generally identified as mould hyphae if they meet the following criteria; walls parallel and of uniform thickness, ends blunt, transverse. Hyphae tubular, appearing identical under the microscope with opposite side walls observed in the same focal plane. (Tomato cell wall debris ribbon-like, opposite walls often lying in two different planes, so that, depending on the actual illumination, the fragments appear variable in thickness.) Mould hyphae show

— parallel walls of uniform thickness, with characteristic granulation;
— parallel walls of uniform thickness, with characteristic branching;
— occasionally, parallel walls of uniform thickness, one end bluntly transverse, the other rounded;
— occasionally, slightly tapering walls of uniform thickness, with characteristic granulation and septa.

The identification of mould hyphae in the course of the Howard counting should always be positive, i.e. if the observer cannot decide whether or not a filament is a fungus, the filament is not counted as a mould fragment.

341

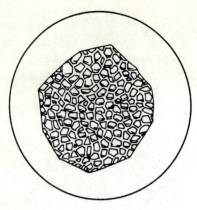

Fig. 65. Epidermal or skin cells (× 200)

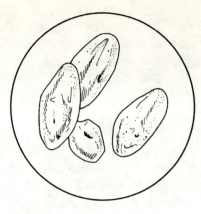

Fig. 66. Flesh cells (× 200)

Fig. 67. Fibrovascular tissues (× 200)

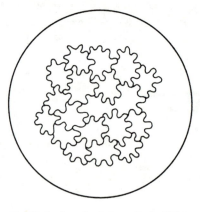

Fig. 68. Seed cavity lining cells (× 200)

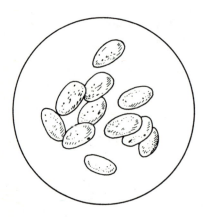

Fig. 69. Isolated cells from core (× 200)

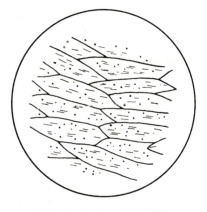

Fig. 70. Seed covering cells (× 200)

342

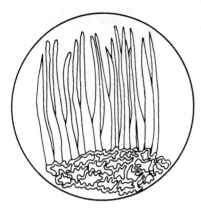

Fig. 71. Seed hairs and seed covering cells (× 200)

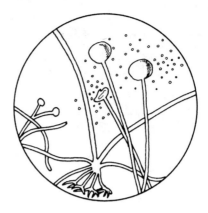

Fig. 72. Rhizopus × 200

Fig. 73. Aspergillus × 200

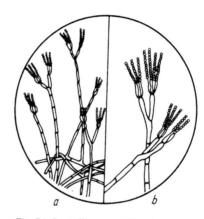

Fig. 74. Penicillium a : × 200; *b :* × 600

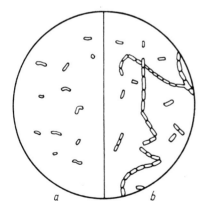

Fig. 75. Oospora a : × 200; *b :* × 500

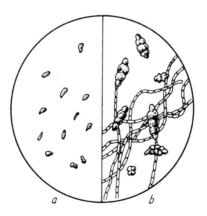

Fig. 76. Alternaria a : × 200; *b :* × 600

343

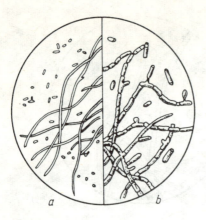
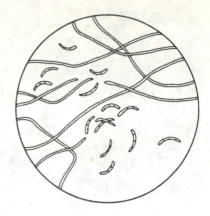

Fig. 77. Colletotrichum a : × 200;
b : × 600

Fig. 78. Fusarium × 200

The following parasitic and saprophytic microscopic fungi occur most frequently in the tomato:

Alternaria, Colletotrichum, Fusarium, Mucor, Rhizopus, Oospora, Penicillium, Aspergillus, Botrytis, Phytophthora infestans, Stemphylium and *Cladosporium.*

Illustrations of these fungi and tomato particles are shown in *Figures 65–78.*

8.4.2.5   Quantitative examination of the microflora

Table 57 lists the methods most widely used in food microbiology for examining the microflora of canned products. Further details are given elsewhere in this book.

8.5   Examination of ingredients and additives

The microbiological contamination of additives may play an important role in the spoilage of canned foods. Of those saprophytes that occur in additives, yeasts, moulds, and thermophilic bacteria (*viz.* the flat-souring *Bacillus stearothermophilus,* the gas-forming and hydrogen–sulphide-producing *Clostridium thermosaccharolyticum,* and *C. nigrificans* which also causes hydrogen–sulphide spoilage), present the greatest problems. In the production of canned foods, stress should be laid on the eradication of spores of the mesophilic putrefying bacteria, some of which are extraordinarily thermoresistant. In the following, several methods are described which have been recommended by the National Canners Association (NCA, USA, 1968) for the examination of microbial contamination in food additives.

344

Table 57

Quantitative examination of the microflora occurring in products of the canning industry

Type of product	Purpose of examination; enumeration of:	Culture medium and method of inoculation	Incubation	
			temperature °C	period
Fruit juice, dried fruit	total yeasts, moulds and bacteria	Malt agar (pH 5·4), spreading	25	5 days
	yeasts and moulds	Czapek's agar (pH 3·5), spreading;	25	5 days
	osmophiles	60% fructose, 0·5% yeast extract, poured plate agar	25	2 weeks
Soured products	total yeasts, moulds and bacteria	Malt agar (pH 5·4)	25	5 days
	yeasts and moulds	Malt–salt agar (pH 3·5), spreading	25	5 days
	salt-tolerant, halophilic microbes	Pour-plates; medium and diluent containing 15% salt	30	5 days
	lactobacilli	Rogosa's acetate agar	32	3 days
	coliforms	MPN method	35	24 and 48 h
Dried vegetables, dehydrated soup, vermicelli, quick-frozen products	mesophilic aerobic bacteria	Plate-count agar, poured-plate method	30	48 h
	coliforms	MPN method	35	24 and 48 h
	sulphite-reducing clostridia	SPS agar, poured plates	35	24 h
	yeasts and moulds	Czapek's agar, spreading	25	5 days

8.5.1　Examination of sugar

8.5.1.1　Sampling and processing of sample

One sample of 200–250 g is taken from each of 5 packaging units in every shipment. The samples are mixed together, and 20 g of the sugar are placed in a 250 ml graduated Erlenmeyer flask, and sterile water is added to the 100-ml mark.

8.5.1.2　Mesophilic, aerobic total count

1 ml of the sugar solution is pipetted into each of 5 sterile Petri dishes, and pour-plates are prepared with glucose–tryptone medium [54]. The plates are inverted after solidification and incubated at 30 °C for 72 h. The total number of colonies on the plates is multiplied by 10 to obtain the total viable count of mesophilic aerobes per 10 g sugar.

8.5.1.3　Yeast and mould counts

1 ml of the sugar solution is pipetted into each of 5 sterile Petri dishes, and pour-plates are prepared with soured potato-glucose agar [17]. The plates are allowed to solidify, and are incubated at 30 °C for 4–5 days. The total number of colonies on the plates is multiplied by 10 to obtain the total viable count of yeasts and moulds per 10 g sugar.

8.5.1.4　Thermophilic spore count

8.5.1.4.A　Heat-activation

The sugar solution (see 8.5.1.1) is boiled rapidly, and simmered for 5 minutes to kill the vegetative cells and spores of low heat tolerance. The flasks are then immediately immersed in cold water.

8.5.1.4.B　Flat-sour count

2-ml portions of the heat-treated sugar solution are pipetted into each of 5 Petri dishes. The solidified plates are placed in an incubator at 50–55 °C for 48–72 h. The total number of colonies on the 5 plates is multiplied by 5 to obtain the thermophilic aerobic spore count per 10 g sugar.

The characteristic colonies of flat-souring bacteria on medium [54] are circular (2–5 mm in diameter), with a characteristic opalescent centre and a yellow halo. The latter may be very narrow or absent around the colonies of weak acid-formers, or it may not be observable if the entire medium has become yellow in the presence of a large number of colonies. The colonies should be counted immediately after removing the plates from the incubator, since the yellow colour tends to fade quickly. If acid formation is difficult to discern after incubation for 48 h, incubation may be continued up to 72 h.

If the density of colonies is high, the counting may be less accurate and the colonies may be atypical in size and shape. In such cases, pour-plating should be repeated with a sample of higher dilution. It should be noted that any sample giving rise to an uncountable number of colonies must be rejected according to the standards recommended by the NCA.

8.5.1.4.C Detection of thermophilic anaerobes that do not produce hydrogen sulphide

Equal volumes (totalling 20 ml) of a sugar solution heat-treated as described in (8.5.1.4.A, are dispensed into test-tubes containing freshly deoxygenated liver broth [88]. The tubes are sealed with sterile agar [3], pre-warmed to 50–55 °C, and incubated at the latter temperature for 72 h.

The presence of thermophilic anaerobes is suggested by gas formation, i.e. disruption of the agar plug, and an odour resembling that of cheese. The method is not quantitative.

8.5.1.4.D Enumeration of thermophilic anaerobes producing hydrogen sulphide (bacteria causing sulphide spoilage)

Equal volumes (totalling 20 ml) of a sugar solution heat-treated as described in 8.5.1.4.A are dispensed into 6 test-tubes containing melted sulphite agar [133]. Freshly-exhausted liver deep agar tubes are inoculated. The tubes are immersed in cold water to solidify the medium, then pre-warmed to 50–55 °C and incubated at this temperature for 48 h.

Bacteria caused sulphide spoilage form characteristic black strips in the sulphite agar. Gas does not accumulate because the hydrogen sulphide becomes fixed as iron sulphide. However, some thermophilic anaerobes may be present that do not produce hydrogen sulphide, but form instead strongly reducing hydrogen which blackens the entire medium. This phenomenon is easily distinguishable from the striped blackening mentioned above.

The blackened strips are counted and multiplied by 2·5 to obtain the number of anaerobic spores of hydrogen-sulphide producing bacteria per 10 g sugar.

8.5.2 Examination of starch for thermophilic spores

8.5.2.1 Sampling and processing of samples

5 samples weighing 20–25 g are taken from each shipment or batch. The samples are mixed together and 20 g are weighed into a dry, graduated Erlenmeyer flask. Sterile cold water is added to the 100-ml mark, and a homogeneous suspension is formed by cautious shaking.

8.5.2.2 Detection of flat-sour spores

10 ml of the starch suspension are delivered from a wide-mouth pipette into a 300-ml flask containing 100 ml of sterile glucose–tryptone agar medium [54] at a temperature of 55–60 °C. The flask is gently shaken for 3 minutes in boiling water to make the starch pasty, and it is then autoclaved at 109 °C for 10 minutes. The autoclaved flask is gently shaken to hasten cooling, taking care to prevent bubbles from forming in the medium. When the mixture of starch and medium has cooled sufficiently, approximately equal portions of the contents are dispensed into 6 Petri dishes. The plates are allowed to solidify, and then 2% agar containing no nutrient is poured over the medium to prevent the spreading of colonies. When the covering layer has solidified, the plates are incubated at 50–55 °C for 48–72 h.

8.5.2.3 Detection of thermophilic anaerobes not producing hydrogen sulphide

Approximately equal volumes (totalling 20 ml) of the cold starch suspension are dispensed into 6 test-tubes containing liver broth [88]. The inoculated tubes are taken three at a time, and rotated between the palms of the hands to mix the contents. The tubes are then placed in boiling water for 15 minutes and during the first 5 minutes of heating, they are spun 4 or 5 times between the fingers. The tubes are cooled in cold water and then pre-warmed to 50–55 °C and incubated at this temperature for 72 h. Owing to the starch content, sealing of the samples as described in 8.5.1.4.D may be omitted.

The results are evaluated as described in 8.5.1.4.C. A positive reaction is indicated by gas formation.

8.5.2.4 Detection of thermophilic anaerobes producing hydrogen sulphide

Equal volumes (totalling 20 ml) of the cold starch suspension are dispensed into 6 test-tubes containing melted sulphite agar [133]. Mixing, heating, and stirring are carried out as described above.

The tubes are incubated at 50–55 °C for 48 h, and the results evaluated as described in 8.5.1.4.D.

8.5.3 Thermophilic spore count in other additives

The thermophilic spore counts in spices, cocoa, condensed milk, flour, molasses, etc. can be determined as described in 8.5.1 (for soluble substances) and 8.5.2 (for insoluble substances).

8.5.4 Counting of the aerobic spore content of flour, giving rise to ropiness in bread and other bakery products

Principle. The number of heat-resistant cells is determined by cultivation in a suspension of flour prepared with 0·5% saline (Vas, 1962).

Procedure. 100 ml of a 0·5% NaCl solution are measured into a sterile, wide-mouthed 250-ml bottle, and 10 ml of the same solution are transferred into each of 9 sterile, narrow-mouthed 50-ml bottles. Some freshly sterilized quartz sand is added to the large bottle. 32 g of a mixed flour sample are weighed out on a sterile watch glass and added to the contents of the 250-ml bottle which are then shaken thoroughly for 2 minutes. 10 ml of the suspension thus obtained are transferred to one of the 50-ml bottles using a sterile pipette with a cut end, and the bottle is shaken thoroughly for 1 minute or more. A twofold dilution series is prepared in the remaining 50-ml bottles. About 10 ml of the stock suspension are pipetted into a sterile test-tube which is placed together with all the bottles of the dilution series in a water bath at 82 °C. After 13 minutes, during which time the vegetative cells are killed, the suspensions are allowed to cool, and 5 aliquots of 1 ml are transferred with a sterile pipette from each suspension into 4–5 ml of sterile nutrient broth (i.e. 5 cultures for each of 10 dilutions). The contents of the tubes are mixed by rolling the tubes between the palms or using a vortex mixer. After incubation at 37 °C for 2 days, the number of tubes showing a surface skin is counted. After incubation for 24 h, those cultures showing no membranous growth should be mixed again. A well-defined skin is indicative of the presence of spores which cause ropiness (*Bacillus subtilis*).

Table 58

Barton–Wright's table for assessing the number of spores of bacteria causing ropiness in flour

y	0·0	0·1	0·2	0·3	0·4	0·5	0·6	0·7	0·8	0·9
0		1224	907	732	615	528	461	406	362	324
1	291	263	239	217	198·2	181·2	166·0	152·3	139·9	128·8
2	118·6	109·4	101·0	93·3	86·3	79·9	73·9	68·5	63·5	58·9
3	54·6	50·7	47·1	45·8	40·7	37·8	35·1	32·7	30·4	28·3
4	26·3	24·5	22·8	21·8	19·77	18·42	17·16	15·98	14·89	13·87
5	12·93	12·05	11·23	10·47	9·75	9·09	9·48	7·90	7·37	6·87
6	6·41	5·98	5·57	5·20	4·85	4·52	4·21	3·93	3·67	3·42
7	3·19	2·97	2·77	2·58	2·40	2·23	2·08	1·931	1·793	1·662
8	1·539	1·423	1·312	1·208	1·108	1·014	0·924	0·838	0·797	0·679
9	0·604	0·532	0·464	0·398	0·334	0·273	0·215	0·1581	0·1036	0·0509
10	0·0									

For the numerical evaluation of these spores, Barton–Wright has drawn up a table (on the basis of the statistical data of Fisher and Yates) which applies to data obtained from 5 cultures for each of 10 dilutions *(Table 58)*.

The number of spores per 1 g of the stock suspension can be calculated by obtaining the value of *y* from the equation

$$y = \frac{A - Y}{n},$$

where A is the total number of inoculated cultures, Y is the number of tubes recorded as positive after incubation, and n is the number of replicate cultures prepared from each dilution. The number of spores is then obtained from *Table 58* by inserting the value of *y*.

An example for the use of Table 58

If the total number of inoculated cultures was 50 and 32 tubes of them were recorded as positive, and 5 replicates were prepared from each dilution, then

$$y = \frac{50-32}{5} = 3·6.$$

On the basis of *Table 58*, the number of spores per 1 ml of stock-suspension (32 g of flour suspended in 100 ml of saline) is 35·1. Therefore, the number of roping spores per 1 g of flour is

$$\frac{35·1 \times 100}{32} = 110.$$

350

The figures in the *Table 58* are the numbers of spores per 1 ml of stock suspension as a function of y.

In this experimental arrangement, the difference in spore count between two flour samples can be regarded as significant if the quotient of the two spore counts exceeds 4 (provided that the value of y lies between 0·1 and 9·3). The results are most reliable for values of y of about 6, and the series of dilutions should be set up accordingly.

9.   CULTURE MEDIA, STAINS AND INDICATORS

[1] *Acetate agar*
(A medium for the isolation and enumeration of lactic acid bacteria.)

Basal medium:	
Peptone (Evans)	10·0 g
Lab-Lemco meat extract	
(Oxoid)	10·0 g
Yeast extract	5·0 g
Agar	15·0 g
Triammonium citrate	2·0 ml
Mineral solution (see below)	5·0 ml
Glucose	10·0 g
Tween 80	0·5 ml
Distilled water	1000·0 ml

$$pH = 5·4$$

Mineral solution:	
$MgSO_4 . 7H_2O$	8·0 g
$MnSO_4 . 4H_2O$	2·0 g
Distilled water	100·0 ml

Dissolve peptone, meat extract, yeast extract and agar in 1 litre of distilled water, mix well at 50 °C for 15 minutes. Add citrate and mineral solution. Adjust the pH to 5·4. Filter if necessary. Add glucose and Tween 80, mix the solution thoroughly and dispense. Sterilize at 121 °C* for 16 minutes.

* For the interconversion of temperatures and pressures see *Table 1*, p. 32.

Preparation of 2M acetic acid–sodium acetate buffer, pH 5·4:

Dissolve 23·3 g sodium acetate (containing water of crystallization) and 1·7 g glacial acetic acid in distilled water, and make up to 100 ml. Dilute the buffer tenfold with distilled water and check the pH (with the undiluted buffer, the sodium ions would attack the glass electrode). Dispense, and sterilize at 115 °C for 20 minutes.

Preparation of complete acetate agar:

Pipette 10 ml of the sterile acetate buffer in 90 ml melted sterile culture medium at 50 °C. Shake cautiously to avoid frothing and prepare pour-plate cultures in the usual way. The final pH is 5·4 ± 0·05.

If lactobacilli are to be examined on this medium, it is advisable to pour a second layer of agar on top of the solidified first layer. In this way, micro-aerophilic conditions are created.

[2] *Adams agar*
(A medium for examining sporulation)

Glucose	0·4 g
Sodium acetate	2·3 g
Agar	20·0 g

Dissolve the ingredients in 1000 ml boiling distilled water. Dispense into test-tubes and sterilize at 112 °C (1·5 atm) for 15 minutes. Solidify as slopes.

[3] *Agar solution, 2%*
(A medium for fermentation tests, instead of paraffin.)

Agar	200·0 g
Distilled water	1000·0 ml

Add the agar to the distilled water and boil until the agar is dissolved. Dispense in flasks as required and sterilize at 121 °C for 15 minutes. The solution can be stored in the refrigerator for long periods.

[4] *Brain-heart nutrient broth*
(A medium for the cultivation of micro-organisms of complex nutrient requirement.)

Dried calf-brain decoction (industrial preparation)	12·5 g

Dried bovine heart decoction (industrial preparation)	5·0 g
Peptone	10·0 g
Glucose	2·0 g
Sodium chloride	5·0 g
Sodium sulphate	2·5 g
Distilled water	1000·0 ml

pH when ready for use, 7·4.

Dissolve the ingredients in boiling distilled water. Cool to room temperature, and check the pH. Dispense 10-ml volumes into test-tubes and sterilize at 121 °C for 15 minutes. The medium can be stored in the refrigerator for several months.

[5] *Culture medium for the demonstration of liquefaction of gelatin and H_2S production by anaerobic bacteria*

Meat extract	4·0 g
Yeast extract	4·0 g
Gelatin	30·0 g
Purified peptone	100·0 ml
Distilled water	1000·0 ml

Purification of peptone:

Peptone	1000·0 g
Distilled water	500·0 ml

Dissolve by heating and add 25 ml of a 40% barium chloride ($BaCl_2 \cdot 2H_2O$) to the solution and filter. Mix the ingredients under boiling conditions, adjust the pH to 7·6 with N NaOH, filter, and sterilize in the autoclave at 121 °C for 20 minutes.

To the basal nutrient medium melted in a water bath, add 10 ml of a 2% sodium thiosulphate solution ($Na_2S_2O_3 \cdot 5H_2O$) and 0·5 g ferrous ammonium sulphate ($FeSO_4/NH_4/_2SO_4$). Boil in the water bath for 20 minutes and dispense 10-ml volumes into sterile test-tubes. Store at +4 °C.

[6] *Andrade nutrient broth*

Nutrient broth:

Peptone	10·0 g
Sodium chloride	5·0 g
Andrade indicator	1·0 ml
Distilled water	1000·0 ml

Dissolve the ingredients, add glucose to give a concentration of 1%, and adjust the pH to 7·4–7·5. Dispense the nutrient in 5-ml volumes, place a Durham tube in each tube and sterilize at 121 °C for 15 minutes. Alternatively, the glucose solution can be sterilized separately. Acid formation is indicated by a change in the red colour of the medium and gas formation is indicated by a gas bubble in the Durham tube.

Andrade indicator:

Acid fuchsin	0·6 g
N NaOH solution	16·0 ml
Distilled water	1000·0 ml

Dissolve the fuchsin and add the NaOH solution. Allow to stand for 18–20 h. Add a further 1–2 ml of the NaOH solution if the lilac colour of the fuchsin has not turned to straw yellow.

[7] *Arginine broth*

D-glucose	0·05 g
Dipotassium hydrogen phosphate	0·2 g
Tryptone	0·5 g
Yeast extract	0·25 g
Arginine HCl	0·3 g
Distilled water	100·0 ml

Final pH: 7·0. Sterilize at 121 °C for 15 minutes.

[8] *Azide-egg-yolk agar*
(A medium for testing for *Staph. aureus*.)

Meat extract	2·0 g
Peptone	10·0 g
Sodium chloride	3·0 g
Disodium hydrogen phosphate	0·2 g
Sodium azide	0·15 g
Distilled water	1000·0 ml

Dissolve all the ingredients except the sodium azide. Adjust the pH to 7·6. Sterilize at 121 °C for 15 minutes. Cool to 50–60 °C, add the azide salt, mix well, and sterilize at 121 °C for 15 minutes. The basal medium can be stored, or the complete medium can be prepared immediately. For storage, dispense the medium into flasks before the second sterilization is carried out.

23*

To prepare the complete medium, add 150 ml sterile egg-yolk emulsion [147] to each melted or still liquid, batch of basal medium. Mix well and pour plates.

Commercially prepared egg-yolk medium should be checked by cultivating coagulase-positive *Staph. aureus* on the medium.

[9] *Modified Baird–Parker medium for the demonstration of glucose fermentation*

Tryptone	10·0 g	
Yeast extract	1·0 g	
Glucose	10·0 g	
Bromcresol purple	0·004 g	(1 ml of a 0·4% solution)
Agar	2·0 g	
Distilled water	1000·0 ml	

pH when ready for use, 7·2.

Dissolve all the ingredients, except the bromcresol purple, by warming in the water. Check the pH. Dispense into flasks or bottles as required and sterilize at 115 °C for 20 minutes.

Add the bromcresol purple solution to the medium immediately before use. Mix thoroughly and dispense 10-ml volumes into sterile test tubes. Sterilize by steaming for 20 minutes. The complete medium can be stored in the refrigerator for 7–10 days.

The bromcresol purple solution is prepared by adding 0·4 g bromcresol purple to 100 ml of 96% ethanol and dissolving with gentle shaking.

[10] *Baird–Parker's selective agar medium*
(A medium for testing for *Staph. aureus*.)

Basal medium:	
Tryptone	10·0 g
Yeast extract	1·0 g
Meat extract	5·0 g
Glycine	12·0 g
Lithium chloride	5·0 g
Sulfamethazine solution	27·5 ml
Agar	20·0 g
Distilled water	1000·0 ml

pH after sterilization, 7·2 (at room temperature).

Add all the ingredients except the sulfamethazine to the water and boil until the ingredients have dissolved. Check the pH. Dispense 90-ml volumes of the solution into flask not larger than 90 ml in volume. Sterilize at 121 °C for 15 minutes. The basal medium can be stored in the refrigerator for 1 month.

Sulfamethazine solution:	
Sulfamethazine	0·2 g
0·1 N sodium hydroxide	10·0 ml
Distilled water	(to make 100 ml of solution)

Dissolve the sulfamethazine in the NaOH solution and make up to volume with distilled water. It is advisable to prepare the solution immediately before use.

The addition of sulfamethazine is necessary if the sample is, or is expected to be, contaminated by *Proteus*.

Complete the medium by adding potassium tellurite solution and egg-yolk emulsion (~20%) before use.

Tellurite solution:	
Potassium tellurite	1·0 g
Distilled water	100·0 ml

Dissolve the tellurite in water of temperature 45 °C, sterilize by filtration and dispense as required. The solution can be stored in the refrigerator for several months.

Preparation of the complete medium without sodium pyruvate:

Melt the basal medium and cool to 50 °C. Add 1 ml of the potassium tellurite solution and 5 ml of egg-yolk emulsion [147] to each 90-ml volume of the medium, mixing well after the addition of each ingredient. The medium thus prepared can be stored in the refrigerator for 1 month.

Preparation of the complete medium with sodium pyruvate:

Sodium pyruvate solution:	
Sodium pyruvate	20·0 g
Distilled water	100·0 ml

Dissolve the pyruvate in a 100-ml volumetric flask. Sterilize by filtration. The solution can be stored in the refrigerator for 1 month.

Pour plates in Petri dishes. Spread 0·5 ml of the sodium pyruvate cautiously over each plate by leaving inverted in a desiccator or in an incubator at 50 °C for 30 minutes. Store not longer than for 48 h.

[11] *Bierbrauer's enrichment nutrient broth*

Basal broth:

Disodium hydrogen phosphate: $(Na_2HPO_4 \cdot 12H_2O)$	2·0 g
Sodium chloride	2·0 g
Witte peptone	2·0 g
Peptone	2·0 g
Yeast extract	2·0 g
Meat extract	4·0 g
Distilled water	1000·0 ml

Dissolve the ingredients in the order given above in constantly warmed distilled water and make up to the required volume. Filter through cotton wool and adjust the pH to 7·2–7·4. Dispense 900-ml volumes into flasks. Sterilize in the autoclave at $121 \pm 1\,°C$ for 20 minutes. The medium can be stored at room temperature for 7–10 days.

Sodium thiosulphate solution, see [146]

Iodine solution, see [146]

Sterile bovine bile:

Measure 100-ml volumes of fresh bovine bile (as obtained at the abattoir) into flasks. Sterilize in the autoclave at 121 °C for 20 minutes. Store for not longer than 10–12 days at room temperature.

Malachite-green solution:	
Malachite green	0·4 g
Distilled water	100·0 ml

Dissolve the dye and allow the solution to stand in a dark bottle for at least a day before use.

Brilliant-green solution see [146].

Final composition of Bierbrauer's enrichment broth medium:

Basal nutrient broth	900·0 ml
Calcium carbonate (heated to glowing immediately before use)	50·0 g
Sodium thiosulphate solution	100·0 ml

Iodine solution	20·0 ml
Sterile bovine bile	100·0 ml
Malachite-green solution	24·0 ml
Brilliant-green solution	5·0 ml

Add the ingredients to the basal broth aseptically in the order given above. Shake thoroughly after each addition. The dye solutions are added in two steps; first of all add 16·0 ml of the malachite-green solution and 1·6 ml of the brilliant-green solution, and incubate the medium at 37 °C for 3 h. Then add the remainder of the solutions. Dispense 1000-ml portions of the medium into 2000-ml flasks. Altogether for 7 days, allow the medium to stand at room temperature for 1 day and at 4–10 °C for a further period of 6 days. The medium can be stored at 4 °C for 7–10 days.

[12] *Brilliant-green–bile broth*

Peptone	10·0 g
Glucose	5·0 g
Disodium hydrogen phosphate anhyd.	6·45 g
Potassium dihydrogen phosphate	2·0 g
Dried bovine bile	20·0 g
Brilliant green	0·015 g
Distilled water	1000·0 ml

Mix the ingredients and boil for not more than 30 minutes. Adjust the pH to 7·2±0·1 at 20 °C. Dispense into sterile test-tubes. Store in the refrigerator, unsterilized, for not more than 1 week.

[13] *Brilliant-green–phenol-red agar*
(Edel–Kampelmacher medium.)

Basal agar medium:	
Meat extract	4·0 g
Peptone	10·0 g
Sodium chloride	3·0 g
Disodium hydrogen phosphate anhyd.	0·8 g
Sodium dihydrogen phosphate	0·6 g

Agar	18·0 g
Distilled water	900·0 ml

pH after sterilization: 7·0 ± 0·1 (at 40 °C).

Dissolve the ingredients while boiling the distilled water. Check the pH. Dispense into flasks not larger than 500 ml in volume and sterilize at 121 °C for 15 minutes.

Sugar-phenol-red solution:	
Lactose	10·0 g
Sucrose	10·0 g
Phenol red	0·09 g
Distilled water (to make up to 100·0 ml)	

Dissolve the ingredients and maintain the solution at 70 °C for 20 minutes. Cool to 55 °C and use immediately.

Brilliant-green solution, see [146].

Preparation of the brilliant-green–phenol-red agar:

Basal agar	900·0 ml
Sugar–phenol-red solution	100·0 ml
Brilliant-green solution	1·0 ml

Add the brilliant-green solution aseptically to the sugar–phenol-red solution at 55 °C. Add the mixture to the basal medium, pre-cooled to 50–55 °C. Mix well.

Preparation of the brilliant-green–phenol-red agar (Edel–Kampelmacher) plate:
Pour about 40 ml of the fresh medium at a temperature of 45 °C into sterile, large-volume Petri dishes. Allow the medium to solidify. Immediately before use, dry the plate opened and inverted in a desiccator or in an incubator at 50 °C for 30 minutes. Do not keep undried plates longer than 4 h at room temperature, or 1 day in a refrigerator.

[14] *Brilliant-green–lactose–bile nutrient solution*

Peptone	10·0 g
Lactose	10·0 g
Bovine bile	20·0 g
Brilliant green	0·0133 g

Dissolve the peptone and the lactose in 500 ml water and add the bovine bile dissolved in 200 ml water. Make up to 975 ml with water and adjust the pH to 7·4. Add 13·3 ml of a 1% aqueous brilliant-green solution, make up to 1 litre and filter if necessary. Dispense 10-ml portions of the clear solution into test-tubes containing a Durham tube. Sterilize at 121 °C for 10 minutes.

[15] *Bromcresol-purple (alcoholic) solution*

Dissolve 0·05 g bromcresol purple in 500 ml 95% ethanol (or denaturated alcohol) and add, drop by drop, 0·1 N sodium hydroxide until the colour of the solution becomes grayish-purple. Adjust the specific gravity to 0·895 (equivalent to 68% ethanol) by adding distilled water.

[16] *Bromcresol-purple–lactose–agar (BLC agar)*
(A medium for the determination of acid- and alkali-producing bacteria.)

Peptone	3·0 g
Yeast extract	2·0 g
Meat extract	3·0 g
Lactose	5·0 g
Skim milk	10·0 ml
Precipitated lime (powdered chalk)	10·0 g
Bromcresol purple (0·04% solution	50·0 ml
Agar	15·0 g
Tap water	1000·0 ml

Disperse the peptone, the yeast extract, the meat extract, and the agar in water, and dissolve while keeping the solution warm. Adjust the pH to 6·8, and add the lactose, the powdered chalk (pre-sterilized by dry heat), and the bromcresol purple solution. Sterilize in the autoclave at 115 °C for 15 minutes.

[17] *Potato-glucose agar*
(A medium for the preparation of Dalmau plates.)

Soak 100 g of washed, peeled, and grated potatoes in 300 ml tap water in the refrigerator overnight, filter through cheese-cloth, and autoclave at 112 °C (1·5 atm) for 1 h. Make up 230 ml of the extract to 1 litre and add 20 g glucose and 20 g agar. Sterilize at 112 °C (1·5 atm) for 15 minutes. Adjust the pH to 5·6.

To prevent bacterial multiplication, add to the cooled medium (at about 50 °C) 10% tartaric acid solution until the pH is reduced to 3·5.

4·0 g of dehydrated potato extract may be used instead of the liquid extract.

[18] *Wheat-meal agar*
(A medium for the detection of spore-formers.)

Soak 1 part of wheat-meal in 2 parts of tap water for 12 h at room temperature. Filter the fluid and add tap water until the original volume is restored. Add 1% tryptone (Difco) and 0·5 g yeast extract. When these have completely dissolved, add 1·5% agar, bring to the boil, dispense, and autoclave at 2 atm for 12 minutes. Final pH, 7·0.

[19] *Cetrimide (GMAC) agar (modified)*
(For testing for *P. aeruginosa*.)

Basal medium:	
Gelysat	0·2 g
Magnesium chloride	
$(MgCl_2 \cdot 6H_2O)$	1·4 g
Potassium sulphate	10·0 g
Cetrimide	0·3 g
Mannitol	5·0 g
Disodium hydrogen	
phosphate	
$(Na_2HPO_4 \cdot 2H_2O)$	
solution	1·7 ml
Potassium dihydrogen	
phosphate solution	3·0 ml
Agar	15·0 g
Distilled water	900·0 ml

The phosphate solutions are prepared by dissolving 71·3 g $Na_2HPO_4 \cdot 2H_2O$ in 1000 ml distilled water, and 54·4 g potassium dihydrogen phosphate in 1000 ml distilled water, respectively.

The basal medium is made up by dissolving the ingredients in boiling distilled water, and adjusting the pH to 7·05. Sterilize at 121 °C for 20 minutes.

Preparation of the complete medium:

Add 100 ml of a 10% acetamide solution pre-sterilized by filtration, and 6 ml phenol-red indicator solution **[41]**, to sterile freshly-prepared, or melted basal medium; then either dispense 6-ml portions aseptically into sterile test-tubes and allow the medium to set as slopes (for the confirmation of *P. aeruginosa* after pre-enrichment), or pour plates.

Gelysat is a preparation obtained from gelatin by enzymatic digestion.

362

[20] *Christensen's urea agar*

Basal medium:
Peptone	1·0
Sodium chloride	5·0 g
Potassium dihydrogen phosphate	2·0 g
Glucose	1·0 g
Agar	20·0 g
20% phenol-red solution	6·0 ml
Distilled water	1000·0 ml

pH 6·8–7·0.

Dispense the basal medium into flasks or test-tubes and sterilize by autoclaving at 121 °C for 20 minutes. Cool to 50 °C and add aseptically to each tube sufficient sterile 20% urea solution (pre-sterilized by filtration) to give a final concentration of 2%. To check the urease activity of a culture, basal medium containing no urea is also inoculated. The ammonia produced from the decomposition of urea changes the colour of the indicator to pink or red.

[21] *Clarke's developer*
(A reagent for the demonstration of liquefaction of gelatin.)

Dissolve 15 g $HgCl_2$ in 20 ml concentrated HCl and make up to 100 ml with distilled water.

[22] *Triple sugar-iron agar (TSI)*

Meat extract	3·0 g
Yeast extract	3·0 g
Peptone	20·0 g
Sodium chloride	5·0 g
Lactose	10·0 g
Sucrose	10·0 g
Glucose	1·0 g
Ferric citrate	0·3 g
Sodium thiosulphate ($Na_2S_2O_3 \cdot 5H_2O$)	0·3 g
Phenol-red	0·024 g
Agar	12·0 g
Distilled water	1000·0 ml

pH 7·4 ± 0·1 at 40 °C and after sterilization.

Dissolve the ingredients in boiling water. Check the pH. Dispense 10-ml portions into tubes of 17–18 mm diameter. Sterilize at 121 °C for 10 minutes, and lay the tubes so as to form slopes.

[23] *Czapek's agar*

Sucrose	30·0 g
Sodium nitrate	3·0 g
Dipotassium hydrogen phosphate	1·0 g
Magnesium sulphate ($MgSO_4 \cdot 7H_2O$)	0·5 g
Ferrous sulphate ($FeSO_4 \cdot 7H_2O$)	0·01 g
Potassium chloride	0·5 g
Agar	20·0 g
Distilled water	1000·0 ml

Dissolve the ingredients, and sterilize by autoclaving at 115 °C for 20 minutes. Adjust the pH of the sterilized medium to 3·5 by adding 10 ml 10% lactic acid. Cool the medium to 50 °C or below before plates are poured.

[24] *Czapek's agar with malt*
(For the culture of species of the Mucorales and Deuteromycetes.)

Add 2% commercially produced malt extract to Czapek's agar.

[25] *Semi-solid agar for the development of flagellar antigens*

Meat extract	3·0 g
Peptone	5·0 g
Agar	8·0 g
Distilled water	1000·0 ml

pH 7·0 ± 0·1 (at 40 °C and after sterilization).

Dissolve the water-free ingredients in boiling water and check the pH. Dispense into flasks not larger than 500 ml in volume. Sterilize at 121 °C for 20 minutes.

[26] *Deoxycholate-citrate medium*

Meat extract	5·0 g
Proteose peptone	5·0 g
Lactose	10·0 g
Sodium citrate	8·5 g

Ferric citrate	1·0 g
Sodium thiosulphate	5·4 g
Sodium deoxycholate	5·0 g
Neutral red	0·02 g
Agar	12·0 g

pH 7·3.

Measure the ingredients into 1000 ml distilled water, heat to boiling and stir to dissolve. Cool to 45–50 °C. Dispense volume of 15–20 ml into Petri dishes. Do not sterilize.

[27] *D-cycloserine blood agar*
(A selective medium for the purification of anaerobic cultures.)

Add 50 ml of sterile, defibrinated bovine blood, and 25 ml of sterile basal broth [32] containing 0·4 g dissolved D-cycloserine bitartarate, to 925 ml of sterile gelatin–basal agar [166], pre-cooled to 50 °C. Mix with circular movements, avoiding frothing at the surface. Pour volumes of about 25 ml into sterile Petri dishes. Place the plates in an incubator at 37 °C to test for sterility, and then store at +4 °C. The medium should always contain exactly 400 µg D-cycloserine per ml.

[28] *Drigalski's agar plate (modified)*
Takács's coloured basal agar

Disodium hydrogen phosphate ($Na_2HPO_4 \cdot 12 H_2O$)	2·0 g
Sodium chloride	2·0 g
Witte peptone	2·0 g
Peptone	2·0 g
Yeast extract	2·0 g
Meat extract	4·0 g
Agar	18·0 g
Distilled water	1000·0 ml

Dissolve the ingredients of the basal broth (except for the agar) in the water, then add the agar. Dissolve by boiling. Adjust the pH to 7·6–7·7. Dispense volumes of 800 ml into 1000-ml flasks. Sterilize at 121 °C for 20 minutes.

Composition of the modified Drigalski's agar plate:
| Lactose (inositol-free) | 8·0 g |
| Bromthymol blue solution (1·5 g bromty- | |

molblue dissolved
in 100 ml 70% v/v
ethanol)   8·0 ml
Crystal violet solution
(1 g dissolved in 100 ml
distilled water)   0·08 ml
Takács's coloured
basal agar   800·0 ml

Add the ingredients to the basal agar medium and maintain the mixture in a boiling water bath for 30 minutes. Dispense into small sterile Petri dishes.

[29] *EC nutrient medium*

Tryptose, or trypticase	20·0 g
Lactose	5·0 g
Bile salt (no. 3)*	1·5 g
Dipotassium hydrogen phosphate	4·0 g
Potassium dihydrogen phosphate	1·5 g
Sodium chloride	5·0 g

Dissolve the ingredients in 1 litre distilled water, mildly warming if necessary. Dispense volumes of 10 ml into test-tubes each containing a Durham tube. Sterilize at 121 °C for 10 minutes. Final pH 6·8.

[30] *Acetic acid–phenol–thionine*

Dissolve 1·0 g thionine and 2·5 g crystalline phenol in 400 ml distilled water, filter, and add 20 ml glacial acetic acid.

[31] *Szita's E-67 agar medium*
(A medium for the culture of *S. faecalis*.)

Basal medium (meat agar):	
Meat extract	6·0 g
Agar	14–18·0 g

* (Oxoid): standardized mixture of sodium glycolate and sodium taurocholate obtained from fresh bovine bile).

366

Peptone	10·0 g
Sodium chloride	3·0 g
Disodium hydrogen phosphate (Na$_2$HPO$_4$ · 12 H$_2$O)	4·0 g
Distilled water	1000·0 ml

Dissolve the meat extract in boiling water, and cool. Draw off the liquid from beneath the fatty layer and make up with water to 1000 ml. Add and dissolve all the other ingredients by boiling, adjust the pH with NaOH to 7·4. Sterilize by autoclaving at 121 °C for 30 minutes and allow the medium to set. Separate off the clear layer and check the pH. Dispense the basal medium into flasks and sterilize at 121 °C for 30 minutes.

Preparation of the complete medium

Add to the basal medium melted in a water bath or in flowing steam 5 g sodium taurocholate, 0·8 ml of a 1% crystal violet (Grübler) solution, and 10 g glucose. Mix thoroughly. Treat in flowing steam for 30 minutes. Cool to 45 °C and add 5 ml of a 1% potassium tellurite solution [10]. Pour plates, using freshly prepared medium wherever possible. Store the tellurite salt in a desiccator before use.

The inhibitory effect of the medium depends largely on the concentrations of the taurocholate and the crystal violet, and efficiency of these two ingredients is strongly influenced by the composition of the basal medium. When a new batch of basal medium is prepared, the inhibitors must be titrated by using a typical *S. faecalis* strain. (The amount of taurocholate and crystal violet solution required may be as much as 5 g and 5 ml, respectively.)

[32] *Universal basal broth (according to Takács)*
(A medium for the culture of aerobic and facultative anaerobic bacteria.)

Meat extract	4·0 g
Disodium hydrogen phosphate (Na$_2$HPO$_4$ · 12 H$_2$O)	2·0 g
Witte peptone	2·0 g
Peptone	2·0 g
Sodium chloride	2·0 g
Glucose*	1·0 g
Yeast extract	2·0 g

* The medium can be used without glucose, e.g. for the pre-cultivation of *Staph. aureus*.

367

Distilled water
(to make up to)   1000·0 ml

pH 7·6.

Dissolve the ingredients in a little boiling water, filter (using filter paper), and make up to volume. Adjust the pH with N NaOH solution, boil and filter again, and sterilize by autoclaving at 121 °C for 20 minutes.

[33] *Endo agar*

Peptone	10·0 g
Lactose	10·0 g
Dipotassium hydrogen phosphate	3·5 g
Sodium sulphite	2·5 g
Basic fuchsin	0·4 g
Agar	20·0 g

Final pH 7·5.

Prepare the medium on the day of use. Dissolve the ingredients, except for the sulphite and basic fuchsin, in 1 litre distilled water in a boiling water bath. Dispense in volumes of 100 ml and sterilize at 115 °C for 15 minutes. Melt immediately before use and add 1 ml of a 4% basic fuchsin solution (in 95% ethanol) and 2·5 ml of a 10% aqueous sodium sulphite solution to each 100 ml of the medium.

Prepare the sulphite solution immediately before use. Mix the medium well and pour into Petri dishes. (Keep away from light.)

[34] *Enterobacteriaceae enrichment medium*

Peptone	10·0 g
Glucose	5·0 g
Disodium hydrogen phosphate ($Na_2HPO_4 \cdot 2\,H_2O$)	8·0 g
Potassium dihydrogen phosphate	2·0 g
Bovine bile	20·0 g
Brilliant green	0·015 g

Dissolve the ingredients in 1 litre distilled water. Dispense 10 ml volumes into test-tubes, or 100 ml volumes into flasks. Sterilize at 121 °C for 5 minutes; heat and cool as quickly as possible.

[35] *Eosin–methylene blue medium*

Peptone	10·0 g
Lactose	10·0 g
Sucrose	5·0 g
Dipotassium hydrogen phosphate	2·0 g
Eosin Y	0·4 g
Methylene blue	0·063 g
Agar	20·0 g

Dissolve the ingredients, except for the dyes in 955 ml distilled water. Add 20 ml of a 2% aqueous eosin Y solution and 25 ml of a 0·25% aqueous methylene blue solution. Boil until all the ingredients have dissolved. Cool to 50–60 °C, mix well, and dispense as necessary. Sterilize by autoclaving at 121 °C for 15 minutes.

[36] *Medium used in testing for aesculin fermentation*
(Test for *S. faecalis*.)

Peptone (Bacto)	15·0 g
Aesculin	1·0 g
Ferric ammonium citrate	0·5 g
Sodium taurocholate	5·0 g
Distilled water	1000·0 ml

Dissolve the ingredients while keeping the solution warmed. Sterilize by Seitz filtration. Dispense 2 ml volumes aseptically into small test-tubes.

[37] *Yeast extract*
(For examining fermentation.)

Mix 200 g baker's yeast in 1 litre of tap water and autoclave at 112 °C (1·5 atm pressure) for 15 minutes. Allow to settle and use the supernatant. (The latter can be clarified by filtration, or by adding egg white before sterilization.)

[38] *Yeast extract–glucose medium*

Yeast extract	5·0 g
Meat peptone	10·0 g
Glucose	5·0 g

Dissolve the ingredients in 1000 ml water. Sterilize at 121 °C for 15 minutes (final pH 7·0).

[39] *Phenylalanine agar* (Ewing et al., 1957)

Yeast extract	2·4 g
β-phenyl-DL-α-	
-alanine, or	2·0 g
β-phenyl-L-α-alanine	1·0 g
Disodium hydrogen	
phosphate	
($Na_2HPO_4 \cdot 7 H_2O$)	5·0 g
Agar	20·0 g
Distilled water	1000·0 ml

pH 7·2.

Dispense into test tubes. Sterilize at 121 °C for 10 minutes, and set as slopes.

[40] *Phenolphthalein-phosphate agar medium*
(A medium for testing for *S. aureus*.)

Nutrient agar [32]	1000·0 g
Phenolphthalein	
phosphate solution	20·0 ml

Dissolve 0·5 g phenolphthalein phosphate in 100 ml distilled water. Sterilize by filtration and store in the refrigerator.

Melt the agar medium in flowing steam and cool to 50 °C. Add the phenolph-thalein phosphate solution aseptically and pour plates immediately.

The medium can be prepared with an addition of Polymyxin B sulphate, as follows. Dissolve 0·5 of a megaunit of the antibiotic in 10 ml saline. Store the solution in the refrigerator. Add 2·5 ml of the antibiotic solution to the nutrient agar simultaneously with the phenolphthalein phosphate solution. Pour plates and dry the solidified plates for 1–2 h at 37 °C. Store the medium in the refrigerator.

[41] *Phenol-red indicator*

Dissolve 1 g phenol-red in 40 ml 0·1 N NaOH and make up to 460 ml with distilled water.

[42] *Carbohydrate broth with phenol-red indicator*
(For testing for carbohydrate fermentation.)

Peptone	10·0 g
Sodium chloride	5·0 g

Phenol red	0·025 g
	(12·5 ml of [41])
Carbohydrate	0·5 g
Distilled water	1000·0 ml

Dissolve the ingredients, except for the carbohydrate, in water of as low as temperature as possible. Cool the solution to room temperature and adjust the pH (the post-sterilization pH should be 7·0). Dispense 5·5 ml volumes of the basal medium into test-tubes and sterilize at 121 °C for 15 minutes. Cool to room temperature. Pipette 0·5 ml of 5% carbohydrate solution, pH 7·0, into each tube. (Sterilize the carbohydrate solution by filtration immediately before use.)

Place a Durham tube in each culture tube for the detection of gas formation.

Preparation of 0·2% phenol-red solution:
Dissolve 1 g phenol red in 40 ml 0·1 N NaOH. Add distilled water up to 500 ml.

[43] *Phenol-red–egg-yolk–polymyxin agar*

Meat extract	1·0 g
Peptone	10·0 g
Mannitol	10·0 g
Sodium chloride	10·0 g
Phenol red	0·025 g
	(12·5 ml of [41])
Agar	15·0 g

Add the ingredients to 1 litre of distilled water, mix thoroughly and boil to dissolve. Adjust the pH (the post-sterilization pH should be 7·2 ± 0·1). Dispense volumes of 90 ml into flasks. Sterilize at 121 °C for 15 minutes. Cool to 45 °C and add 10 ml egg-yolk emulsion and 1 ml of a 0·1% aqueous polymyxin B sulphate solution, pre-sterilized by filtration.

Preparation of the egg-yolk emulsion

Clean hen's eggs with a brush and sterilize the egg shell by immersing the eggs in aqueous 0·1% $HgCl_2$ solution. Open the eggs aseptically and separate the yolks into a sterile measuring cylinder.

Add an equal volume of sterile 0·85% saline and mix aseptically.

[44] *Physiological phosphate buffer (pH 7)*

Dissolve 3·52 g KH_2PO_4, 10·9 g $Na_2HPO_4 \cdot 7 H_2O$, and 1 g peptone in distilled water and make up to 1 litre.

[45] *Saline*
(For use as a diluent.)

Dissolve 9·0 g NaCl in 1000 ml water, dispense 9 ml volumes into test-tubes, and sterilize at 121 °C (2 atm) for 15 minutes.

[46] *Casein agar, according to Frazier and Rupp*
(A medium for cultivating aerobic protein-decomposing bacteria.)

Casein solution

Soak 3·5 g Hammarstein's casein for 15 minutes in about 50 ml distilled water. Add 100 ml saturated lime-water and shake the mixture until the casein has dissolved. Add 200 ml nutrient broth (composition: Liebig's meat extract, 3 g; Witte peptone, 5 g; NaCl, 5 g; 100 ml distilled water; pH 7·4), plus 10 ml of a solution containing 1·5% $CaCl_2$, plus 10 ml of another solution containing 1·05% $Na_2HPO_4 \cdot 2 H_2O$ (Sørensen's phosphate) and 0·35% KH_2PO_4. Make up to 500 ml with distilled water. Dispense volumes of 25 ml into 100 ml Erlenmeyer flasks. Sterilize by steaming for 20 minutes on each of 3 consecutive days.

Agar

Dissolve 30 g agar in 100 ml water, filter, and dispense volumes of 25 ml in 100 ml Erlenmeyer flasks. Sterilize in the autoclave at 121 °C for 20 minutes.

Complete agar medium

Melt the agar before use and add the cold casein solution. The mixture can be used only once, since floccular precipitation occurs on repeated liquefaction.

[47] *Gel–phosphate diluent*

Gelatin	2·0 g
Disodium hydrogen phosphate	4·0 g
Distilled water	1000·0 ml

Dissolve the ingredients with mild warming. Sterilize by autoclaving at 121 °C for 20 minutes. Final pH 6·2.

[48] *Tellurite enrichment medium*
(after Giolitti and Cantoni)
(For testing for *S. aureus*.)

Basal medium:

Tryptone	10·0 g
Meat extract	5·0 g
Yeast extract	5·0 g
Lithium chloride	5·0 g
Mannitol	20·0 g
Sodium chloride	5·0 g
Glycine	1·2 g
Sodium pyruvate	3·0 g
Distilled water	1000·0 ml

pH after sterilization, 6·9 (measured at room temperature).

Boil the ingredients until completely dissolved. Check the pH, taking account of the required post-sterilization value. Dispense volumes of 19 ml into 20 mm × 200 mm test-tubes, and sterilize at 121 °C for 20 minutes. The basal medium can be stored in the refrigerator for two weeks.

Preparation of the complete medium

Steam the test-tubes containing the basal medium for 20 minutes, cool to room temperature, and transfer 0·1 ml potassium tellurite solution into each tube. (The potassium tellurite solution is prepared as described in the recipe for medium [10].)

[49] *Glucose agar*

Yeast extract	1·5 g
Tryptone	10·0 g
Glucose	10·0 g
Sodium chloride	5·0 g
Bromcresol red	0·015 g
Agar	15·0 g
Distilled water	1000·0 g

Dissolve the ingredients in the distilled water in a warm water bath. Adjust the pH to 7·0 ± 0·1 at 40 °C. Dispense volumes of 15 ml into test-tubes, and sterilize at 121 °C for 20 minutes.

Stab-inoculate the medium from suspect colonies, and incubate at 37 °C for 24 h. A yellow colour throughout the medium is taken as a positive reaction.

[50] *Glucose–yeast extract–peptone water (and solid medium)*

Glucose	20·0 g
Peptone (Bacto)	10·0 g
Yeast extract (Bacto)	5·0 g

Dissolve the ingredients in 1000 ml distilled water. Sterilize at 121 °C for 15 minutes.

Add 20 g agar per litre for the solidified form of the medium.

[51] *Glucose broth and glucose gelatin*

Dissolve 10 g glucose in 1000 ml nutrient broth. Sterilize the dispensed medium on each of three consecutive days for 15 minutes at 121 °C. Use 3% agar or 15% gelatin to prepare the solid medium.

[52] *Semi-liquid deep glucose agar* (according to Takács)
(A medium for testing for glucose fermentation and motility.)

Basal agar:	
Disodium hydrogen phosphate	
($Na_2HPO_4 \cdot 12\,H_2O$)	2·0 g
Sodium chloride	2·0 g
Glucose	4·0 g
Witte peptone	2·0 g
Peptone	2·0 g
Yeast extract	2·0 g
Agar	1·8 g
Meat extract	4·0 g
Distilled water	1000·0 ml

Dissolve the ingredients in boiling water, adjust the pH to 7·2, and filter the medium through cotton wool.

Neutral red solution:	
Neutral red indicator	10·0 g
Distilled water	1000·0 ml

Add 5 ml of 1% aqueous neutral red solution to 1000 ml filtered basal agar medium. Dispense volumes of 10–12 ml into test-tubes. Sterilize by autoclaving at 121 °C for 20 minutes. The medium can be stored at room temperature for 7–10 days.

[53] *Glucose nutrient solution*

Protease peptone	5·0 g
Glucose	5·0 g
Dipotassium hydrogen phosphate	5·0 g

374

Add the ingredients to 1 litre distilled water, and warm cautiously until completely dissolved. Dispense 5 ml volumes into test-tubes. Sterilize at 121 °C for 10 minutes.

[54] *Glucose–tryptone broth/agar*
(For the culture of aerobic and facultative anaerobic bacteria, and for the enumeration of living cells.)

Bromcresol purple*	0·04 g
Tryptone (Bacto)	10·0 g
Glucose	5·0 g
Distilled water	1000·0 ml

For the solidified medium, add 15 g agar. Sterilize by autoclaving at 121 °C for 20 minutes, then adjust the pH to 6·8–7·0.

[55] *Solutions required for the Gram stain*

Crystal violet solution
Dissolve 2 g crystal violet in 20 ml 95% ethanol. Mix the solution with a solution of 0·8 g ammonium oxalate in 80 ml distilled water. Allow the mixture to stand for 24 h before use.

Lugol's solution
Dissolve 1 g iodine and 2 g potassium iodine in a little distilled water by grinding in a mortar. Transfer the solution to a volumetric flask and make up to 300 ml with distilled water.

Acetone–ethanol mixture
Mix 700 ml 95% ethanol with 300 ml acetone.

Safranin solution
Dissolve 0·25 g safranin in 10 ml 95% ethanol. Add 100 ml distilled water and filter.

[56] *Griess–Ilosvay reagent (nitrate-reduction test)*

Solution A
Dissolve 100 g sulphanylic acid in 100 ml 5 N acetic acid.

Solution B
Dissolve 0·5 g alpha-naphthalamine in 100 ml 5 N acetic acid.

---

* Before using the bromcresol purple indicator, test the solution at the concentration in the medium, for its inhibitory effect against a sample containing known numbers of bacteria.

Store the reagents in brown bottles in the dark. Mix solution A with an equal volume of solution B before use. Solutions A and B can be stored separately for long periods, but only for a few days as a mixture.

[57] *Hartley's broth*

Clean some veal or horse rumpsteads, removing tendon and fat tissue, and mince twice in a mincer. Place 1 kg of the minced meat, 1 litre of distilled water, and 90 g sodium carbonate in a vessel. Mix well and keep the mixture refrigerated overnight. Warm up to 85 °C over a naked flame while continually stirring, and when this temperature has been reached, add 1150 ml distilled water, adjusting the temperature to 45 °C. Add 25 ml distilled water and again adjust the temperature to 45 °C. Add 25 ml pig pancreatic extract (see below) and 25 ml chloroform.

Place the medium in a water bath set at 56 °C and allow to equilibrate. Keep the medium at 56 °C in the water bath for 4 h. Then carry out the biuret reaction to assess the degree of digestion. If the reaction is negative, continue the digestion until a bright red colour is obtained, indicating that the digestion is complete.

When digestion is complete, take the preparation out of the water bath, add 20 ml concentrated HCl, mix well, and refrigerate the mixture overnight. Then bring the mixture to boil over a naked flame, and allow it to settle for about 2 h. Decant the supernatant or filter, and make up with distilled water to 1400 ml. Adjust the pH to 8·0–8·2. Dispense into flasks and sterilize at 115 °C for 30 minutes.

Preparation of pig pancreatic extract

Mix well 500 g fat-free minced pig pancreas with 500 ml 95% ethanol and 1500 ml distilled water. Allow the mixture to stand at room temperature for 24 h, then store in the refrigerator.

Biuret reaction

Add 5 ml N NaOH to a 5 ml test sample, mix well, and add a few drops of a 5% copper sulphate solution. Complete digestion is indicated by a bright red colour.

[58] *Holmann's medium, modified by Takács*
(A medium for the demonstration of obligate anaerobes.)

Grind 1 kg of horse meat which has been freed from bones, tendon and fatty tissue. Cook the ground meat in sufficient water just to cover the meat. Filter, and dry the meat. Weigh out 0·6–0·7 g of meat into test-tubes, and add broth of the following composition.

376

Disodium hydrogen phosphate:	
(Na$_2$HPO$_4$ · 12 H$_2$O)	2·0 g
Yeast extract	0·8 g
Sodium chloride	2·0 g
Glucose	1·0 g
Peptone	4·0 g
Meat extract	4·0 g
Starch (water-soluble)	1·0 g
Distilled water	1000·0 ml

Boil the water together with the ingredients, adjust the pH with N NaOH to 7·8–8·0. Filter and dispense the solution into the tubes containing the meat, 10–12 ml per tube. Sterilize by autoclaving at 121 °C for 20 minutes. Store at +4 °C.

[59] *Hugh and Leifson medium*

Peptone	2·0 g
Sodium chloride	5·0 g
Dipotassium hydrogen phosphate	0·3 g
0·2% bromthymol blue solution	15·0 ml
Agar	3·0 g
Distilled water	1000·0 ml

Dissolve the ingredients in the water by heating to 115 °C in the autoclave for 10 minutes. Adjust the pH to 7·1. Dispense the medium in 5 ml volumes into test-tubes and sterilize at 121 °C for 15 minutes. Immediately before use, melt the medium and add 0·5 ml of a 10% glucose (or other carbohydrate) solution already sterilized by filtration. Inoculate the medium and add sterilized liquid paraffin to cover.

[60] *Meat infusion*

Prepare 500 g meat (beaf, veal or horse meat) so that it is free of fat and tendon tissue, mince and grind, and then mix with 1000 ml water. Allow to stand at room temperature overnight, or at 37 °C for 1 h. Filter through a piece of linen, squeezing the residue well, heat-treat at 121 °C for 1 or 2 minutes or take briefly up to 130 °C, or steam for 30 minutes on each of three consecutive days.

[61] *Nutrient broth*

Grind 10 g peptone and 5 g sodium chloride with a little (sterile or non-sterile) meat infusion and dilute to 1000 ml with the same meat infusion. Adjust the pH to 7·2. Filter through cotton wool under vacuum. Dispense and sterilize at 121 °C for 20 minutes.

Add 2% agar to solidify the medium.

Note. Another method of preparing nutrient broth is to dissolve 10 g peptone and 4 g meat extract in 1000 ml water, adjusting the pH to 7·2. Filter and dispense as above.

[62] *Meat–glucose–yeast extract–peptone–gelatine agar*
(A medium for determination of the total viable count in foods of animal origin.)

Glucose	1·0 g
Meat extract	4·0 g
Yeast extract	2·0 g
Bacto peptone	3·0 g
Sodium chloride	2·0 g
Disodium hydrogen phosphate	2·0 g
Gelatine	10·0 g
Agar	13·0 g
Distilled water	1000·0 ml

pH 7·4 ± 0·1.

The yeast extract is prepared by suspending commercial yeast in an equal volume of distilled water. Sterilize by autoclaving at 121 °C for 20 minutes. Allow the cells to settle and use the supernatant.

Dissolve all the ingredients in 1000 ml water. Dispense and sterilize at 121 °C for 15 minutes.

[63] *Meat suspension medium*

Meat suspension

Homogenize 20 g meat in 80 ml saline. Pasteurize at 80 °C for 10 minutes and homogenize again, aseptically.

Agar substrate

Dissolve 30 g agar in 1000 ml water. Dispense in 45 ml volumes and sterilize at 120 °C for 20 minutes.

Buffer- and
nutrient solution:

Potassium dihydrogen phosphate	0·3 g
Disodium hydrogen phosphate	0·65 g
Ammonium hydrogen phosphate	0·1 g
Pyridoxine	0·1 g
Ferric chloride	0·001 g
Magnesium sulphate $(MgSO_4 \cdot 7 H_2O)$	0·05 g

Dissolve each ingredient in 45 ml water and sterilize by filtration and mix 45 ml of the solution (buffer and nutrient solutions) with 45 ml melted agar substrate and 10 ml meat suspension and pour plates 2 mm thickness.

Inoculate the surface of the plate.

[64] *Potassium cyanide broth* (Müller, 1954; Ewing and Edwars, 1960), *modified basal medium*

Peptone "Bacto proteose": Difco No. 3	3·0 g
Sodium chloride	5·0 g
Potassium dihydrogen phosphate, Sörensen	0·225 g
Disodium hydrogen phosphate, Sörensen	5·64 g
Distilled water	1000·0 ml

pH 7·6.

Sterilize the basal medium for 15 minutes at 121 °C.

Cool down it above 0 °C. Add 15 ml of a 0·5% KCN solution (sterile) to the cool medium in sterile 10 × 100 mm tubes. Seal the tubes by a cork previously immersed in heat-sterilized paraffin.

The medium can be stored in sealed tubes in the refrigerator for 2 weeks. The final dilution of the KCN is 1 : 13,300.

[65] *Starch–agar medium*
(A medium for testing for starch fermentation.)

Meat extract	3·0 g
Peptone	5·0 g
Soluble starch	2·0 g
Agar	15·0 g
Distilled water	1000·0 ml

pH when ready for use, 7·2.

Dissolve the ingredients, except for the starch, in boiling water. Add the starch in several small portions while continuously stirring. Boil for a further 3–4 minutes until the starch is dissolved. Cool the mixture to 50–60 °C and adjust the pH, taking into account that the pH should be 7·2 when the medium is in use. Sterilize at 121 °C for 10 minutes. Pour plates.

[66] *Kessler and Swenarton's gentian violet–lactose–peptone–bile broth*
(A medium for the determination of coliforms.)

Bovine bile	50·0 g
Peptone	10·0 g
Lactose	10·0 g
Distilled water	1000·0 ml

Add the bile and the peptone to 1 litre of boiling distilled water. Mix thoroughly and steam for 1 h. Dissolve 10 g lactose in the broth, adjust the pH to 7·2, and filter through cotton wool moistened with boiling water. Add 4 ml of a 1% gentian violet solution. Dispense into test-tubes each containing a Durham tube. Steam for 30 minutes on each of 3 consecutive days.

[67] *King's media*
(Media for testing for strains of *Pseudomonas*.)

Solution A:	
Peptone	20·0 g
Glycerol	10·0 g
Disodium sulphate	10·0 g
Magnesium chloride	1·4 g
Agar	20·0 g
Distilled water	1000·0 ml
Solution B:	
Peptone	20·0 g
Glycerol	10·0 g
Dipotassium hydrogen phosphate	1·5 g

Magnesium sulphate	1·5 g
Agar	20·0 g
Distilled water	1000·0 ml

Final pH 7·0–7·2.

Dissolve the ingredients in boiling water and steam for 1 h. Allow to settle, and filter if necessary. Check the pH. Dispense into flasks as desired, and sterilize by steaming for 1 h. The medium can be stored in the refrigerator for long periods.

The selectivity of King's media can be improved by adding 2·5 ml NF (nitrofurantoine) stock solution to the melted medium before use, and pouring plates after thorough mixing.

The preparation of NF stock solution is detailed in the description of broth medium [105].

[68] Klimmer's medium

Yeast agar	1000·0 ml
Bromthymol blue	10·0 ml
Trypaflavine	10·0 ml
Lactose (dissolved in 15 ml water)	10·0 g

Preparation of yeast agar:	
Peptone	10·0 g
Sodium chloride	5·0 g
Yeast extract	40·0 ml
Agar	25·0 g
Distilled water	1000·0 ml

Dissolve all the solid ingredients except for the agar, adding the agar when the other ingredients are in solution. Autoclave at 121 °C for 10 minutes. Adjust the pH to 7·2–7·6. Decant if a sediment is present.

Add the lactose solution, the bromthymol blue and the trypaflavine to the melted agar and steam for 30 minutes on each of 3 consecutive days.

Yeast extract

Suspend 100 g baker's yeast in 100 ml water and steam for 30 minutes. Maintain at 100 °C for 15 minutes. Allow to stand for 2–3 days.

Bromthymol blue solution

Dissolve 1 g bromthymol blue in 100 ml ethanol.

Dissolve 1 g trypaflavine in 100 ml distilled water. (The suspension to be tested should be diluted in saline—8·5 g sodium chloride in 1000 ml distilled water.)

[69] *Congo red stain for staining capsules*
(Gebhardt, 1970)

Solution 1:	
Congo red	1·0 g
Distilled water	100·0 g
Solution 2 (stored in a dropping bottle):	
Ferric chloride solution	4·0 ml
Glacial acetic acid	4·0 ml
Phenol (5%)	30·0 ml
1% aqueous fuchsin solution	2·0 ml

Filter the solutions a few days after preparation, using medium filter paper.

[70] *Koser's citrate medium*

Sodium ammonium phosphate	1·5 g
Potassium dihydrogen phosphate	1·0 g
Magnesium sulphate $MgSO_4 \cdot 7 H_2O$	0·2 g
Sodium citrate 2 H_2O	2·5 g
Bromthymol blue	0·016 g

Dissolve the ingredients, except for the bromthymol blue, in 990 ml distilled water. Dissolve the bromthymol blue in 10 ml distilled water, filter, and add to the medium. Mix thoroughly. Dispense 5-ml volumes into test-tubes and sterilize at 121 °C for 15 minutes. Final pH, 6·8.

[71] *Crystal violet–neutral red–bile agar*
(A modified MacConkey agar.)

This medium is similar to medium [102]. The ingredients are dissolved by boiling, the pH is adjusted, and the medium is sterilized at 121 °C for 15 minutes. It has been suggested that 5 ml water agar at a temperature of 50 °C may be poured over the

MacConkey plate to prevent spreading of colonies, especially if the medium is to be used for the enumeration of coliforms. Coli-aerogenes colonies appear dark, and are usually 1–2 mm in diameter when covered by the agar layer.

[72] *Crystal violet–blood agar*
(A medium for the purification of anaerobic bacterial cultures.)

Cool 500 ml of gelatin-basal agar to 50 °C, and add 0·2 ml of a 1% aqueous crystal violet solution and 25 ml sterile defibrinated bovine blood. Pour plates and store at +4 °C.

[73] *Corn meal agar*

Mix 12·5 ml yellow corn meal with 300 ml tap water. Place the mixture in a water bath set at 60 °C for 1 h. Filter, make up to 300 ml, add 3·8 g agar, and sterilize by autoclaving at 112 °C. Filter through adsorbent cotton wool. Dispense and autoclave at 1·5 atm pressure for 15 minutes. Plates of this medium cannot be stored, and must be poured just before used. Dry the Petri plates, uncovered and inverted at 50 °C for 30 minutes. When the surface of the medium has dried, draw a line diametrically on the plate and add a few drops of antitoxin to one half of the plate. Allow to dry.

[74] *Kurmann's medium*
(A medium for the detection of propionibacteria.)

Ammonium carbonate	10·0 g
Ammonium lactate	5·0 g
Yeast extract	10·0 g
Distilled water	1000·0 ml
Agar	10·0 g

Dissolve the ingredients and sterilize at 121 °C for 20 minutes. Final pH, 6·8–7·0.

[75] *Litmus milk*

Add to 1000 ml fresh skim milk 1 g azolitmin dissolved in 10 ml water plus a few drops of 1 N NaOH. Dispense, and steam for 20 minutes at 100 °C on each of 3 consecutive days.

[76] *Litmus milk*

Powdered skim milk	100·0 g
Litmus	5·0 g

Weigh the ingredients into a 200-ml flask. Add 1 litre distilled water while slowly stirring. Adjust the pH to 6·8. Filter through muslin or cotton wool. Dispense after sterilizing at 121 °C for 5 minutes. (The cooled medium has a pale mauve colour.)

[77] *Lactose broth*

Meat extract	2·0 g
Peptone	3·0 g
Peptone (Witte)	3·0 g
Lactose	5·0 g

Dissolve the ingredients in 1000 ml distilled water. Adjust final pH 6·7–6·9. Dispense volumes of 225 ml into flasks. Sterilize for 15 minutes at 121 °C.

[78] *Lactose–egg yolk–milk agar* (Willis and Hobbs)

Basal medium:	
Meat extract	4·0 g
Peptone	10·0 g
Sodium chloride	5·0 g
Agar	12·0 g
Distilled water	1000·0 ml
Lactose	1·2 g
Neutral red (1% aqueous solution)	3·0 ml

pH (after sterilization, and at 20 °C), 7·4.

Dissolve the ingredients, except for the lactose and the neutral red, by heating in a boiling water bath. Add the lactose and the neutral-red solution. Check the pH. Sterilize at 121 °C for 20 minutes. The basal medium can be stored at +4 °C for 4 weeks.

Preparation of sterile milk

Centrifuge cow's milk at 3000 rpm for 10 minutes and use the fat-free layers. Dispense volumes of 15 ml into test-tubes. Sterilize at 121 °C for 10 minutes. Sterile milk can be stored at +4 °C for 1 week.

Preparation of sterile egg-yolk emulsion

Wipe the shell of a fresh hen's egg with 96% ethanol and allow to dry. Open the egg and separate the white and the yolk into Petri dishes. Transfer the yolk to a sterile graduated cylinder using a wide-tipped Pasteur pipette or a 20-ml syringe. Add an equal volume of 0·85% saline and mix thoroughly. The emulsion can be stored for 1 week at +4 °C.

Antitoxin

Use *C. perfringens* A antitoxin or therapeutic polyvalent *C. perfringens* antitoxin.

Preparation of the complete medium:	
Basal medium	100·0 ml
Sterile milk	15·0 ml
Egg-yolk emulsion	3·75 ml

Melt some of the basal medium and cool to 50 °C. Add the milk and the egg-yolk emulsion. Mix well. Dispense 15-ml volumes into sterile Petri dishes.

[79] *Laurylsulphate–tryptose medium*

Tryptose, tryptone or trypticase	20·0 g
Lactose	5·0 g
Dipotassium hydrogen phosphate	2·75 g
Potassium dihydrogen phosphate	2·75 g
Sodium chloride	5·0 g
Sodium laurylsulphate	0·1 g

Dissolve the ingredients in 1 litre distilled water. Dispense 10-ml volumes into test-tubes containing Durham tubes. Sterilize by autoclaving at 121 °C for 10 minutes. Final pH, 6·8.

[80] *Littman's bile agar*

Peptone	10·0 g
Glucose	10·0 g
Bovine bile	15·0 g
Crystal violet	0·01 g
Agar	20·0 g
Distilled water	1000·0 ml

Dissolve the ingredients and autoclave the solution at 121 °C for 15 minutes. Adjust the pH to about 7·0. Cool to 50 °C and pour into Petri dishes. Allow to stand at room temperature for 6–8 h before use.

[81] *Litsky and Mallmann's medium*
(For testing for *Streptococcus faecalis*.)

Peptone	20·0 g
Glucose	15·0 g
Sodium chloride	5·0 g
Dipotassium hydrogen phosphate	2·7 g
Potassium dihydrogen phosphate	2·7 g
Sodium azide	0·4 g
Ethyl violet (0·12 g dissolved in 100 ml distilled water)	0·1 ml
Distilled water	1000·0 ml

Dissolve the salts and the peptone in the distilled water in the autoclave. Adjust the pH to 6·9–7·0. Filter and add the remaining ingredients. Dispense into test-tubes and sterilize et 121 °C for 15 minutes.

[82] *Medium for carrying out the lysine decarboxylase test* (Carlquist, 1956)

Hydrochloric-acid hydrolysate of casein	15·0 g
Dipotassium hydrogen phosphate	2·0 g
Glucose	1·0 g
Distilled water	1000·0 ml

pH 6·7–6·8.

Dispense 5 ml volumes into test-tubes and sterilize by autoclaving at 121 °C for 15 minutes.

[83] *Lugol's iodine solution*

Dissolve 1 g iodine and 2 g potassium iodide in 100 ml distilled water. Stir well and allow to stand for 24 h. Add a few more KI crystals if necessary for complete dissolution of the iodine.

[84] *MacConkey's broth*
(For the identification of coliforms.)

Peptone	20·0 g
Bile salt	
(sodium	
taurocholate)	5·0 g
Sodium chloride	5·0 g
Lactose	10·0 g
Bromcresol purple	
(1% solution in	
ethanol)	1·0 ml
Distilled water	1000·0 ml

Dissolve the peptone, the bile salt, and the sodium chloride in water and steem for 1–2 h. Dissolve the lactose while warming the solution for 15 minutes. Cool and filter. Adjust the pH to 7·4. Add the bromcresol purple solution. Dispense into test tubes containing Durham tubes. Steam for 30 minutes on each of 3 consecutive days.

Double-strength medium is similarly prepared; the ingredients are dissolved in half the volume of distilled water.

[85] *MacConkey broth*
(For the cultivation of *E. coli* I and coliforms.)

Peptone	20·0 g
Lactose	10·0 g
Sodium chloride	5·0 g
Bile salt	
(sodium	
taurocholate)	5·0 g
Distilled water	1000·0 ml

Dissolve the ingredients by boiling. Adjust the pH with N NaOH to 7·6. Filter, and add 10 ml of an aqueous 0·75% neutral red solution. Dispense 12-ml volumes into test-tubes containing Durham tubes. Sterilize by autoclaving at 121 °C for 20 minutes.

Note. If a considerable amount of precipitate is formed, it is necessary to add one tea-spoonful of active carbon 5–10 minutes after the solution has been taken off the boil. The neutral red solution should be added after thoroughly mixing the solution and filtering.

[86] *MacConkey's agar*

Peptone	17·0 g
Proteose peptone	
or polypeptone	3·0 g
Lactose	10·0 g
Bile salt No.3	1·5 g
Sodium chloride	5·0 g
Agar	13·5 g
Neutral red	0·03 g
Crystal violet	0·001 g

pH (after autoclaving) 7·0.

Suspend the ingredients in 1000 ml distilled water and mix well. Heat to boiling with continuous stirring until completely dissolved. Cool to 50–60 °C. Check the pH. Dispense as required and sterilize at 121 °C for 15 minutes.

[87] *McClary's agar*
(For sporulation tests.)

Glucose	1·0 g
Potassium chloride	1·8 g
Yeast extract	2·5 g
Sodium acetate	8·2 g
Agar	15·0 g

[88] *Liver broth (and glucose liver broth)*
(A medium for the culture of anaerobic bacteria, OXOID CM 77 and CM 78.)

Remove fatty tissue from bovine liver and cut 500 g of the liver into cubes of about 1 cm. Cover the cubes with 1 litre of distilled water and boil for 1 h. Adjust the pH with NaOH to 8·5. Filter the decoction through cheese cloth and make up with distilled water to 1 litre. Remove the solidified fat by repeated filtration and add the following ingredients to the fat-free filtrate:

Tryptone	10·0 g
K_2HPO_4	1·0 g
Starch (water-soluble)	1·0 g

Before dispensing the liver broth into tubes, fill the tubes to a depth of about 25 mm with the cooked liver pieces retained in the filter cloth.

Autoclave the test-tubes and flasks containing the medium at 121 °C for 20 and 25 minutes, respectively. Final pH, 7.

Before use, place the test-tubes containing the medium in boiling water for 20 minutes to expel oxygen. Cool quickly, taking care to minimize oxygen uptake.

(Glucose liver broth contains 10 g of glucose per litre, added to the liver broth before it is dispensed. Steam the dispensed medium for 15 minutes on each of 4 consecutive days.)

[89] *Liver broth modified by Takács*
(A medium for the detection of obligate anaerobic bacteria.)

Meat extract	20·0 g
Yeast extract	4·0 g
Peptone (Witte)	4·0 g
Glucose	3·0 g
Disodium hydrogen phosphate $Na_2HPO_4 \cdot 12H_2O$	3·0 g
Starch (water-soluble)	1·0 g
Distilled water	1000·0 ml

Cook bovine liver and cut into small cubes. Place cubes in test-tubes, 2 or 3 in each.

Dissolve the ingredients by boiling and adjust the pH to 8·0 with N NaOH. Filter through filter paper and dispense 10-ml volumes into the test-tubes containing the liver cubes. Sterilize by autoclaving at 112 °C for 40 minutes.

[90] *Malt extract broth and malt extract agar*

Malt extract (powdered)	20·0 g
Peptone	3·0 g
Distilled water	1000· ml
Agar	20·0 g

Sterilize at 115 °C for 10 minutes.

[91] *Malt extract salted agar*

1. Malt extract (powdered)	20·0 g
Agar	20·0 g
Distilled water	750·0 ml
2. Sodium chloride	74·0 g
Distilled water	300·0 ml

Sterilize solutions *1* and *2* at 121 °C for 15 minutes.
Mix and cool to about 45 °C before dispensing. Final pH 5.

[92] *Malonate broth (modified by Leifson)*

Yeast extract	0·8 g
Ammonium sulphate	2·0 g
Dipotassium hydrogen phosphate	0·6 g
Potassium dihydrogen phosphate	0·4 g
Sodium chloride	2·0 g
Sodium malonate	3·0 g
Glucose	0·25 g
Bromthymol blue indicator (1 : 500)	12·5 ml
Distilled water	1000·0 ml

pH 6·8.

Sterilize at 121 °C for 15 minutes.

Bromthymol blue indicator:	
Bromthymol blue	1·0 g
0·1 N NaOH	25·0 ml
Distilled water	475·0 ml

[93] *Mannitol–yolk (MY) agar*
(A medium for the culture of *B. cereus*.)

Meat extract	1·0 g
Peptone	10·0 g
Mannitol	10·0 g
Sodium chloride	10·0 g
Phenol red	25·0 mg
(12·5 ml of a 0·2% solution)	
Agar	18·0 g
Distilled water	900·0 or 1000·0 ml

pH (when ready for use), 7·1.

Dissolve the ingredients by warming the solution. Check the pH. Sterilize at 121 °C for 15 minutes. Cool to 49–50 °C. Add 10 ml of a 20% egg-yolk emulsion per 90 ml of the cooled preparation (see details for medium [147]). Pour plates. For the preparation of phenol red indicator, see the details for medium [41].

The medium can be made more selective by adding 10 ml Polymyxin B sulphate solution per litre of medium.

Preparation of Polymyxin B sulphate solution

Dissolve 50 mg of the antibiotic in 50 ml distilled water. Sterilize by filtration. Use freshly prepared medium whenever possible.

[94] *Methylene blue stain*

Dissolve 0·3 g methylene blue in 30 ml 96% ethanol (or industrial alcohol). Add 100 ml water.

[95] *Methyl red solution*

Methyl red	0·25 g
Ethanol (96%)	100·0 ml

[96] *Chalk agar (or glucose–chalk agar)*

Add 50 g glucose, 5 g of precipitated calcium carbonate, and 20 g agar to 1 litre of yeast infusion. (Alternatively, add 5 g of precipitated calcium carbonate to 10-fold diluted yeast autolysate.) Stir vigorously, cool to 50 °C and dispense.

Incubate cultures at 25–28 °C. A positive reaction is indicated by clearing of the opalescence of the medium due to acid production.

[97] *Medium for the demonstration of nitrate reduction and motility*

Tryptone	20·0 g
Disodium hydrogen phosphate	2·0 g
Glucose	1·0 g
Potassium nitrate	1·0 g
Agar	3·0 g
Distilled water	1000·0 ml

Melt the complete medium in a water bath. Adjust the pH to $7·2 \pm 0·1$ at 20 °C.

Transfer the medium to U-shaped tubes (200×10 mm) in 10–12-ml volumes. Sterilize at 121 °C for 15 minutes. The medium can be stored at $+4$ °C for a week. When it has been stored for 2 days or more, the medium should be boiled for 2 minutes before use.

[98] *Sodium azide–TTC agar*
(A medium for demonstrating the presence of *S. faecalis*.)

Peptone	20·0 g
Cellamin (HUMAN, Budapest)	5·0 g
Disodium hydrogen phosphate ($Na_2HPO_4 \cdot 12H_2O$)	4·0 g
Sodium azide	0·4 g
Glucose	2·0 g
Agar	18·0 g
Distilled water	1000·0 ml

Dissolve the ingredients by heating. Adjust the pH to 7·2. Sterilize at 121 °C for 15 minutes. Allow the medium to settle, and dispense the clear supernatant into flasks. Sterilize at 121 °C for 15 minutes. Add 1% triphenyltetrazolium chloride (TTC) to the medium and prepare plates. (The TTC is pre-sterilized by heating to 100 °C.)

[99] *Nessler's reagent*

Potassium iodide	7·0 g
Mercuric iodide	10·0 g
Potassium hydroxide	10·0 g
Distilled water	100·0 ml

Dissolve the potassium iodide and the mercuric iodide in 40 ml of distilled water. Dissolve the potassium hydroxide in 50 ml of distilled water. Mix the two solutions and make up with distilled water to 100 ml. Allow the precipitate to settle and use the clear supernatant.

[100] *Newman's stain*

Methylene blue	1·0 g
Ethanol 96%	54·0 ml
Tetrachloro-ethane	40·0 ml
Glacial acetic acid	6·0 ml

Mix the ethanol and the tetrachloro-ethane and heat to 70 °C. Add the methylene blue and dissolve by cautiously shaking. Cool, add the glacial acetic acid slowly and filter.

[101] *Newman and Charlett stain*

Add 0·25 g basic fuchsin and 50 ml 1% ethanol to 100 ml of Newman's stain [**100**]. The mixture can be stored for a long time.

[102] *Neutral red-bile glucose medium*

Yeast extract	3·0 g
Peptone	7·0 g
Sodium chloride	5·0 g
Bile salt	1·5 g
Glucose	10·0 g
Neutral red*	0·03 g
Crystal violet*	0·002 g
Agar	15·0 g

Add the ingredients to 1 litre of distilled water, boil with continuous stirring until all the ingredients have dissolved. The medium must not be further sterilized. Prepare plates in 15-cm Petri dishes. Store in the refrigerator if the plates are not used immediately.

[103] *Nitrate–peptone water*
(For use in the nitrate-reduction test.)

Dissolve 10 g casein–peptone and 1·0 g of potassium nitrate in 1 litre of distilled water. Sterilize at 121 °C for 15 minutes. Final pH, 7·2.

[104] *Medium for the demonstration of nitrate reduction* (Kauffmann, 1954)

Potassium nitrate (nitrite-free)	0·2 g
Peptone (Bacto)	5·0 g
Distilled water	1000·0 ml

pH 7·4.

Dispense 5-ml volumes of the medium into test-tubes and autoclave at 121 °C for 5 minutes on each of 3 consecutive days. Inoculate from agar slope culture and incubate at 37 °C for 4 days. Add 0·1 ml of a mixture of Solution A and Solution B of Griess–Ilosvay reagent [**56**] (mixed immediately beforehand).

*Filtered, 1% aqueous solution.

[105] *Nitrofurantoin (NF) broth*
(A medium for testing for *P. aeruginosa*.)

Gelysat	7·5 g
Trypticase	7·5 g
Sodium chloride	5·0 g
Distilled water	1000·0 ml

Dissolve the ingredients, by boiling if necessary. Adjust the pH to 7·2. Sterilize at 121 °C for 20 minutes, and cool to about 40 °C. Add, aseptically, 50 ml of a 0·2% NP solution, mix well, and dispense 9-ml volumes into sterile test-tubes, or 90-ml volumes into sterile flasks.

Preparation of NF stock solution

Mix 1 g NF with a few millilitres of polyethylene glycol 300 and make up with glycol to 500 ml. Store the solution in dark flasks. Do not sterilize; auto-sterilization takes place in 3 months. The solution can be stored for 6 months or longer.

[106] *Nitrogen-free basal agar medium*
(A medium for testing for nitrogen assimilation.)

Glucose	20·0 g
Potassium dihydrogen phosphate	1·0 g
Magnesium sulphate	0·5 g
Agar (washed)	20·0 g

Dissolve the ingredients in 1000 ml of distilled water and dispense in 18–20 ml amounts. Sterilize at 112 °C for 15 minutes.

[107] *Nógrády and Rodler's modified double-layered polytropic medium*

Basal medium:	
Yeast extract	4·0 g
Meat extract	4·0 g
Purified Richter peptone solution	100·0 ml
Agar	20·0 g
Distilled water	1000·0 ml

Dissolve the ingredients by heating. Adjust the pH to 7·6. Dispense 200-ml volumes into 250-ml flasks and sterilize by autoclaving at 121 °C for 20 minutes. The basal medium can be stored for 7–10 days at room temperature.

Purified Richter peptone solution

Richter peptone	100·0 g
Distilled water	500·0 ml
40% barium chloride solution	25·0 ml

Dissolve the peptone in the water by boiling, and add the barium chloride to the boiling solution. Filter through cotton wool.

The bottom layer of the medium

Sterilized basal medium	100·0 ml
2% sodium thiosulphate solution ($Na_2S_2O_3 \cdot 5H_2O$)	10·0 ml
Ferrous (II) ammonium sulphate ($Fe/NH_4/_2/SO_4/_2 \cdot \cdot 6H_2O$)	0·5 g

Mix the ingredients together and place the mixture in a boiling water bath for 20 minutes. Dispense 3-ml volumes into sterile test-tubes with plugs. Let them solidify.

The upper layer of the medium

Sterilized basal medium	200·0 ml
Lactose	2·0 g
Sucrose	0·2 g
Glucose	0·1 g
Phenol-red solution	1·0 ml
Urea solution, 50%	4·0 ml

Phenol-red solution	
Phenol-red	0·5 g
N NaOH	2·0 ml
N HCl	1·5 ml
Distilled water	1000·0 ml

Allow to stand for 24 h, filter through a G5 glass fibre filter, and store at room temperature.

Add the ingredients aseptically to the melted basal medium. Mix well and hold in boiling waterbath for 20 minutes, cool at 50 °C and add the urea solution to it. Then pour the medium 3 ml volumes onto the bottom layer, but making a slope agar deep part of which is at least 10 mm. (So the upper layer consists of two parts: a deep or butt region and a slope region.) Store at 4 °C for 7–10 days.

[108] *ONPG (β-galactosidase) reagent*

ONPG (o-nitrophenyl β-D-galacto- pyranoside)	80·0 mg
Sterile distilled water	15·0 ml
1/15 M phosphate buffer, pH 7·0	5·0 ml

Take up the solution on strips of filter paper* and hang up the strips in a clean room. Do not sterilize them before use. Cut out pieces the same size as standard antibiotic discs, and store in a dark cool place in a desiccator.

Preparation of 1/15 M phosphate buffer

Solution A	
KH_2PO_4	9·078 g
Distilled water	1000·0 ml
Solution B	
Na_2HPO_4	11·876 g
Distilled water	1000·0 ml

A mixture made up of 39% solution A and 61% solution B has a pH of 7·0.

[109] *Peptone solution, for use as a diluent*

Dissolve 1 g of peptone in 1 litre of distilled water and adjust the pH to $7·0 \pm 0·1$. Dispense the solution into test-tubes or flasks used for preparing dilutions. The volume of the diluent should be the same as the volume required for the dilution, $\pm 2\%$, after sterilization at 121 °C for 15 minutes. If the deviation in volume is greater than 2%, add diluent aseptically if necessary.

[110] *Peptone water*

Dissolve 10 g of peptone and 5 g of NaCl in 1000 ml distilled water by steaming. Adjust the pH to 7·2. Dispense, and sterilize at 121 °C for 15 minutes.

---

*Macherey–Nagel No. 214. Cut strips 4 cm wide by 50 cm long from each filter paper. A strip of this size takes up 20 ml of the reagent; the paper is bright-white, containing practically no chemical impurities. Sterility cannot be guaranteed, but a high degree of purity is ensured; the use of forceps is essential. A slightly yellowish paper can be used if it does not stain the distilled water.

[111] *Peptone water, buffered*

Peptone	10·0 g
Sodium chloride	5·0 g
Disodium hydrogen phosphate ($Na_2HPO_4 \cdot 12 H_2O$)	9·0 g
Potassium dihydrogen phosphate	1·5 g
Distilled water	1000·0 ml

pH (after sterilization and at 20 °C) 7·0 ± 0·1.

Dissolve the ingredients in boiling water. Check the pH. Dispense 225-ml volumes into 500-ml flasks. Sterilize at 121 °C for 20 minutes.

[112] *Polónyi's milk agar*

Nutrient agar	75·0 ml
Sterile skim milk (mixed at 45 °C)	25·0 ml

Nutrient agar

Peptone (Witte)	10·0 g
Agar	20·0 g
Sodium chloride	3·0 g
Disodium hydrogen phosphate ($Na_2HPO_4 \cdot 12 H_2O$)	2·0 g
Meat infusion*	1000·0 ml

Soak the agar overnight and then add the NaCl and the disodium hydrogen phosphate. Dissolve by warming in a water bath. Adjust the pH to 7·5–7·6. Filter and dispense the filtrate as required. Autoclave until a pressure of 2 atm (121 °C) is reached.

Preparation of skim milk

Dispense 100-ml volumes of skim milk (acidity not less than pH 6·0) into flasks. Sterilize by steaming for 30 minutes on each of three consecutive days.

Before use, melt the nutrient agar in flowing steam and cool to 45 °C. Heat the skim milk to the same temperature and mix well with the nutrient agar.

* Meat infusion may be substituted by 3 g of Liebig's meat extract dissolved in 1000 ml distilled water, or by 7 g of Yestor pulp dissolved in 1000 ml of distilled water. In the latter case, NaCl is omitted.

[113] *Proteose–peptone (PP) media*
(For the examination of *S. faecalis*.)

Basal medium:		
Peptone, proteose	10·0	g
Glucose	1·0	g
Sodium chloride	5·0	g
Thiamine hydrochloride	0.005	g
Distilled water	1000·0	ml

pH (after sterilization) 6·9–7·0.

Dissolve the ingredients by boiling. Cool to 50–60 °C and check the pH. Dispense into flasks as required. Sterilize at 121 °C for 15 minutes.

a) *PP agar slopes*
Add 15 g of agar to 1 litre of basal medium. Sterilize at 121 °C for 15 minutes and dispense into test-tubes. Allow the medium to gel in a sloping position.

b) *PP 40% bile broth*
Add 400 g of bovine bile to 1 litre of basal medium and adjust the pH to 6·9–7·0. Dispense 5-ml volumes into test-tubes. Sterilize for 121 °C for 15 minutes.

c) *6·5% sodium chloride PP broth*
Add sodium chloride to the basal medium at 65 g per litre. Dispense, and sterilize at 121 °C for 15 minutes.

d) *PP broth, pH 9·6*
Adjust the pH of the basal medium to 9·6. Dispense 5-ml volumes into tubes. Sterilize at 121 °C for 15 minutes.

e) *TTC–PP agar*
Add 15 g of agar and 0·1 g of triphenyltetrazolium chloride per litre of basal medium. Sterilize at 121 °C for 15 minutes. Pour plates or prepare agar slopes.

f) *PP agar containing 0·05% tellurite*
Add 0·5 g of potassium tellurite to 1 litre of the basal medium. Sterilize at 121 °C for 15 minutes. Prepare plates or agar slopes in tubes.

[114] *Resazurin solution*
(For the resazurin reduction test.)

Dissolve 1 tablet containing 2·5 mg of resazurin in 50 ml pre-boiled water. Keep in brown bottles. Use the freshly-prepared solution.

398

[115] *Ringer's solution*

Sodium chloride	9·00 g
Potassium chloride	0·42 g
Calcium chloride	
($CaCl_2 \cdot 2 H_2O$)	0·32 g
Sodium hydrogen	
carbonate	0·20 g
Distilled water	1000·0 ml

To prepare quarter-strength Ringer's solution, dilute one volume of the solution with three volumes of distilled water. Sterilize at 121 °C for 20 minutes.

[116] *Robertson's cooked meat medium*

Cut 500 g of fresh, fat-free bovine heart into small pieces and simmer for 20 minutes in 500 ml boiling 0·05 N NaOH. After cooking, adjust the pH to 7·4. Strain off the liquid and drain the meat pieces on filter paper. Place a few pieces (1–2-cm cubes) in test-tubes and add 10-ml of peptone water [103] or nutrient broth [138]. Sterilize by autoclaving at 121 °C for 20 minutes.

[117] *Robertson's modified medium*
(A medium for testing for *S. aureus*.)

Preparation of boiled meat:	
Bovine heart, cleaned and cut into	
pieces	1000·0 g
Sodium hydroxide, 0·05 N	1000·0 ml

Pour the NaOH solution into a stainless steel or intact enamelled container. Add the minced bovine heart. Mix cautiously and simmer for 20 minutes with frequent stirring. Adjust the pH to 7·5.

Remove superfluous liquid by straining through a dense gauze and squeezing the meat if necessary. The pieces are further dried on filter paper.

Preparation of the complete medium

Place the dried meat pieces to a depth of about 2·5 cm, in test-tubes of the required size. Add a medium of the following composition (sufficient to form a liquid column about 4·5 cm in height):

Beef extract "Lab-Lemco"	10·0 g
Peptone	10·0 g
Sodium chloride	100·0 g
Distilled water	1000·0 ml

Dissolve the ingredients by warming, adjust the pH to 7·5, and add the liquid medium to the meat as above. Sterilize at 121 °C for 20 minutes.

[118] *Thiocyanate enrichment medium*
(A medium used in testing for *S. aureus*.)

Add 4 g of potassium thiocyanate (or sodium thiocyanate) per 100 ml sterile nutrient broth **[32]**. Warm the broth to 70 °C, dissolve the thiocyanate and dispense 9-ml volumes into test-tubes. Sterilize at 121 °C for 20 minutes.

The medium can be stored at room temperature for 3–4 days or in the refrigerator for a longer time.

[119] *Rogosa's agar*
(For the isolation and enumeration of lactobacilli.)

Casein peptone or tryptone	10·0 g
Yeast extract	5·0 g
Glucose	20·0 g
Tween 80	1·0 g
KH_2PO_4	6·0 g
Ammonium citrate	2·0 g
Sodium acetate	
(containing water of crystallization)	25·0 g
Glacial acetic acid	1·32 ml
$MgSO_4 \cdot 7 H_2O$	0·575 g
$MnSO_4 \cdot 2 H_2O$	0·12 g
$FeSO_4 \cdot 7 H_2O$	0·034 g
Agar	15·0 g
Distilled water	1000·0 ml

Final pH 5·4.

Add all the ingredients to the water and dissolve by steaming in the autoclave. Dispense into sterile vessels and steam for a further 50 minutes. Store in the refrigerator.

Modified Rogosa agar

Use trypsin-treated milk instead of trypticase, and increase the overall concentration of the medium by 10%.

[120] *Simmon's citrate agar*

Magnesium sulphate ($MgSO_4 \cdot 7 H_2O$)	0·2 g
Ammonium dihydrogen phosphate	1·0 g
Dipotassium hydrogen phosphate	1·0 g
Sodium citrate (dehydrated)	2·0 g
Sodium chloride	5·0 g
Bromthymol blue	0.08 g
Agar	25·0 g

Final pH 6·8–7·0.

Add the ingredients to 1000 ml of distilled water, and dissolve by boiling. Dispense in 10-ml volumes. Sterilize at 121 °C for 15 minutes. Set as short agar slopes.

[121] *Medium of Slanetz and Bartley*
(A medium for the detection and determination of enterococci in stuffed meat preparations.)

Basal broth:	
Peptone	10·0 g
Meat extract	3·0 g
Yeast extract	5·0 ml
Tween 80	0·5 ml
Ethyl violet	0·00012 g
Distilled water	1000·0 ml

pH 6·2.

Add the ingredients to the water. Sterilize at 1·5 atm for 20 minutes.

Additives

Glucose (10% aqueous solution)	10·0 ml
Sodium carbonate (10% aqueous solution)	4·0 ml
Sodium azide (10% aqueous solution)	0·8 ml

Sodium taurocholate	
(10% aqueous solution)	2·5 ml
Bromcresol purple	
(0.4% ethanolic solution)	2·0 ml

Dispense the basal broth in 180-ml volumes. Add the additives except for the indicator solution. Sterilize by steaming on each of three consecutive days.

[122] *Salt agar*

Ammonium nitrate	1·0 g
Ammonium sulphate	1·0 g
Dipotassium hydrogen phosphate	4·0 g
Potassium dihydrogen phosphate	2·0 g
Sodium chloride	1·0 g
Glucose	10·0 g
Yeast extract	1·0 g
Agar	15·0 g
Distilled water	1000·0 ml

Dissolve the ingredients. Dispense, and sterilize at 115 °C for 15 minutes. Before use, melt and cool to 70 °C, adding 57 ml of a sterile 10% citric acid solution; pH 3·5.

[123] *Salted agar*
(A medium for testing for *S. aureus*.)

Add 65 g sodium chloride, 10 g mannitol, and 10 ml of a 2% phenol-red solution to 1 litre of melted gelatin basal agar [166]. Dispense as required and sterilize at 121 °C for 20 minutes. Pour plates or store the medium (for 7–10 days at room temperature, or for a longer time in the refrigerator).

The phenol-red and the mannitol may be omitted.

[124] *Salted milk agar*
(A medium for testing for *S. aureus*.)

Basal medium:	
Meat extract	3·0 g
Peptone	5·0 g
Sodium chloride	65·0 g
Agar	15·0 g
Distilled water	1000·0 ml

pH (after sterilization) 7·4.

Dissolve the ingredients in the water by boiling with frequent stirring. Cool to 50–60 °C and check the pH. Dispense into flasks as usual and sterilize at 121 °C for 15 minutes.

Preparation of the complete medium

Melt the basal medium and cool to 50 °C. Add 10 ml of sterile milk per 100 ml of the medium. Mix well and pour plates. The complete medium can be used immediately or it can be stored in the refrigerator for 2–3 days.

Preparation of the sterile milk

Centrifuge cow's milk (acidity not less than pH 6·4) at 2500–3000 rpm. Remove the fat from the surface and filter the skim milk through cotton wool. Dispense as required into test tubes or flasks. Sterilize for 1 h on each of 3 consecutive days by steaming or in a boiling water bath. The sterile milk thus prepared can be stored for up to 1 month.

[125] *Salted blood agar*
(A medium for testing for *S. aureus*.)

Melt 1 litre of medium [123], cool to 45–48 °C and add 50 ml sterile, defibrinated, bovine blood, moving the vessel continuously as the blood is added. Pour plates, measuring 20 ml into each Petri dish. The medium can be stored in the refrigerator for 2 days. Plates showing no haemolysis, no darkening, and no bacterial contamination, may be used beyond 2 days.

[126] *Malt extract agar*

Malt extract (brown malt)	900·0 ml
Tap water	720·0 ml
Agar	25·0 g

Mix the ingredients and melt the agar by steaming. Filter the medium through a moistened cotton wool filter. Dispense into flasks, or transfer 10-ml volumes to test-tubes. Plug with cotton wool, steam for 30 minutes on each of 3 consecutive days.

To examine a sample by inoculation of malt extract, place in a Petri dish 0·5 ml of a sterile 5% lactic acid solution (to adjust the pH) together with the sample, and then add 10–11 ml of malt agar from a test-tube.

[127] *Streptococcus enriching (SE) medium*
(A medium used in testing for *S. faecalis*.)

Tryptone	20·0 g
Yeast extract	5·0 g

Dried bovine bile	10·0 g
Sodium chloride	5·0 g
Sodium citrate	1·0 g
Aesculin	1·0 g
Ferric ammonium citrate	0·5 g
Sodium azide	0·25 g
Distilled water	1000·0 ml

Dissolve the ingredients by heating. Adjust to pH 7·0 and dispense 9-ml volumes into test-tubes. Sterilize at 121 °C for 20 minutes.

[128] *Streptococcus confirmatory (SC) agar medium*

The composition of this medium is the same as that of medium **[127]**, except that the SC medium contains 15 g agar. Complete dissolution of the ingredients requires continuous boiling for 15 minutes. Sterilize at 121 °C for 20 minutes. Caution is required while pouring plates to avoid contamination. Final pH 7·0.

[129] *Selenite–brilliant green broth*
(Stokes–Osborn)

Basal broth:	
Peptone	5·0 g
Yeast extract	5·0 g
Sodium taurocholate	1·0 g
Mannitol	5·0 g
Sodium hydrogen selenite	4·0 g
Distilled water	900·0 ml

Dissolve the first 4 ingredients by heating in a boiling water bath for 5 minutes. Cool and add the selenite. Adjust the pH to $7· \pm 0·1$ at 20 °C. Store at $+4$ °C in the dark for not longer than a week.

Buffer solution:	
Solution A	
Potassium dihydrogen phosphate	34·0 g
Distilled water	1000·0 ml
Solution B	
Dipotassium hydrogen phosphate	43·6 g
Distilled water	1000·0 ml

Mix 2 volumes of solution A with 3 volumes of solution B. Adjust the pH to $7·0 \pm 0·1$ at 20 °C.

Brilliant-green solution [146]
Composition of the Stokes–Osborn medium

Basal broth	900·0 ml
Buffer solution	100·0 ml
Brilliant-green solution	1·0 ml

Add the buffer to the basal broth, heat to 80 °C, cool, and add the brilliant-green solution. Dispense 100-ml volumes of the complete medium into 500-ml sterile flasks. Use the medium on the day of preparation.

[130] *Selenite–cystine enrichment medium*

Selenite basal broth:

Tryptose	5·0 g
Lactose	4·0 g
Disodium hydrogen phosphate	
$(Na_2HPO_4 \cdot 12 H_2O)$	10·0 g

Dissolve 4 g of sodium hydrogen selenite in 900 ml of distilled water by heating and dissolve the tryptose, the lactose and the disodium hydrogen phosphate in this solution. Final pH is approx. 7·0.

L-cystine solution:

L-cystine	0·1 g
N NaOH	15·0 ml
Distilled (to make up to 100 ml)	

Dissolve the L-cystine in the N NaOH and make up to the final volume. Filter the solution through a bacteriological filter, e.g. a G5 glass filter.

Composition of the complete medium

Dispense 225-ml volumes of the basal broth into flasks, or transfer 10-ml volumes to test-tubes. Maintain at 100 °C for 10 minutes. Cool, and add L-cystine solution, 1 volume per 100 volumes of the basal medium.

Note. Use the complete medium on the day of preparation.

[131] *Basal agar carbon-free medium*
(A medium for testing for sugar assimilation.)

Ammonium sulphate	5·0 g
Potassium hydrogen phosphate	1·0 g

Magnesium sulphate	0·5 g
Agar (washed)	20·0 g

Dissolve the ingredients in 1000 ml of distilled water. Dispense 18–20 ml volumes into flasks. Autoclave at 1·5 atm for 15 minutes.

Preparation of washed agar

Soak 20 g agar in 1 litre of distilled water. Change the water daily for 8 days, then rinse the agar several times.

[132] *Medium for examining the decomposition of organic acids* (according to Kauffmann, 1961)

Basal medium:
Bacto peptone (Difco)	10·0 g
NaCl	5·0 g
Distilled water	1000·0 ml

Add 12 ml of 1 : 500-diluted bromthymol blue as an indicator. Prepare the indicator solution as described in [92].

[133] *Sulphite agar*
(OXOID CM79 and CM 80)
(A deep agar medium for the enumeration of thermophilic anaerobe bacteria producing hydrogen sulphide.)

Tryptone	10·0 g
Sodium sulphite	1·0 g
Agar	20·0 g
Distilled water	1000·0 ml

Place an iron wire or an iron nail in each test-tube of the medium. The wires of nails should be cleaned in diluted hydrochloric acid and thoroughly rinsed before use. The wire or nail may be substituted by 10 ml of 5% ferric citrate.

It has been suggested that 5 g of sodium thioglycolate, or 10 ml of a 5% solution of this should be added to the medium. Adjustment of pH is not essential.

Sterilize at 121 °C for 20 minutes. Store for up to a week.

[134] *Sulphite–polymyxin medium (Mossel–De Waart)*

Basal medium:
Meat extract	3·0 g
Pancreatin-digested casein	15·0 g

Yeast extract	10·0 g
Water	1000·0 ml

pH (after sterilization and at 20 °C) 7·0 ± 0·1.

Add the ingredients to the water and heat to boiling. Check the pH.
Dispense 20-ml volumes of the basal medium into 20×200-mm test-tubes. Sterilize at 121 °C for 20 minutes. Store in the refrigerator for up to 4 weeks.

Solutions
Ferrous sulphate solution

Ferrous sulphate ($FeSO_4 \cdot 7 H_2O$)	0·4 g
Water	100·0 ml

Dissolve the iron sulphate in the water. Sterilize by filtration. Store the solution in the refrigerator for up to a week at 4 °C.

Sodium sulphite solution

Sodium sulphite ($Na_2SO_3 \cdot 7 H_2O$)	1·0 g
Water	100·0 ml

Dissolve and sterilize by filtration. Do not store the solution.

Polymyxin B-sulphate solution

Polymyxin B sulphate (not less than 6000 IU per 1 mg)	0·050 g
Sterile water	50·0 ml

Dissolve the Polymyxin B in the water in a sterilized flask. The solution can be stored in the refrigerator for up to a week.

Composition of the medium

Basal medium	20·0 ml
Ferrous sulphate solution	1·0 ml
Sodium sulphite solution	1·0 ml
Polymyxin B sulphate solution	0·2 ml

Preparation

Dispense 20-ml or 10-ml volumes of the basal medium into test-tubes. Add immediately before use the ferrous sulphate solution, the sodium sulphite solution, and the Polymyxin B sulphate solution. To avoid precipitation, do not mix the ingredients together before adding to the basal medium.

407

[135] *Sulphite–Polymyxin–sulphadiazine agar*

Tryptone	15·0 g
Yeast extract	10·0 g
Ferric citrate	0·5 g
Agar	15·0 g
Distilled water	1000·0 ml

Dissolve the ingredients by steaming. Adjust the pH to 7·0. Dispense 100-ml volumes of the medium into flasks or screw-capped bottles. Sterilize at 121 °C for 15 minutes. Before pouring plates melt the medium, cool to 45 °C, and add aseptically the following solutions (each pre-sterilized by filtration) per 100 ml of the medium.

Freshly-prepared	
10% sodium sulphite	0·5 ml
0·1% Polymyxin B solution	1·0 ml
1·2% sodium sulphadiazine	
solution	1·0 ml

Mix well and pour plates.

[136] *Semi-liquid agar medium for testing for sulphite reduction (Takács)*
(A medium for demonstrating the motility and sulphite-reducing ability of obligate and facultative anaerobes.)

Meat extract	4·0 g
Peptone	5·0 g
Yeast extract	8·0 g
Glucose	1·0 g
Agar	1·8 g
Distilled water	1000·0 ml

Dissolve the ingredients by boiling. Adjust the pH to $7·2 \pm 0·1$ by adding 0·1 N NaOH. Filter, and dispense 1000-ml volumes into flasks. Sterilize by autoclaving at 121 °C for 20 minutes. Store at room temperature. Immediately before use, melt the medium in a boiling waterbath and add 20 ml of the sterile supplementary solution.

Supplementary solution	
Ferric citrate (Fe $C_6H_5O_7$)	5·0 g
Sodium sulphite ($Na_2SO_3 \cdot 7H_2O$)	10·0 g
Distilled water (to make up to 200·0 ml)	

Heat the ingredients in a water bath to dissolve. Filter through a G5 glass filter. Add the freshly-prepared solution to the basal medium and dispense aseptically into plugged sterile test-tubes, 10–12 ml per tube.

[137] *Egg-yolk agar for the sulphite-reduction test*

Melt 865 ml of sterile gelatine agar, pH 7·6–7·7 [166], by boiling. Cool to 50 °C. Add a mixture of 50 ml egg yolk, 50 ml of saline, and 25 ml of basal broth [32] in which 0·40 g of D-cycloserine bitartarate, 1·0 g ferric citrate ($FeC_6H_5O_7$) and 2·0 g sodium sulphite ($Na_2SO_3 \cdot 7H_2O$) has been dissolved. Mix by circular movements of the vessel, taking care to avoid foaming. Pour plate, about 25 ml per Petri dish. Incubate at 37 °C for 24 h as a test for sterility, and store in the refrigerator. Care must be taken to ensure that the cycloserine concentration in this medium is 400 μg per ml exactly.

[138] *Nutrient broth*
(OXOID CM67, CM68)

Meat extract ("Lab-Lemco")	10·0 g
Peptone	10·0 g
Sodium chloride	5·0 g
Agar	15·0 g

Dissolve the ingredients in 1 litre of distilled water. Check the pH. Dispense, and sterilize by autoclaving at 121 °C for 15 minutes. pH (after sterilization) 7·5.

[139] *Nutrient gelatine for the enumeration of microcolonies*

Gelatine	15·0 g
Tryptone	1·0 g
Yeast extract	0·5 g
Glucose	0·2 g

Mix the ingredients, dissolve in 100 ml water and sterilize by autoclaving at 121 °C for 15 minutes. Cool quickly and store in the refrigerator.

[140] *Thronley's semi-solid arginine medium*

Peptone	0·1 g
Sodium chloride	0·5 g
Dipotassium hydrogen phosphate	0·03 g
Arginine HCl	1·0 g
Phenol red	0·001 g
Distilled water	100·0 ml
Agar	0·3 g

pH 7·2.

Sterilize at 121 °C for 15 minutes.

[141] *Sterile milk*

Skim milk with a low microbe count to use. Dispense 10-ml volumes amounts into test-tubes and plug with cotton wool. Sterilize by steaming at 100 °C for 15 minutes on each of 3 consecutive days.

[142] *Lactose–phenol-red nutrient solution*
(A medium for water analysis.)

Yeast broth	1000·0 ml
Lactose	15·0 g
Phenol-red indicator [41]	75·0 ml

Dispense volumes of 2·5 ml, 5 ml, and 25 ml into test-tubes, and into flasks, resp., calibrated to 25 ml each containing a Durham tube. Sterilize by autoclaving at 100 °C for 30 minutes.

Yeast broth:

Tap water	1000·0 ml
Peptone	15·0 g
Sodium chloride	7·5 g
Yeast extract	3·0 g

Dissolve and autoclave for half an hour. Adjust the pH to 7·3–7·4 at 30–40 °C. Filter through a fluted filter paper. Dispense, and sterilize by autoclaving at 121 °C for 15 minutes. Store in the refrigerator.

[143] *Lactose broth*
(OXOID CM137 and CM139)

Meat extract	3·0 g
Peptone	5·0 g
Lactose	5·0 g
Distilled water	1000·0 ml

pH (after sterilization) 6·4–7·0.

Heat slightly to dissolve the components and distribute into tubes filled with Durham tubes. Sterilize at 121 °C for 15 minutes.

[144] *Lactose-broth agar*

Liebig's meat extract	3·0 g
Peptone (Witte)	5·0 g

Lactose	10·0 g
Sodium chloride	5·0 g
Agar	20·0 g
Distilled water	1000·0 ml

Mix the ingredients, except for the lactose, and steam until dissolved. Adjust the pH, using bromthymol blue indicator, to 6·8 (grass-green colour). Sterilize for 15 minutes at 2 atm pressure. Check and, if necessary, adjust the pH. Filter the hot medium through cotton wool moistened with boiling water. Add the lactose. Dispense the medium into test-tubes or flasks. Sterilize by steaming at 100 °C for 30 minutes on each of 3 consecutive days.

[145] *Thermoacidurans agar*
(National Canners Association, 1968)
(A selective medium for cultivating some of the acid-tolerant bacteria, especially *Bacillus coagulans (B. thermoacidurans)*.)

Solution 1

Proteose–peptone (Bacto)	5·0 g
Yeast extract	5·0 g
Glucose	5·0 g
Dipotassium hydrogen phosphate	4·0 g
Distilled water	500·0 ml

Adjust the pH to 5·0 with HCl.

Solution 2

Agar	20·0 g
Distilled water	500·0 ml

Sterilize Solution 1 and Solution 2 separately at 121 °C for 25 minutes. Melt and mix immediately before use.

[146] *Tetrathionate broth (Müller–Kauffmann)*

Basal medium:

Meat extract	2·0 g
Peptone	10·0 g
Sodium chloride	3·0 g
Calcium carbonate	45·0 g
Distilled water	1000·0 ml

pH (after sterilization and at 20 °C) $7·0 \pm 0·1$.

Add the ingredients of the basal medium to the water and dissolve by boiling. Check the pH. Sterilize at 121 °C for 20 minutes.

Sodium thiosulphate solution

Sodium thiosulphate ($Na_2S_2O_3 \cdot 5H_2O$)	50·0 g
Distilled water (to make up to 100·0 ml)	

Dissolve the thiosulphate in a little water and make up to 100 ml. Sterilize at 121 °C for 20 minutes.

Iodine solution

Iodine (sublimated)	20·0 g
Potassium iodide	25·0 g
Distilled water (to make up to 100·0 ml)	

Dissolve the potassium in a little water, add the iodine, and shake gently until the iodine is completely dissolved. Make up to 100 ml. Keep the solution in a tightly-closed dark bottle, and protect from the light.

Brilliant-green solution

Brilliant green	0·5 g
Distilled water	100·0 ml

Dissolve the brilliant green in the water. Allow to stand for at least a day in the dark so that auto-sterilization can take place.

Bovine bile solution

Bovine bile (dried)	10·0 g
Distilled water	100·0 ml

Dissolve the dried bile by boiling. Sterilize at 121 °C for 20 minutes.

Composition of the Müller–Kauffmann medium

Basal broth	900·0 ml
Sodium thiosulphate solution	100·0 ml
Iodine solution	20·0 ml
Brilliant-green solution	2·0 ml
Bovine bile solution	50·0 ml

Add the supplementary solutions to the basal broth in the order given above. Shake vigorously. Dispense 100-ml volumes aseptically into sterile 500-ml flasks.

Store in the refrigerator for up to a week after preparation.

[147] *Egg-yolk agar medium*
(A medium for the lipase reaction.)

Soak 4 fresh eggs in 96% ethanol (sufficient to cover the eggs). Transfer 50 ml of sterile saline aseptically into a sterile 250-ml flask. Take the eggs one by one, ignite the ethanol, and break open the top of the egg with a sterile scalpel and sterile forceps. Make only a small opening, and flame around the edges. Separate the egg-white by rotating the egg. Flame again, and transfer the yolk with the aid of a sterile pipette into the flask containing 50 ml saline. Mix by circular movement of the flask, taking care to prevent frothing. Add the yolk emulsion aseptically to 1000 ml of sterile gelatin basal agar [166], melted in a water bath and cooled to 50 °C. Final pH 7·6–7·7. Pour plates, 20 ml per Petri dish. Incubate at 37 °C for 24 h as a test for sterility, and store in the refrigerator.

[148] *Toluidine blue (TB)–deoxyribonucleic acid (DNA) agar (TDA)*
(A medium for the identification of *S. aureus*.)

Agar (granulated, Difco)	10·0 g
Calcium chloride, 0·01 m	1·0 ml
Sodium chloride	10·0 g
Toluidine blue, 0·1 M	3·0 ml
DNA, Difco	0·3 g
TRIS buffer, 0·05 M, pH 9·0	1000·0 ml

Add the ingredients, except for the toluidine blue, to the buffer solution. When the ingredients are completely dissolved, add the TB solution and mix well. Dispense the TDA medium into rubber-stoppered tubes and store at room temperature. Melt the mixture before use.

The TDA medium can be stored at room temperature for 4 days, without the need for pre-sterilization. It can be melted and solidified repeatedly. TB inhibits the growth of Gram-negative spore-formers. The TB–DNA complex is thermostable.

[149] *Tributyrin agar*

Add 5 drops of tributyrin to 100 ml of melted broth agar just before pouring plates. The medium should be uniformly turbid owing to the emulsified fat droplets.

[150] *Trypsin-treated milk*

Add 5 g trypsin and 10 ml chloroform to 1000 ml skim milk or reconstituted skim milk powder. Steam, and allow to stand at 37 °C for 24 hours, then steam for 20 minutes and filter before cooling. Adjust the pH with glacial acetic acid to 6·65 ± 0·02.

Nutrient solution

Yeast extract (Difco)	6·0 g
Diammonium hydrogen citrate	2·4 g
Potassium dihydrogen phosphate	7·2 g
Glucose	24·0 g
Tween 80	1·2 g
Saline	6·0 ml
Distilled water	100·0 ml

Saline

Magnesium sulphate ($MgSO_4 \cdot 7H_2O$)	11·5 g
Manganese sulphate ($MnSO_4 \cdot 4H_2O$)	2·8 g
Ferrous sulphate ($FeSO_4 \cdot 7H_2O$)	0·08 g
Distilled water	100·0 ml

Dissolve the ingredients by mild heating. Add 60 ml of sodium acetate–acetic acid buffer, pH 5·37 ± 0·02. Add distilled water to make up to 200 ml. Final pH 5·0. Do not sterilize. Keep in the refrigerator until use.

Preparation of the complete medium

Dissolve 19 g of agar in 700 ml trypsin-treated milk by autoclaving at 121 °C for 20 minutes. Add the nutrient solution (pre-warmed to 50 °C) to the boiling agar mixture. Make up to 1000 ml with trypsin-treated milk.

Dispense 10-ml amounts of the medium and store in the refrigerator. Note that overheating may lead to the formation of a black precipitate.

[151] *Trypticase–peptone–glucose–yeast extract broth with added sterile trypsin (TPGY broth)*

Trypticase	50·0 g
Bacto peptone	5·0 g
Yeast extract	20·0 g
Glucose	4·0 g
Sodium thioglycolate	1·0 g
Distilled water	1000·0 ml

Dissolve the ingredients and dispense in quantities ready for use (15-ml amounts in 20-ml test-tubes). Final pH 7·0 ± 0·1. Sterilize at 121 °C for 6–12 minutes. Use the medium within 2 weeks of preparation.

414

Prepare a 1·5% trypsin solution (Difco 1: 250). Sterilize by filtering through a "Millipore" or similar filter. Add the trypsin solution to the TPGY medium and cool to room temperature; the amount of sterilized trypsin solution added is that required to give a final concentration of 0·1% trypsin. Mix cautiously to avoid bubbling. Do not store the medium after the trypsin has been added.

[152] *Trypticase–soybean–salt broth*
(A medium for testing for *S. aureus*.)

Trypticase	17·0 g
Phytone	3·0 g
Sodium chloride	100·0 g
Dipotassium hydrogen phosphate	2·5 g
Glucose	2·5 g
Distilled water	1000·0 ml

Dissolve the ingredients by heating in the water. Dispense 9-ml volumes into test-tubes. Final pH 7·2. Phytone is a plant peptone prepared by papain digestion.

[153] *Tryptophan broth, modified by Ljutov*
(A medium for carrying out the indole test.)

Tryptose	10·0 g
Sodium chloride	5·0 g
Dl-tryptophan	1·0 g
Distilled water	500·0 ml

Dissolve the ingredients in boiling water and filter. Adjust the pH to 7·2–7·4. Dispense into short length test-tubes. Sterilize by autoclaving at 121 °C for 20 minutes.

Kovács's reagent
(for the demonstration of indole production)

p-dimethylaminobenzaldehyde	5·0 g
Amyl alcohol (tertiary)	75·0 ml
Hydrochloric acid	
(specific weight 1·19)	25·0 ml

Dissolve the p-dimethylaminobenzaldehyde in the amyl alcohol in a water bath set at 50–60 °C; use a glass-stoppered dark bottle. Cool, and add the hydrochloric acid slowly, with continuous stirring. The final reagent is pale yellow.

The reagent can be stored for a long time if it is kept away from light.

[154] *Tryptone–glucose–yeast-extract agar (TGE agar, Plate Count Agar)*
(A medium for the determination of the total microbial count.)

Tryptone	5·0 g
Yeast extract	2·5 g
Glucose	1·0 g
Agar	15·0 g
Distilled water	1000·0 ml

Dissolve the ingredients in the water by heating. Adjust the pH to 7·0. Dispense, and sterilize at 121 °C for 15 minutes.

[155] *Tryptone–glucose–meat extract–skim milk agar (TGEM agar)*
(A medium for the determination of the total microbial count.)

Tryptone	5·0 g
Glucose	1·0 g
Meat extract	3·0 g
Fresh skim milk	10·0 ml
Agar	15·0 g
Distilled water	1000·0 ml

Dissolve the tryptone, the glucose, and the meat extract in cool distilled water. Add the agar and heat by steaming. Add the skim milk and steem for 5 minutes. Sterilize by autoclaving at 2 atm for 20 minutes. Do not heat for longer than this time.

Skim milk cannot be added if it is undiluted or if tenfold diluted milk is being tested.

The IDF (International Dairy Federation) Standard prescribes the use of 2·5 g of yeast extract instead of 3 of meat extract.

[156] *TTC broth*
(A medium for examination of the contaminating flora of milk.)

Boil 500 g of beef, 10 g peptone, and 5 g sodium chloride in 1000 ml distilled water for 2·5–3 h. Filter, and sterilize in a steamer. Add 11 ml of an aqueous 2% TTC solution per 100 ml of the broth (pH 6·2–6·4). Dispense 5-ml or 2-ml quantities into test-tubes and store in a cool, dark room.

[157] *Tween agar*

Peptone	10·0 g
Sodium chloride	5·0 g

Calcium chloride	0·1 g
Tween 80	10·0 g
Distilled water	1000·0 ml
Agar	15·0 g

Dissolve the ingredients by steaming. Final pH 7·0–7·4. Sterilize at 115 °C for 15 minutes.

[158] *Universal medium*

Sweet milk whey	200·0 ml
Yeast decoction (1: 10)	100·0 ml
Meat extract	4·0 g
Peptone	2·0 g
Water	700·0 ml

Add the white of one egg to 4 litres of the medium. Mix the ingredients thoroughly in an enamelled vessel, cover, and autoclave briefly up to a pressure of 2 atm. Filter through cotton wool. Make up with water to the original volume. Adjust the pH with NaOH to 7·2. Dispense into 1-litre or 2-litre flasks and sterilize as above. Allow to stand for 1 day or more. Filter before use to remove the precipitated protein. Dissolve 10 g glucose per 1000 ml of the medium. Dispense into test-tubes and sterilize at 1·5 atm for 30 minutes.

Preparation of sweet milk whey

Warm skim milk to 35 °C and add 1 g of rennet per 5 litres. Maintain at 35 °C for 8. minutes. Filter, and freeze in suitable sized portions. Store in the frozen state.

Preparation of the yeast decoction

Add 100 g yeast to 1000 ml of water and mix. Autoclave briefly up to a pressure of 2 atm. Allow to settle for several days. Separate off the clear liquid, dispense into flasks and sterilize as above.
Solidify the nutrient solution by adding 2% agar. Dissolve, and dispense the medium. Sterilize at 1·5 atm for 30 minutes.

[159] *Blood agar medium*
(A medium for carrying out haemolysis tests and for the detection and determination of bacteria requiring native protein.)

Add 20 ml of sterile defibrinated bovine blood to 400 ml gelatin basal agar medium [166] with a pH of 7·6–7·7, cooled to 50 °C. Mix by circular movement of the vessel, taking care to prevent frothing. Transfer 20-ml aliquots aseptically into sterile Petri dishes. Place the solidified plates in an incubator at 37 °C for 24 h as a test for sterility. Store in the refrigerator.

[160] *Vitamin-free yeast medium*

Dissolve 16·7 g of vitamin-free basal medium in 100 ml of distilled water with continuous heating. Dilute tenfold with distilled water. Sterilize by filtration.

Composition of the Bacto vitamin-free basal medium

Ammonium sulphate	5·0 g
Bacto glucose	10·0 g
L-histidine HCl	0·01 g
DL-methionine	0·02 g
DL-tryptophan	0·02 g
Boric acid	0·00050 g
Copper sulphate	0·00004 g
Potassium iodide	0·00010 g
Ferric chloride	0·00020 g
Manganese sulphate	0·00040 g
Sodium molybdate	0·00020 g
Zinc sulphate	0·00040 g
Potassium dihydrogen phosphate	0·85 g
Dipotassium hydrogen phosphate	0·15 g
Magnesium sulphate	0·50 g
Sodium chloride	0·10 g
Calcium chloride	0·10 g

[161] *Water (sterile)*

9·5-ml volumes of tap water in test-tubes or flasks are sterilized at 2 atm for 30 minutes.

[162] *Vogel and Johnson's agar*
(For examining *S. aureus*.)

Tryptone	10·0 g
Yeast extract	5·0 g
Mannitol	10·0 g
Dipotassium hydrogen phosphate	5·0 g
Lithium chloride	5·0 g

Glycine	10·0 g
Agar	16·0 g
Phenol-red	
(12·5 ml of a 0·2% solution)	0·025 g
Potassium tellurite	0·2 g
Distilled water	1000·0 ml

Final pH 7·2.

Dissolve the ingredients, except for the tellurite, in the water by heating with careful stirring. Boil for 1 minute. Dispense 100-ml amounts into flasks and sterilize at 121 °C for 15 minutes. Cool to 45–50 °C. Add 2 ml of a 1% potassium tellurite solution per 100 ml of the medium. Check the pH. Pour plates.

For the preparation of phenol-red solution, see medium [41]. The 1% tellurite solution is prepared separately and sterilized at 121 °C for 15 minutes.

[163] *Voges–Proskauer modified reagent*

| α-Naphthol | 5·0 g |
| Ethanol (absolute) | 100·0 ml |

Prepare the solution on the day of use. Add 0·6 ml of the solution and 0·2 ml of a 40% KOH solution to the culture under examination.

[164] *Wilson and Blair modified medium*
(For the enumeration of clostridia in foods of animal origin and canned food.)

Basal medium
Meat extract	4·0 g
Peptone	5·0 g
Yeast extract	0·5 g
Distilled water	1000·0 ml

pH (before sterilization) 7·6–7·7; pH (after sterilization) 7·4 ± 0·1.

To create the appropriate degree of anaerobiosis, add 2·5% agar. Sterilize at 121 °C for 20 minutes.

Add 10 ml of 0·1% sodium sulphite ($Na_2SO_3 \cdot 7H_2O$) solution and 10 ml 0·05% ferric citrate ($FeC_6H_5O_7 \cdot 5H_2O$ solution), each freshly sterilized by filtering through a G-2 or H-5 glass filter, to 1000 ml of the basal medium. Dispense 10-ml volumes into sterile test-tubes. Before inoculation, expel the oxygen from the medium by boiling, and then cool to 45–48 °C. Incubate without sealing (i.e. aerobically) at 37 °C.

[165] *Solution required for the Ziehl–Neelsen stain*

Carbol fuchsin
Add 100 ml of a 5% aqueous phenol solution to 10 ml 3% fuchsin dissolved in absolute alcohol. Mix, allow to stand overnight, then filter.

Decolourizing acid–alcohol
Add 0·01 g potassium hydroxide dissolved in 100 ml of distilled water to a mixture of 97 ml 96% alcohol and 3 ml concentrated hydrochloric acid.

[166] *Gelatin Basal Agar (Takács)*

(A medium for: the detection of obligate aerobes and facultative anaerobes; the determination of the total viable count by Koch's plating procedure; the detection of proteolytic bacteria by Clark's developer.)

Basal broth [32]	1000·0 ml
Agar	15·0 g
Gelatin	5·0 g

pH 7·6.

Mix the ingredients of the basal broth together with the agar and the gelatin and boil to dissolve. Adjust the pH to 7·6 and sterilize by autoclaving at 121 °C for 20 minutes.

[167] *Gelatin medium*

Meat extract	3·0 g
Peptone	5·0 g
Gelatin	50·0 g

pH (after autoclaving) 6·9 ± 0·1.

Add the ingredients to 1 litre of distilled water, mix thoroughly, and boil to dissolve. Check the pH. Dispense into test-tubes to a depth of 2–3 cm. Sterilize by autoclaving at 121 °C for 15 minutes.

[168] *Yeastrel agar*
(A medium for the determination of the total viable microbial count.)

Yeastrel yeast extract (Brewers Food Supply Co., England)	3·0 g

Peptone	5·0 g
Agar	15·0 g
Fresh whole milk or skim milk	10·0 ml
Distilled water	10000 ml

Final pH (at room temperature) 7·2.

Dissolve the yeastrel and the peptone by heating. Cool, and adjust the pH to 7·4. Add the agar and the milk to the broth. Autoclave at 121 °C for 20 minutes and then filter the hot medium, check the pH at 50 °C and adjust to 7·0 if necessary. Dispense, and sterilize at 121 °C for 15 minutes.

[169] *Coretti's medium for the isolation and identification of bacteria causing greening in meat products*

Tryptone	5·0 g
Yeast extract	2·5 g
D-glucose	1·0 g
Agar	15·0 g

Dissolve the ingredients in 1000 ml of distilled water. Adjust the pH to 6·0–6·2. Cool the basal medium and add black manganese oxide (MnO_2) suspended in gelatin (6 ml suspension per 100 ml of the medium).

Preparation of the suspension

Add 5% (w/v) very finely ground black manganese oxide, suspended in a little water to a 10% aqueous solution of gelatin. Sterilize. Spread on the inoculated, solidified plates and incubate at 30 °C for 48 h. The production of hydrogen peroxidase is indicated by a transparent halo around the colonies, seen against the background of the blackish-gray agar.

[170] *Gardner's medium for the isolation and enumeration of Microbacterium thermosphactum* (Gardner, 1966)

Peptone	20·0 g
Yeast extract	2·0 g
Glycerol	15·0 g
K_2HPO_4	1·0 g
$MgSO_4 \cdot 7H_2O$	1·0 g
Agar	13·0 g

Dissolve the ingredients in 1000 ml of distilled water. Adjust the pH to 7·0. Dispense in 100-ml volumes, and sterilize at 121 °C for 15 minutes. Cool to 45 °C. Add the following solutions, each made up with sterile distilled water.

Streptomycin sulphate	500 μg per ml*
Actidion	50 μg per ml*
Thallium acetate	50 μg per ml*

Enumerate the colonies after incubation at 20–22 °C for 48 h.

[171] *Gardner's medium for the determination of the total viable count in liquid brine and in meat products cured over a long period* (Gardner, 1968)

Peptone	10·0 g
Lab-Lemco	10·0 g
Yeast extract	2·0 g
Sodium chloride	40·0 g**
D-glucose	1·0 g
Agar	13·0 g

Final pH 7·0.

Dissolve the ingredients in 1000 ml distilled water. Sterilize at 121 °C for 15 minutes. Incubate at 20–22 °C for 4 days.

[172] *Gardner's medium for the demonstration of salt-tolerant Vibrio in liquid brine and in cured products* (Gardner, 1973)

Peptone	10·0 g
Lab-Lemco	10·0 g
Yeast extract	2·0 g
Sodium chloride	40·0 g
D-glucose	1·0 g
Agar	13·0 g

Dissolve the ingredients in 1000 ml distilled water. Adjust the pH to 7·0. Dispense, and sterilize at 121 °C for 15 minutes. Cool, and add a solution of crystal violet. The latter is pre-sterilized at 121 °C for 15 minutes, and sufficient is added to give a final concentration of 10 μg ml^{-1}.

Mix the solutions, pour plates, and dry the plates at 37 °C overnight. Do not store for more than 24 h after preparation.

[173] *R.C.M. broth (Reinforced Clostridial Medium)*
(For the cultivation of anaerobic spore-formers.) (OXOID CM 149)

Yeast extract	3·0 g
"Lab-Lemco" beef extract	10·0 g

---

* Final concentrations.
** The final salt concentration of the medium depends on the salt concentration of the brine (see text).

Peptone	10·0 g
Starch (soluble)	1·0 g
Glucose	5·0 g
Cysteine hydrochloride	0·5 g
Sodium chloride	5·0 g
Sodium acetate	3·0 g
Agar	0·5 g
Distilled water	1000·0 ml

Dissolve the ingredients (or 38 g of the Oxoid powdered medium) in distilled water at a temperature of 40–50 °C, shaking the solution for 15 minutes. Adjust the pH to 6·8. Mix well, dispense and autoclave at 115 °C for 10 minutes.

[174] *Stern broth*

Peptone	2·0 g
Sodium chloride	0·05 g
Meat extract	1·0 g
Distilled water	100·0 ml

Dissolve the ingredients by boiling. Adjust the pH to 7·8–8·0 and filter through qualitative filter paper. Add, per 100 ml of the filtrate, 5–6 drops of oversaturated fuchsin solution, followed by 2 ml of freshly-prepared 10% sodium sulphite solution and 0·8 ml of glycerol.

Dispense into flasks. Autoclave at 121 °C for 20 minutes. Store at + 4 °C. Dispense 3-ml volumes into sterile, plugged test-tubes immediately before use.

REFERENCES

Albertz, R.–Rechelmann, H.–Leistner, L. (1972): Erfahrungen mit einer Gerät (PETRIMAT 90) zum automatischen und sterilen abfüllen von Agar-Nährböden in Petrischalen. *Die Fleischwirtschaft,* **52,** pp. 1027–1028.

Alford, J. A.–Smith, J. L.–Lilly, H. D. (1971): Relations of Microbial Activity to Changes in Lipids of Foods. *J. Appl. Bact.,* **34,** pp. 133–146.

American Meat Institute Foundation (1960): *The Science of Meat and Meat Products.* Freeman, San Francisco, Calif.

American Public Health Association, Inc. (1960): *Standard Methods for the Examination of Dairy Products Microbiological and Chemical.* (11th Ed.) Amer. Publ. Health Assoc., Inc., New York.

Angelotti, R.–Hall, H. E.–Foter, M. J.–Lewis, K. H. (1962): Quantitation of *Clostridium perfringens* in Foods. *Appl. Microbiol.,* Baltimore, **10,** p. 193.

Anon. (1964): Az állati eredetű élelmiszer-nyersanyagok, félkész- és késztermékek higieniai szabályzata (Hygienic Regulation Statutes for Food Raw Materials, Semi-finished and Finished Products). *Élelmiszeripari Értesítő,* Budapest, **13,** pp. 45–80.

Anon. (1966): *Microbiological Examination of Precooked Frozen Foods.* Association of Food and Drug Officials of the United States. Quarterly Bulletin. The Editorial Committee Berkeley, Calif.

Anon. (1969): *Sartorius Membranfilter* G.m.b.H., Göttingen. (A leaflet.)

Anon. (1970): Microbiological Specifications and Testing Methods for Irradiated Food. *Technical Reports,* Series, No. 104, International Atomic Energy Agency, Vienna.

A.P.H.A. (1966): *Recommended Methods for the Microbiological Examination of Foods.* (2nd Ed.) Amer. Publ. Health Assoc. Inc., New York.

Árokszállásy, Z.–Bánhegyi, J.–Boros, Á.–Gallé, L.–Hortobágyi, T. (1968): *Növényhatározó. I. kötet.* Baktériumok — Mohák (Plant Identification Manual. Vol. I. (Bacteria–Mosses). Tankönyvkiadó, Budapest.

Árpai J. (1967): *Baktériummeghatározó kulcs* (Bacterium Identification Key). Műszaki Könyvkiadó, Budapest.

Ayres, J. C. (1960): The Relationship of Organisms of the Genus *Pseudomonas* to the Spoilage of Meats, Poultry and Eggs. *J. Appl. Bact.,* **23,** p. 471.

Bacteriological Analytical Manual for Foods. FDA, Washington D.C. (1972)

Baibel, R. H.–Seeley, H. W. Jr. (1974): *Genus I. Streptococcus.* In: Buchanan, R. E.–Gibbons, N. E., pp. 490–509.

Bailey, J. M.–Meymandi-Nejad, A. (1961): *J. Lab. Clin. Med.,* New York, **58,** pp. 667–672.

Baird-Parker, A. C. (1966): *Methods for Classifying Staphylococci and Micrococci.* In: Gibbs, B. M.–Skinner, F. A., pp. 59–64.

Baird-Parker, A. C. (1972): *Classification and Identification of Staphylococci and their Resistance to Physical Agents.* In: Cohen Jay, O., pp. 1–20.

Baird-Parker, A. C. (1974): *Genus II. Staphylococcus.* In: Buchanan, R. E.–Gibbons, N. E., pp. 483–489.

Baird-Parker, A. C. (1975): *Micrococcaceae*. In: Buchanan, R. E.–Gibbons, N. E.: Bergey's Manual of Determinative Bacteriology, pp. 478–483.

Bánhegyi, J.–Tóth, S.–Ubrizsy, G., Vörös, J. (1976): *Magyarország mikroszkopikus gombáinak határozó kézikönyve* (Manual for Identification of Microscopic Fungi Occurring in Hungary). Akadémiai Kiadó, Budapest.

Barabás, J.–Vadász, J. (1966): *Mikroszkópos fényképezés* (Microscopic Photography). Műszaki Könyvkiadó, Budapest.

Barraud, C.–Kitchell, A. G.–Labots, H.–Reuter, G.–Simonsen, B. (1967): Standardization of the Total Aerobic Count of Bacteria in Meat and Meat Products. *Die Fleischwirtschaft*, **47**, pp. 1313–1319.

Barron, G. L. (1968): *The Genera of Hyphomycetes from Soil*. The Williams and Wilkins Co., Baltimore.

Bartholomew, J. W.–Mittwer, T. (1950): A Simplified Bacterial Spore Stain. *Stain Technol.*, **25**, p. 153.

Baumgart, J. (1973): Der "Stomacher" ein neues Zerkleinerungsgerät zur Herstellung von Lebensmittel-suspensionen für die Keimzahlbestimmung. *Die Fleischwirtschaft*, **53**, p. 1600.

Beech, F. W.–Carr, J. G. (1960): Selective Media for Yeasts and Bacteria in Apple Juice and Cider. *J. Sci. F. Agric.*, London, **11** (1), pp. 38–40.

Bergdoll, M. S. (1972): *The Enterotoxins*. In: Cohen Jay, O., pp. 301–332.

Bergdoll M. S. (1973): Enterotoxin Detection. In: Hobbs, P. C.–Christien, J. H. B., pp. 287–292.

Bogdanov, V. M.–Bashirova, R. S.–Kirova, K. A.–Korneev, I. P.–Kostrova, E. I.–Petrzhikovskaya, L. M.–Pankratov, A. Ya.–Svitich, K. A. (1968): Tekhnicheskaya Mikrobiologiya Pishchevih Produk-tov. *Izd. Pishchevaja Promishlennosti*, Moscow.

Breed, R. S. (1911): The Determination of the Number of Bacteria by Direct Microscopic Examination. *Zbl. Bakt. Abt. II. Naturwiss. Abt.*, Jena, **30**, pp. 337–340.

Brewer, J. H.–Allgeier, D. L. (1966): Safe Self-Contained Carbon Dioxide-hydrogen Anaerobic System. *Appl. Microbiol.*, Baltimore, **14**, pp. 958–988.

Brown, W. R. L.–Ridout, C. W. (1960): An Investigation of Some Sterilization Indicators. *Pharm. J.*, **184**, p. 5.

Buchanan, R. E.–Gibbons, N. E. (Eds.) (1975): *Bergey's Manual of Determinative Bacteriology* (8th Ed.). The Williams and Wilkins Co., Baltimore.

Burri, R. (1928): The Quantitative Smear-Culture: A Simple Means for the Bacteriological Examination of Milk. *Proc. World's Dairy Congr.*, London, 690 p.

Cantoni, C.–Molnár, M.–Renon, P.–Giolitti, G. (1967a): Lipolytic Micrococci in Pork Fat. *J. Appl. Bact.*, **30**, p. 190.

Cantoni, C.–Molnár, M.–Renon, P.–Giolitti, G. (1976b): Untersuchung über die Lipide von Dauerwürsten. *Nahrung*, **11**, p. 341.

Carlquist, P. R. (1956): A Biochemical Test for Separating Paracolon Groups. *J. Bact.*, **71**, pp. 339–341.

Cassel, E. A. (1965): Rapid Graphical Method for Estimating the precision of Direct Microscopic Counting Data. *Appl. Microbiol.*, Baltimore, **13**, pp. 293–296.

Catsaras, M.–Gulistani, A. W.–Mossel, D. A. A. (1974): Contaminations superficielles des Caroasses réfrigérées de bovins et de chevaux. *Rec. Med. Vet.*, **150**, p. 287.

Charlett, S. M. (1954): An Improved Staining Method for the Direct Microscopical Counting of Bacteria in Milk. *Dairy Ind.*, **19**, p. 632.

Charley, V. L. S. (1959): The Prevention of Microbiological Spoilage in Fresh Fruit. *J. Sci. Food Agric.*, **10**, p. 349.

Chatgny, M. A. (1961): Protection Against Infection in the Microbiological Laboratory: Devices and Procedures. *Adv. Appl. Microbiol.*, New York–London, **3**, p. 131.

Clark, D. S.–Thatcher, F. S. (1974): *Microorganisms in Food*. Vol. 2. University of Toronto Press, Toronto.

Clegg, L. F. L.–Thomas, S. B.–Cox, C. P. (1951): A Comparison of Roll–Tube and Petri Dish Colony Counts on Raw Milk. *Proc. Soc. Appl. Bact.*, **14**, p. 171.

Cochran, W. G. (1950): Estimation of Bacterial Densities by Means of the "Most Probable Number". *Biometrics*, Washington, **6**, pp. 105–116.

425

Cohen Jay, O. (1972): *The Staphylococci*. Wiley J. and Sons Inc.,

Collins, C. H. (1967): *Microbiological Methods*. Buttererworth, London.

Coretti, K. (1958): Technische Verbesserung des Mangandioxyd–Nährbodens zum Nachweis bakteriel-ler Peroxyde, *Jahresbericht 1958 des Bundesforschungsanstalt für Fleischwirtschaft, Kulmbach 36–37*.

Costin, I. D.–Grigo, J. (1974): Bioindikatoren zur Autoklavirungskontrolle. Einige theoretische Aspekte und praktische Erfahrungen bei der Entwicklung und Anwendung. *Zbl. Bakt. Hyg.; I. Abt. Orig. A*, **227**, pp. 483–521.

Costin, I. D.–Grigo, J. (1975): Bioindikatoren für die Überwachung des Autoklavierprozesses: Anforderungen und Leistungen. *Archiv für Lebensmittelhygiene*, No. 1, pp. 34–35.

Coulter, W. H. (1953): *U.S. Patent* No. 2. 656. 508.

Cowan and Steel's Manual for the Identification of Medical Bacteria (1974). (2nd ed., revised by S. T. Cowan.) Cambridge University Press.

Cowell, N. D.–Morisetti, M. D. (1969): Microbiological Techniques, Some Statistical Aspects. *J. Sci. Food Agric.*, London, **20**, pp. 573–579.

Cross, T. (1970): The Diversity of Bacterial Spores. *J. Appl. Bact.*, **33**, pp. 95–102.

Cruickshank, R. (1965): *Medical Microbiology*. (11th ed.) Livingstone, Edingburgh.

Csaba, K. (1961): *Útmutatás az élelmiszerek bakteriológiai és parazitológiai vizsgálatához*. Sokszoro-sított kézirat (Guide in Bacterial and Parasitological Testing of Foodstuffs. A manifold.) OÉTI, Budapest.

Csiszár, J. (1935): Eine Methode zur bakteriologischen Überflächenuntersuchung. *D. Molk.–Ztg.*, **56**, p. 1404.

Csiszár, V. (1964): *Húsvizsgálat és húshigiéne* (Meat Control and Meat Hygiene). Mezőgazdasági Kiadó, Budapest.

Darmady, E. M.–Hughes, K. E. A.—Jones, J. D.–Prince, D.–Tuke, W. (1961): Sterilization by Dry Heat. *J. Clin. Path.*, **14**, p. 38.

Deibel, R. H.–Seeley, H. W. Jr. (1975): *Streptococcaceae*. In: Buchanan, R. E.Gibbons, N. E. pp. 217–226.

De Man, J. C.–Bindschedler, O.–Moussa, R. S. (1974): Economic Aspects and Rationalization of Conventional Microbiological Methods for the Examination of Foods. *IAMS Kiel, 16th–20th September, 1974*.

Demeter, K. J. (1939): Über des Abklatschverfahren–Vorschlag einer verbesserten Methode. *D. Molk.–Ztg.*, **60**, pp. 1155–1156.

Demeter, K. J. (1967): *Bakteriologische Untersuchungsmethoden der Milchwirtschaft*. Verlag Eugen Ulmer, Stuttgart.

Denny, C. B. (1970): Collaborative Study of Procedure for Determining Commercial Sterility of Low–Acid Canned Foods. *J. ADAC*, **53**, pp. 713–715.

Difco Manual of Dehydrated Culture Media and Reagents for Microbiological and Clinical Laboratory Procedures. (9th Ed.), 1953. Difco Laboratories Inc., Detroit, Laboratories Inc. Detroit, Mich.

Donnelly, C. B.–Black, L. A.–Lewis, K. H. (1960): An Evaluation of Simplified Methods for Determining Viable Counts of Raw Milk. *J. Milk Fd. Technol.*, **14**, pp. 201–205.

Doudoroff, M.–Palleroni, N. J. (1975): *Genus I. Pseudomonas*. In: Buchanan, R. E.–Gibbons, N. E.: Bergey's Manual of Determinative Bacteriology, pp. 217–226.

Dowell, V. R. S.–Hawkins, T. M. (1968): *Laboratory Methods in Anaerobic Bacteriology*. Publ. 1803. Governmt. Printing Office Washington, D. C.

Drews, G. (1968): *Mikrobiologisches Praktikum für Naturwissenschaftler*. Springer Verlag, Berlin–Heidelberg–New York.

Drion, E. F.–Mossel, D. A. A. (1977): The Reliability of the Examination of Foods, Processed for Safety, for Enteric Pathogens and Enterobacteriaceae: a Mathematical and Ecological Study. *J. Hyg. Camb.*, **78**. pp. 301–324.

Duncan, A. J. (1965): *Quality Control and Industrial Statistics*. Homewood, III, Illinois.

Edwards, P. R.–Ewing, W. H. (1972): *Identification of Enterobacteriaceae.* (3rd ed.), Minneapolis, Burgess Publ. Co. Atlanta, USA.

Elliott, F. (1975): Method for Preserving Mini-Cultures of Fungi under Mineral Oil. *Labor. Practice,* p. 751.

Ellis, M. B. (1971): *Dematiaceous Hyphomycetes.* Commonwealth Mycological Institute, Kew.

Engelbrecht, E. (1974): The BIOREACTOR and BIODILUTOR as Tools for the Automation of Tests to be Performed on Serial Dilutions of Samples. In: *Automation of Microbiological Food Analysis.* IAMS Committee on Food Microbiology and Hygiene, IX. International Symposium, Kiel, 16–20 September, 1974, Abstr. 4 12.

Eörsi, M. (1947): Népegészségügy (Public Health). Supplement of *Orvosok Lapja,* **30,** p. 1172.

Ewing, W. H.–Edwards, P. R. (1960): The Principal Divisions and Groups of Enterobacteriaceae and their Differentiation. *Int. Bull. Bact. Nomencl. Taxon.,* **10,** p. 1.

Ewing, W. H.–Tatum, H. W.–Davis, B. R. (1957): *Publ. Health Lab.,* **15,** p. 118.

Fábri I. (1972): *A konzervgyártás mikrobiológiai vizsgálata. Konzervipari Zsebkönyv* (Microbiological Control of Canning. A Canned-Food-Industrial Vade-mecum). Kardos, E.–Szenes, E. (Eds), Mezőgazdasági Kiadó, Budapest, pp. 586–606.

Farkas, J. (1962): Elektronikus részecskeszámláló alkalmazási lehetőségei a kutatásban és az iparban (Application of Electronic Particle Counter in Research and Industry). *Élelmezési Ipar,* Budapest, **16,** pp. 365–367.

Farkas, J. (1971): Baktérium spóraszám meghatározási eljárások (Methods of Bacterial Spore Counting). *Élelmiszervizsgálati Közlemények,* Budapest, **17,** pp. 183–190.

Favero, M. S.—Berquist, K. R. (1968): Use of Laminar Air-Flow Equipment in Microbiology. *Appl. Microbiol.,* Baltimore, **16,** pp. 182–183.

Fish and Shellfish Hygiene. Technical Report Series No. 550, WHO (1974).

Food-Borne Disease: Methods of Sampling and Examination in Surveillance Programmes. Technical Report Series No. 543, WHO.

Foster, E. M. (1974): Interpretation of Analytical Results for Bacterial Standard Enforcement. Association of Food and Drug Officials of the United States. *Quarterly Bulletin,* **38,** No. 4, pp. 267–276.

Frazier, W. C.–Marth, E. H.–Deibel, R. H. (1968): *Laboratory Manual for Food Microbiology.* Burgess Publ. Co., Minneapolis, Minn.

Frost, W. D. (1921): Improved Technic for the Micro or Little Plate Method of Counting Bacteria in Milk. *Infect. Diseases,* **28,** pp. 176–184.

Fulghum, R. S. (1971): Mobile Anaerobe Laboratory. *Appl. Microbiol.,* Baltimore, **21,** pp. 769–770.

Fung, D. Y. C. (1974): Semi-Automated Viable Cell Count Techniques and Automated Staining Procedure. In: *Automation of Microbiological Food Analysis.* IAMS Committee on Food Microbiology and Hygiene. IX. International Symposium, Kiel, 16–20 September, 1974, Abstr. 4. 13.

Gainor, G.–Wegemer, D. E.– (1954): Studies on a Psychrophilic Bacterium Causing Ropiness in Milk. I. Morphological and Physiological Considerations. *Appl. Microbiol.,* Baltimore, **2,** p. 95.

Gardner, G. A. (1966): A Selective Medium for the Enumeration of *Microbacterium Thermosphactum* in Meat and Meat Products. *J. Appl. Bact.,* **29,** pp. 455–460.

Gardner, G. A. (1968): Effects of Pasteurization or Added Sulphite on the Microbiology of Stored Vacuum Packed Baconburgers. *J. Appl. Bact.,* **31,** pp. 462–478.

Gardner, G. A. (1973): A Selective Medium for Enumerating Salt Requiring *Vibrio* spp. from Wiltshire Bacon and Curing Brines. *J. Appl. Bact.,* **36,** pp. 329–333.

Gardner, M.–Martin, W. J. (1971): Simplified Method of Anaerobic Incubation. *Appl. Microbiol.,* Baltimore, **21,** p. 1092.

Gebhardt, L. P. (1970): *Microbiology Laboratory Manual.* The C. V. Mosby Co., Sant Louis.

Gibbs, B. M.–Shapton, D. A. (1968): *Identification Methods for Microbiologists. Part B.* Academic Press, London, New York.

427

Gibbs, B. M.–Skinner, F. A. (1966): *Identification Methods for Microbiologists. Part A*. Academic Press, London, New York.

Gibson, T.–Gordon, R. E. (1974): *Genus I. Bacillus*. In: Buchanan, R. E.–Gibbons, N. E. pp. 529–545.

Gibson, T.–Gordon, R. E. (1975): *Genus I. Bacillus*. In: Buchanan, R. E.–Gibbons N. E.: Bergey's Manual of Determinative Bacteriology, pp. 529–545.

Gilbert, R. J. (1972): A Comparative Assessment of Medium for the Isolation and Enumeration of Coagulase Positive Staphylococci from Foods. *J. Appl. Bact.*, **35**, pp. 673–679.

Gilbert, R. J.–Stringer, M. E.–Peace, T. C. (1974): The Survival and Growth of *Bacillus cereus* in Boiled and Fried Rice in Relation of Outbreak of Food Poisoning. *J. Hyg. Camb.*, **73**, pp. 433–444.

Gilbert, R. J.–Taylor, A. J. (1975): Das Auftreten von *Bacillus cereus* Lebensmittelvergiftungen in Grossbritannien. *Archiv für Lebensmittel-Hygiene*, No. 1.

Gilbert, R. J.–Wieneke, A. A. (1973): *Staphylococcal Food Poisoning with Special Reference to the Detection of Enterotoxin in Food*. In: Hobbs, B. C.–Christian, J. H. B. pp. 273–286.

Goepfert, J. M.–Spira, W. M.–Glatz, B. A.–Kim, H. U. (1973): Pathogenicity of *Bacillus cereus*. In: Hobbs, B. C.–Christian, J. H. B., pp. 69–76.

Görög, J. (1961): *Ipari mikrobiológiai gyakorlatok. I. rész*. (Exercises in Industrial Microbiology. Part I). Tankönyvkiadó, Budapest.

Gray, W. D. (1959): *The Relation of Fungi to Human Affairs*. Henry Holt and Co., Inc., New York.

Guthy, Kl. (1972): Zur Verwendung von elektronischen Partikelzählgeräten (Coulter Counter) in der Milchwirtschaft. *Milchwirtschaft*, **23** (13), pp. 10–11.

Hadlok, R. (1960): Bakteriologische Untersuchungen von Roh und Fertigerzeugnis bei der Konservenherstellung. *Die Fleischwirtschaft*, **12**, p. 1035.

Hammer, B. W.–Babel, F. J. (1957): Dairy Bacteriology. (4th ed.) Wiley, New York.

Harrigan, W. F.–McCance, M. F. (1976): *Laboratory Methods in Food and Dairy Microbiology*. Academic Press, London–New York–San Francisco.

Hartman, P. A. (1968): *Miniaturized Microbiological Methods*. Academic Press, New York–London.

Heeschen, W.–Reichmuth, J.–Tolle, A.–Zeidler, H. (1969): Zur Bestimmung der psychrotrophen Keimflora in der Anlieferungsmilch. *Milchwissenschaft*, Nürnberg, **24**, pp. 721–726.

Heeschen, W.–Reichmuth, J.–Tolle, A.–Zeidler, H. (1969): Die Konservierung von Milchproben zur bakteriologischen, zytologischen und hemmstoffbiologischen Untersuchung. *Milchwissenschaft*, Nürnberg, **24**, pp. 729–734.

Heller, C. L. (1954): A Simple Method for Producing Anaerobiosis. *J. Appl. Bact.*, **17**, p. 202.

Hendrie, M. S.–Shewan, J. M. (1966): *The Identification of Certain Pseudomonas Species*. In: Gibbs, B. M.–Skinner, F. A., pp. 1–8.

Henis, Y.–Gould, J. R.–Alexander, M. (1966): Detection and Identification of Bacteria by Gas Chromatograph. *Appl. Microbiol.*, **14**, p. 613.

Hermán G.–Hoch, V. (1971): Phage Typing of D-group Streptococci. II. Isolation of Supplementary Phages for Classification of Enterococci Untypable with Roumanian Phages. *Acta Microbiol. Acad. Sci. Hung.*, Budapest **18**, pp. 101–104.

Hess, E.–Lott, G. (1970): Kontamination des Fleisches während und nach der Schlachtung. *Die Fleischwirtschaft*, **50**, p. 47.

Hobbs, B. C.–Christien J. H. B. (1973): *The Microbiological Safety of Food*. Academic Press, London and New York.

Hobbs, B. C.–Clifford, W.–Gosh, A. C.–Gilbert, R. J.–Kendall, M.–Roberts, D.–Wieneke, A. A. (1973): Sampling of Food and Other Materials for Bacteriological and Ecological Studies. In: *Sampling — Microbiological Monitoring of Environments*. (Board, R. G.–Lovelock, D. W. Eds.) Academic Press, London, p. 233.

Hobson, P. N.–Mann, N. O.–Summers, R. (1968): A Small Mobile Laboratory for Microbiological Field Work. *Lab. Pract.*, Manchester, **17**, pp. 599–602.

428

Hoch, V., Hérmán, G. (1971): Phage Typing of D-group Streptococci. I. Typing of Enterococci with Roumanian Phages. *Acta Microbiol. Acad. Sci. Hung.*, Budapest, **18**, pp. 95–99.

Hoskins, J. K. (1934): Most Probable Numbers for the Evaluation of Coli-aerogenes Tests by Fermentation Tube Methods. *Pub. Hlth Rep.*, **49**, pp. 393–405.

Hungate, R. E. (1969): A Roll–Tube Method for Cultivation of Strict Anaerobes, pp. 117–132. In: Morris, J. R.–Ribbons, D. W. (Eds.) *Methods in Microbiology.* Vol. 38, Acad. Press, New York–London.

Hylsop, N. St. G. (1961): *Nature*, London, **191**, p. 305.

Hyatiäinen, M.–Pohja, M. S. and Niskanen, A. (1975): Über mikrobiologische Untersuchungsmethoden und über Qualitätsbeurteilung des Fleisches. *Die Fleischwirtschaft*, **55**, pp. 549–552.

Informal Consultation on the Principles of Organization and Management of Food Hygiene Programmes. WHO (1971).

Ingold (1961): *The Biology of Fungi*, Hutchinson Educational (Publishers), Ltd., London.

Ingram, M. (1957): The General Microbiology of Bacon Curing Brines with Special References to Methods of Examination, *Proc. 2nd Intern. Symp. of Food Microbiol. of Fish and Meat Curing Brines*, Cambridge, 1957, pp. 121–136.

Ingram, M. (1969): Sporeformers as Food Spoilage Organisms. The Bacterial Spore. Gould, G. W.–Hurst, A. (Eds), Academic Press, London–New York, pp. 548–610.

Ingram, M.–Barnes, E. M. (1956): A Simple Modification of the Deep Shake Tube for Counting Anaerobic Bacteria. *Lab. Pract. Manchester*, **5**, p. 145.

Ingram, M.–Roberts, T. A. (1976): The Microbiology of the Red Meat Carcass and the Slaughterhouse. *Royal Soc. Hlth. J. Dec.* (A. R. C. Meat Res. Inst. Memoir No. 775).

Ingram, M.–Bray, D. F.–Clark, D. S.–Dolman, C. E.–Elliot, R. F.—Thatcher, F. S. (Eds) (1974): *Microorganisms in Foods. II. Sampling for Microbiological Analysis.* University of Toronto Press, Toronto.

Insalta, N.F.—Schulte, S. J.–Berman, J. H. (1967): Immunofluorescence Technique for Detection of Salmonellae in Various Foods. *Appl. Microbiol.*, **15**, p. 1145.

International Dairy Federation (1963): International Standard FILIDF 18: 1962: Standard Capacity Test for the Evalution of the Disinfectant Activity of Dairy Disinfectants. *Dairy Ind.*, **28**, p. 610.

International Dairy Federation (1964): International Standard FILIDF 19: 1962: Standard Suspension Test for the Evaluation of the Disinfectant Activity of Dairy Disinfectants. *Dairy Ind.*, **29**, pp. 34–38.

ISO/DIS 3811 Meat and Meat Products — Detection and Enumeration of Presumptive Coliform Bacteria and Presumptive *Escherichia coli* (Reference method).

ISO/TC 34/Sc 6/WG2 Meat and Meat Products — Detection and Count of Enterobacteriaceae.

Janke, A.–Dickscheit, R. (1967): *Handbuch der mikrobiologischen Laboratoriumstechnik.* Verl. T. Steinkopf, Dresden.

Janke, A.—Janke, E. G. (1959): Über das Pseudomycel und die Formgattungen *Candida* und *Mycoderma. Sidowia*, Beih. I, p. 193.

Jasper, D. E.–Dellinger, J. (1966): Variations in Volume of Milk Delivered by a Standard 0·01 ml Loop. *J. Milk Food Technol.*, **29**, pp. 199–200.

Jenei, E.–Váczi, L. (1966): *Alkalmazott bakteriológia és elméleti alapjai* (Applied Microbiology and its principles). Medicina Könyvkiadó, Budapest.

Jennison, M. W.–Wadsworth, G. P. (1940): Evaluation of the Errors Involved in Estimating Bacterial Numbers by the Plating Method. *J. Bact.*, **39**, pp. 389–397.

Johns, C. K. (1954): Relation Between Reduction Times and Plate Counts of Milk before and after Pasteurization. *J. Milk Food Technol.*, **17**, pp. 369–371.

Johst, F. (1970): Elektronische Zellzahlung bei der Anlieferungsmilch als Grundlage für ein neues System der Qualitätsermittlung. *Milchforschung–Milchpraxis*, **12**, pp. 156–161.

Kandler, O. (1961): Die Anwendung der Nitratreduktion zur Qualitätskontrolle der Rohmilch. *D. Molk. Ztg.*, **82**, pp. 1283–1285.

Kaneko, T.–Holder-Franklin, M. and Franklin, M. (1975): Multiple Syringe Inoculator for Agar Plates. *Appl. Environm. Microbiol.*, **33**, pp. 892–985.

Karlikanova, S. N. (1968): O metode opredeleniya nalichiya antibiotikov v moloke. *Moloch. Prom.*, Moszkva, **29** (6), pp. 22–25.

Katona, F. and Pusztai, S. (1975): *Élelmiszer-higiéniai vizsgálatok állatorvosi laboratóriumokban* (Food-Hygienic Investigations in Veterinary Laboratories). Mezőgazdasági Kiadó, Budapest, 92. p.

Kauffmann, F. (1954): *Enterobacteriaceae.*Ejnar Munksgaard Publisher, Copenhagen.

Kauffmann, F. (1960): *Acta path. microbiol. Scand.*, **49**, p. 393.

Kauffmann, F. (1961): *Die Bakteriologie der Salmonella-Species.* Munksgaard, Kopenhagen.

Kauffmann, F.–Edwards, P. R.–Ewing, W. H. (1956): *Int. Bull. Bact. Nomencl. Taxon.*, **6**. p. 29.

Kelch, F. (1960): Die bakteriologische Untersuchung von Fleisch- und Fleischkonserven. *Die Fleischwirtschaft*, **12**, p. 354.

Kelch, F.–Hadlok, R. (1960a): Untersuchungsmethodik für Fleischkonserven. *Die Fleischwirtschaft*, **12**, p. 157.

Kelch, F.–Hadlok, R. (1960b): Zur Methodik der bakteriologischen Untersuchung von Fleischkonserven ohne Bombageerscheinung. *Die Fleischwirtschaft*, **12**, p. 915.

Kelsey, J. C. (1958): *The Testing of Sterilisers.* Lancet, II. p. 306.

Ketting, F. (1959): *Laboratóriumi gyakorlatok, III* (Laboratory Exercises, III). Műszaki Könyvkiadó, Budapest.

Kiermeier, F. (1968): *Handbuch der Lebensmittelchemie III/1, III/2, Tierische Lebensmittel.* Springer-Verlag Berlin–Heidelberg–New York.

King, W. L.–Hurst, A. (1963): A Note on the Survival of Some Bacteria in Different Diluents. *J. Appl. Bact.*, London–New York, **26**, p. 504.

Kitchell, A.G. (1957): The Micrococci of Pork and Bacon and of Bacon Brines. *Proc. 2nd International Symp. of Food Microbiol. The Microbiol. of Fish and Meat Curing Brines*, Cambridge, pp. 191–196.

Kubitschek, H. E. (1969): Counting and Sizing Micro-organisms with the Coulter Counter. In: Norris, J. R.–Ribbons, D. W. (Eds), p. 593.

Labots, H.–Galesloot, Th. E. (1965): Laboratory Evaluation of Dairy Disinfectants. Examination of Surface-Active Disinfectants by Means of the Capacity Test. *Nederlands Melken Ziveltijdach*, Wageningen, **19**, pp. 139–147.

Leistner, L. (1970): Vorkommen und Bedeutung von Clostridien in Fleischkonserven. *Arch. Lebensmittelhyg.*, **21**, p. 145.

Leistner, L.–Hechelman, H. (1974): *Mechanisierung und Automatisierung mikrobiologischer Arbeitsmethoden.* Institut für Bakteriologie und Histologie der Bundesanstalt für Fleischforschung, Kulmbach.

Lerche, M. (1957): *Lehrbuch der tierärztlichen Lebensmittel-Überwachung.* M. H. Schaper, Hannover.

Live I. (1972): Staphylococci in Animals: Differentiation and Relationship to Human Staphylococcosis. In: Cohen Jay, O., pp. 443–456.

Ljutov V. (1961): Technique of Methyl Red Test. *Acta Pathol. Microbiol. Scand.*, **51**, p. 369.

Ljutov, V. (1963): Technique of Voges–Proskauer Test. *Acta Pathol. Microbiol. Scand.*, **58**, p. 325.

Lodder, J. (Ed.) (1970): *The Yeasts.* North-Holland Publ. Co., Amsterdam–London.

Lorenz, W. (1953): Zur Bestimmung der Milchkeimzahl mit Hilfe "Kleinen Plattenmethode". *Öst. Milchw.*, Wien, **21**, pp. 345–346.

Losonczy, S.-né (1969): Gyors csíraszám-meghatározási módszer kidolgozása és alkalmazása a húsiparban (A Rapid Method for Determination of Viable Count and its Application in Meat Industry). *Húsipar,* Budapest, **18**, pp. 161–166.

Losonczy, S.-né (1970): Gyors csíraszám-meghatározási módszer gyakorlati alkalmazása a húsiparban, II. (Application in Meat Industry of a Rapid Method for Assessing Viable Count, II). *Húsipar,* Budapest, **19**, pp. 228–231.

430

Lötzsh, R.–Tauchmann, F.–Meyer, W. (1973): Einsatz des "Stomacher" in der Mykotoxin-Analytik. *Die Fleischwirtschaft,* **54,** pp. 943–945.

Lowry, O. H.–Rosenbrough, N. J.–Farr, A. L.–Randall, R. J. (1951): *J. Biol. Chem.,* Baltimore, **193,** pp. 265–275.

Marton, A.–Nagy-Dani É. (1976): *Staphylococcus aureus* törzsek antibiotikum rezisztenciája és enterotoxin termelése közötti összefüggés vizsgálata (Correlation between the Resistance to Antibodies of, and Enterotoxin Production by *Staphylococcus areus*). *Egészség-tudomány,* **20,** pp. 150–155.

Matuszewski, T.–Supinska, J. (1939): Studies on the Methylene Blue Reduction Test. I. *Zbl. Bakt. Abt. II. Naturwiss. Abt.,* Jena, **101,** pp. 45–64.

McDade, J. J.–Philips, G. B.–Sihinski, N. D.–Waitfield, W. J. (1969): Principles and Applications of Laminar-Flow Devices. In: Norris, J. R.–Ribbons, D. W. (Eds) Methods in Microbiology, Vol. 1, pp. 137–168.

McGinnis, M. R.–Padhye, A. A.–Ajello, L. (1974): Storage of Stock Cultures of Filamentous Fungi, Yeasts, and Some Aerobic Actinomycetes in Sterile Distilled Water. *Applied Microbiology,* **28,** pp. 218–222.

Meynell, G. G.–Meynell, E. (1965): *Theory and Practice in Experimental Bacteriology.* Cambridge Univ. Press, Cambridge.

Microbial Aspects of Food Hygiene. Technical Report Series, No. 399, WHO (1968).

Microbiological Aspects of Food Hygiene. Technical Report Series, No. 598, WHO (1976).

Ministry of Agriculture, Fisheries and Food (1968): Bacteriological Techniques for Dairy Purposes. *Technical Bulletin,* **17,** Her Majesty's Stationery Office, London.

MNOSZ 3602-49: Tartósított élelmiszerek mintavétele (Sampling from Preserved Food).

MNOSZ 3646-54: Élelmiszerek mikrobiológiai vizsgálata, mintavétel mikrobiológiai vizsgálat céljára (Microbiological Examination of Foodstuffs. Sampling for Microbiological Testing).

MNOSZ 3644-54: Élő csírák számának megállapítása (Determination of Viable Count).

Møller, V. (1954): Simplified Tests for Some Amino Acid Decarboxylases and for the Arginine Dihydrolase System. *Acta Pathol. Microbiol. Scan.,* **36,** p. 158.

Möller, O.–Reimoser, J. (1969): Eine modifizierte Nitratreduktionsprobe zur Qualitätskontrolle der Rohmilch. *D. Molk.-Ztg.,* **90,** pp. 977–984.

Monod, J. (1949): The Growth of Bacterial Culture. *A Rev. of Microbiology, Stanford,* **3,** p. 371.

Mortimer, P. R.–McCann, G. (1974): Food-poisoning Episodes Associated with *Bacillus cereus* in Fried Rice. *Lancet,* pp. 1043–1045.

Moss, C. W.–Lewis, V. J. (1967): Characterization of Clostridia by Gas Chromatography. *Appl. Microbiol.,* **15.** p. 390.

Mossel, D. A. A. (1956): The Kluyver Fermentation Test for Detecting Preservatives in Foods. *Food Manufact.,* **31,** London, pp. 190–197.

Mossel, D. A. A. (1975): *Microbiology of Foods and Dairy Products. Occurrence, Prevention and Monitoring of Hazards and Deterioration.* The University of Utrecht, Faculty of Veterinary Medicine.

Mossel, D. A. A. (1975): *Microbiology of Foods and Dairy Products.* The University of Utrecht.

Mossel, D. A. A.–Ingram, M. (1955): The Physiology of the Microbial Spoilage of Foods. *J. Appl. Bact.,* **18,** pp. 232–268.

Mossel, D. A. A.–Zwart, H. (1959): Die Quantitative Bakterioskopische Bewertung von Gemüse/Fleisch-konserven. *Arch. Lebensmittelhyg.,* **10,** p. 229.

Mossel, D. A. A.–Mengerink, W. H. J.–Scholts, H. H. (1962): Use of Modified Mac Conkey Agar Medium for the Selective Growth and Enumeration of Enterobacteriaceae. *J. Bacteriol.,* **84,** p. 381.

Mossel, D. A. A.–Koopman, M. J.–Cornelisseu, A. M. (1963): The Examination of Foods for Enterobacteriaceae Using a Test of the Type Generally. *J. Appl. Bact.,* London–New York, A. **26,** p. 444.

431

Mossel, D. A. A.–Koopman, M. J.–Jongerius, E. (1967): Enumeration of *Bacillus cereus* in Foods. *Appl. Microbiol.*, Baltimore, **15**, pp. 650–653.

Mossel, D. A. A.–Krol, B.–Moerman, P. C. (1972): Bacteriological and Quality Perspectives of Salmonella Radicidation of Frozen Boneless Meats. *Alimenta,* **11**, p. 51.

MSZ 3640/1–74. Húsok és húsalapú élelmiszerek mikrobiológiai vizsgálata. Általános előírások (Microbiological Testing of Meats and Foodstuffs Containing Meat. General Prescriptions).

MSZ 3640/8-74. Húsok és húsalapú élelmiszerek mikrobiológiai vizsgálata. Szalmonellák kimutatása (Microbiological Testing of Meats and Foodstuffs Containing Meat. Detection of *Salmonella*).

MSZ 3640/10-75. Húsok és húsalapú élelmiszerek mikrobiológiai vizsgálata. Enterobaktériumok kimutatása és számának meghatározása (Microbiological Testing of Meats and Foodstuffs Containing Meat. Detection and Counting of Viable Cells of Enteric Bacteria).

MSZ 3640/12-75. Húsok és húsalapú élelmiszerek mikrobiológiai vizsgálata. *Clostridium perfringens* kimutatása és számának meghatározása (Microbiological Testing of Meats and Foodstuffs Containing Meat. Detection and Counting of Viable Cells of *Clostridium perfringens*).

MSZ 3640/12-75. Húsok és húsalapú élelmiszerek mikrobiológiai vizsgálata. A feltételezetten koliform baktériumok és a feltételezetten *Escherichia coli* kimutatása és számának meghatározása (Microbiological Testing of Meats and Foodstuffs Containing Meat. Detection and Counting of Viable Cells of Presumptive Coliforms and Presumptive *E. coli*).

MSZ 3643-59. Összes csírák számának megállapítása (Determination of Total Viable Count).

MSZ 3641-62. Élelmiszerek mikrobiológiai vizsgálata. Hőkezeléssel csírátlanított készítmények tartóssági próbája (Microbiological Control of Foodstuffs. Test for Keeping Quality of Products Sterilized by Heating).

MSZ 3742-70. Tej és tejtermékek mikrobiológiai vizsgálata (Microbiological Testing of Milk and Dairy Products.)

MSZ 22901-71. Ivóvíz bakteriológiai vizsgálata (Bacteriological Testing of Drinking-Water).

Müller, G. (1974): *Mikrobiologie pflanzlicher Lebensmittel.* VEB Fachbuch-Verlag, Leipzig.

Narayan, K. G. (1965): *The Incidence and Significance of Clostridia in Meat and Meat Animals.* Kandidátusi értekezés. (Thesis.) Budapest.

Narayan K. G. (1966): The Sulphite-Reducing Ability of Clostridia in a Modified Sulphite Agar Medium. *Acta Vet.,* **16**, pp. 45–52.

Narayan, K. G. (1967): Culture, Isolation and Identification of Clostridia. *Zbl. Bakteriol.,* I. Orig., **202**, p. 212.

National Canners Association, Research Laboratories (1968): Laboratory Manual for Food Canners and Processors. Vol. 1. Microbiology and Processing. The AVI Publishing Co., Inc., Westport, Connecticut.

Nickerson, J. T.–Sinskey, A. (1972): *Microbiology of Foods and Food Processing.* American Elsevier Publishing Co., New York.

Niven, C. F. Jr. (1951): Sausage Discoloration of Bacterial Origin. *Amer. Meat Inst. Found. Bull.,* No. 13.

Norris, J. R.–Ribbons, D. W. (Eds) (1969): *Method in Microbiology, Vol. 1.* Academic Press, London–New York.

Norris, J. R.–Swain, H. (1971): Staining Bacteria, In: Norris, J. R.–Ribbons, D. W. (Eds), Vol. 5 A, pp. 105–134.

Nowak, J. (1927): *Documenta Microbiologica. 1. Bakterien.* Gustav Fischer Verl., Stuttgart.

Nygren, B. (1962): Phospholipose C-producing Bacteria and Food Poisoning. *Acta Pathol. Microbiol. Scand.,* Suppl., **160**, p. 1.

Olson, J. C. (1973): Some Approaches to Regulatory Control of Microbiological Hazards in Foods. *Dev. Ind. Microbiol.,* **14**, p. 169.

Ormay, L. (1962): *Az orvosi laboratóriumi asszisztensek kézikönyve, I–II* (Manual for Medical Laboratory Technicians. Vols I–II). Medicina Könyvkiadó, Budapest.

432

Ormay L. (1969): *Szakmai irányelvek, az élelmiszerek termelésének és forgalomba hozatalának higiéniai–mikrobiológiai ellenőrzésére* (Professional Principles in Hygienic–Microbiological Testing and Issuing of Food). Országos Élelmezés- és Táplálkozástudományi Intézet, Budapest.

Ormay L. (Ed.) (1970): *Élelmiszerbakteriológiai vizsgálatok* (Food-Bacteriological Tests). Orvosto-vábbképző Intézet, Budapest.

Országos Közegészségügyi Intézet (OKI) (1969): *Módszertani Útmutató. A közegészség-ügyi–járványügyi állomások járványügyi bakteriológiai laboratóriumainak egységesített módszerei* (A Methodological Guide. Standardized Methods for Hygienic–Bacteriological Laboratories of Public Health Stations). OKI Nyomda, Budapest.

Osváth-Marton, A.—Domján, J. (1974): Enterotoxin Production by *Staphylococcus aureus* Strains in Hungary. *J. Hyg. Epid. Microbiol. Immunol.*, Prague, pp. 289–293.

Pankhurst, E. S. (1967): A Simple Culture Tube for Anaerobic Bacteria. *Lab. Pract.*, Manchester, **16,** pp. 58–59.

Partmann, W.–Montfort, L. (1952): Ein neuer Impfkasten. *Zbl. Bakt. Abt. I. Ref.*, Stuttgart, **157,** pp. 611–619.

Paton, A. M.–Jones, S. M. (1971): Techniques Involving Optical Brightening Agents. In: Norris, J. R.–Ribbons D. W. (Eds), Vol. 5A, pp. 135–144.

Pelczar, M. J. Jr.–Reid, R. D. (1965): *Microbiology.* McGraw-Hill, New York.

Perkins, J. J. (1956): *Principles and Methods of Sterilization.* Charles C. Thomas, Springfield, Illinois.

Pijper, A. (1947): Methylcellulose and Bacterial Motility. *J. Bact.*, **53,** p. 257.

Pratt, G. B. (1950): *Statistical Significance of the Howard Mould Count.* Memorandum of the Research Division, American, Can. Co.

Proszt, G.–Varga B. (1965): *Általános mikrobiológiai gyakorlatok.* Jegyzet (Exercises in General Microbiology. Lecture Notes). Kertészeti és Szőlészeti Főiskola Élelmiszertechnológiai és Mikrobiológiai Tanszék, Budapest.

Pulay, G. (1969): A tej fermentációs folyamatait gátló anyagok kimutatási módszerei. Kézirat (Detection of Substances Inhibiting Fermentation Processes in Milk. Manuscript). Tejipari Tröszt TEÁ, Budapest.

Pulay, G.–Z. Bittera, R. (1959): A bacto-strip eljárás használhatósága a tej és a tejszín koliform fertőzöttségének kimutatására (Applicability of the Bacto-Strip Procedure in Detection of Coliforms in Milk and in Sweet Cream). *Tejipar,* Budapest, pp. 49–54.

Pulay, G.–Z. Bittera, R. (1960): A kóliform csírák tejhigiénés tejipari jelentősége, gyors kimutatási módszereinek bírálata (The Importance of Coliforms in Milk Hygiene and Dairy Industry. Criticism of their Detection). *Tejipar,* Budapest, pp. 59–63. ¨

Pulay, G.–Zsinkó, M. (1969): Mezofil savanyítók aktivitásának vizsgálata (Testing of the Activity of Mesophilic Acidifiers). *Tejipar,* Budapest, **18,** pp. 25–32.

Pusztai, S. (1969): Enterococcusok jelentősége és kimutatása töltelékes húskészítményekből (The Importance and Detection of *Enterococcus* in Stuffed Meat Products). *Magyar Állatorvosok Lapja,* Budapest, **24,** pp. 593–596.

Pusztai, S.–Vetési, F.–Hoch, V. (1972): Toxicity Testing of *Enterococcus* Strains Responsible for Food Poisoning. *Acta Vet. Acad. Sci. Hung.*, Budapest **22** (3), pp. 299–306.

Put, H. M. C.–Van Doren, H.–Warner, W. R.–Kruiswijk, J. Th. (1972): The Mechanism of Microbiological Leaker Spoilage of Canned Foods: a Review. *J. Appl. Bact.*, **35,** pp. 7–27.

Quesnel, L. B. (1971): Microscopy and Micrometry. In: Norris, J. R.–Ribbons D. W. (Eds), Vol. 5A, pp. 1–103.

Ramming, G.–Linke, H.–Leistner, L. (1971): Einfluß verschiedener Techniken der Durchmischung der Verdünnungsflüssigkeit auf die Ausbeute und Streuung der Keimzahlbestimmung. *Die Fleischwirt-schaft,* **51,** pp. 1652–1653.

Raper, K. B. and Fennel, D. I. (1965): *The Genus Aspergillus.* Williams and Wilkins Co., Baltimore.

Raper, K. B. and Thom, C. (1949): *A Manual of the Penicillia.* Williams and Wilkins Co., Baltimore.

Reuter, G. (1970): Mikrobiologische Analyse von Lebensmitteln mit selektiven Median. *Arch. Lebensmittelhyg.*, Hannover, **2**, pp. 30–35.

Roeder, G. (1954): Grundzüge der Milchwirtschaft und des Molkereiwesens. Paul Parey, Hamburg–Berlin.

Rogosa, M.–Mitchell, J. A.–Wiseman, R. F. (1951): A Selective Medium for the Isolation and Enumeration of Oral and Fecal Streptococci. *J. Bact.*, Baltimore, **62**, p. 132.

Ross, K. F.A (1967): *Phase Contrast and Interference Microscopy for Cell Biologists*. Edward Arnold, London.

Royce, A.–Bowler, C. (1959): *J. Pharmac.*, London, **11**, p. 294.

Schönherr, W. (1965): *Tierärztliche Milchuntersuchung*. S. Hirzel Verlag, Leipzig.

Schulz, M. E.–Voss, E.–Kay, H.–Siegfried, H.–Mrowetz, G. (1955): Betriebsuntersuchungen mit Hilfe transportable Untersuchungs-geräte. *Kieler Milchw. Forschber.*, Hildesheim, **7**, pp. 207–224.

Sedlák, J.–Rische, H. (1961): *Enterobacteriaceae Infektionen*. VEB Georg Thieme, Leipzig.

Seidler, M. (1976): Welche Vorteile bietet die Verwendung quadratischer Plastikpetrischalen im bakteriologischen Routinebetrieb? *Die Fleischwirtschaft*, **56**, pp. 517–518.

Sharpe, A. N. (1973): Automation and Instrumentation Developments for the Bacteriology Laboratory. In: Board, R. G.–Lowelock, D. W. (Eds): *Sampling–Microbiological Monitoring of Environments*. Academic Press, London–New York.

Sharpe, A. N.,–Jackson, A. K. (1972): Stomaching: a New Concept in Bacteriological Sample Preparation. *Appl. Microbiol.*, **24**, pp. 175–178.

Sharpe, A. N.–Kilsby, D. C. (1970): Ultrasound and Vortex Stirring as Bacteriological Sampling Methods for Foods. *J. Appl. Bact.*, London–New York, **3**, pp. 351–357.

Sharpe, M. E. et al. (1966): Identification of the Lactic Acid Bacteria. In: Gibbs, B. M.–Skinner, F. A., pp. 65–79.

Sierra, G. (1964): Hydrolysis of triglicerides by a bacterial proteolytic enzyme. Can. 3. *Microbiol.*, **10**, p. 926.

Simmons, J. S. (1926): A Culture Medium for Differentiating Organisms of the Typhoid-colon-aerogenes Groups and for the Isolation of Certain Fungi. *J. Infect. Dis.*, **39**, p. 209.

Sirockin, G.–Cullimore, S. (1969): *Practical Microbiology*. McGraw Hill, London.

Skerman, V. B. D. (1959): *Genera of Bacteria*. The Williams-Wilkins Comp., Baltimore.

Skorodumova, A. M. (1949): *A tej és tejtermékek műszaki mikrobiológiájának gyakorlati kézikönyve* (Practical Manual of Technical Microbiology of Milk and Dairy Products [in Russian]). Ogizh. Sel'khozgizh., Moscow.

Sobeck-skal, E.–Neuer, B. (1969): Untersuchungen zur Keimzahlbestimmung nach der Kleinplattenmethode unter Berücksichtigung der Psychrotrophenzählung in Rohmilch. *Öst. Milchw. Wiss. Beil.*, Wien, **24**, pp. 7–18.

Spicher, G. (1969): Beitrage zur Vereinheitlichung der Ermittlung das Keimgehaltes von Getreideprodukten. 2. Über den Einfluss der Vorbereitung des Getreides auf das Ergebnis der Ermittlung der Bakterien — und Schimmelpilzkeimzahl. *Brot Gebäck, Bochum*, **23**, pp. 61–69.

Stamer, J. R.–Albury, M. N.–Pederson, C. S. (1964): Substitution of Manganese for Tomato Juice in the Cultivation of Lactic Acid Bacteria. *Appl. Microbiol.*, **12**, p. 165.

Steiner, E. H. (1967): Statistical Methods in Quality Control. In: Herschdoerfer, S. M.: *Quality Control in the Food Industry, Vol. 1*. Academic Press, London–New York, pp. 121–231.

Stirling, A. C.–Stevens, M. K.–Lawley, D. N. (1950): Colony Counts on Strips of Agar in Tubes. *J. Gen. Microbiol.*, London–New York, **4**, pp. 339–344.

Straka, R. P.–Stokes, J. L. (1957): Rapid Destruction of Bacteria in Commonly used Diluents and its Elimination. *Appl. Microbiol.*, Baltimore, **5**, p. 21.

Strong, D. H.–Woodburn, M. J.–Mancini, M. M. (1961): Preliminary Observations on the Effect of Sodium Alginate on Selected Nonsporing Organisms. *Appl. Microbiol.*, **9**, pp. 213–218.

434

Stumbo, C. R. (1973): *Thermobacteriology in Food Processing.* (2nd Ed.) Academic Press, New York–London.

Sváb, J. (1967): *Biometriai módszerek a mezőgazdasági kutatásban* (Biometric Methods in Agricultural Research). Mezőgazdasági Kiadó, Budapest.

Sykes, G. (1969): Methods and Equipment for Sterilization of Laboratory Apparatus and Media. In: Norris, J. R.–Ribbons, D. W. (Eds), Vol 1.

Szegő, M. (1969): *Javaslat a termelői tej bakteriológiai tisztaság szerinti minősítésének módszerére és szervezetére. A termelői tej bakteriológiai tisztaság szerinti minősítése tárgyú tudományos ankét előadásai* (Recommendation for the Method and Organization of Grading by Purity of Producers' Milk. Papers Delivered at a Conference on Grading of Producers' Milk on Bacteriological Purity). pp. 13–22.

Takács, J. (1960): Standard táptalaj az állati eredetű élelmiszerek összes élő csíráinak megszámlálására (Standard Nutrient Medium for Counting of Viable Counts in Foodstuffs of Animal Origin). *Magyar Állatorvosok Lapja,* Budapest, **15,** pp. 416–418.

Takács, J. (1964a): *Az Enterobacteriaceae-családba tartozó baktériumok elkülönítése a laboratóriumi húsvizsgálatban, különös tekintettel a salmonellákra* (Isolation of Bacteria of the Enterobacteriaceae Family in Laboratory Testing of Meat, with Special Regard to Salmonellae). (Thesis.) Budapest.

Takács, J. (1964b): Vizsgálatok az O_1 és R-salmonella fágpróba felhasználására a kiegészítő laboratóriumi húsvizsgálatban (Attempts to Apply the O_1 and R *Salmonella* Phage Test in Supplementary Laboratory Meat Control). *Magy. Állatorvosok Lapja,* Budapest, **19,** pp. 127–131.

Takács, J. (1968): Coli és coliform csírák elkülönítése a víz és az élelmiszerek vizsgálatában (Isolation of *E. coli* and Coliforms in the Frame of Food and Water Control). *Magyar Állatorvosok Lapja,* Budapest, **23,** pp. 38–43.

Takács, J. (1971): Comparative Examinations for the Effectiveness of Media to Detect Clostridia from Canned Meat Products. *Arch. Lebensmittelhyg.,* **22,** pp. 101–104.

Takács, J. (1972): Összehasonlító vizsgálatok a clostridiumok legeredményesebb kimutatását célzó táptalajok hatásfokára, különös tekintettel a konzervek vizsgálatára (Comparative Studies Aimed at Finding the Most Efficient Detection of Clostridia by Culturing, with Special Regard to Examination of Canned Food). *Magyar Állatorvosok Lapja,* Budapest, **26,** pp. 275–281.

Takács, J.–Bné Nagy, Gy., (1973): Az O_1 és R-Salmonella fágpróba értéke laboratóriumi gyakorlatban. (Evaluation of the O_1 and R *Salmonella* Phage Test in the Laboratory Practice). *Magyar Állatorvosok Lapja,* Budapest, **28,** pp. 161–168.

Takács, J.–Narayan, K. G. (1965a): Összehasonlító vizsgálatok a módosított szulfit-agar hatásfokára és a clostridiumok szulfit érzékenységére (Comparative Studies on the Efficiency of the Modified Sulphite Nutrient Agar and the Sensitivity of Clostridia to Sulphite). *Magyar Állatorvosok Lapja,* Budapest, **20,** pp. 564–567.

Takács, J.–Narayan, K. G. (1965b) A clostridiumok szulfit-redukáló képessége módosított táptalajban (The Sulphite-Reducing Capacity of Clostridia in a Modified Nutrient Medium). *Magyar Állatorvosok Lapja,* Budapest, **20,** pp. 211–214.

Takács, J.–Zné, Imreh, E. (1973): Selective Propagation of Clostridia in Media Containing D-cycloserine Tartarate. *Acta Microbiol. Acad. Sci. Hung.,* Budapest, **20,** p. 64.

Takács, J.–Zné, Imreh, E. (1975): Use of D-cycloserine Bitartarate for Selective Isolation of Clostridia. *Acta Vet. Acad. Sci. Hung.,* Budapest, **25,** pp. 283–289.

Tanner, F. W. (1950): *Laboratory Manual and Work Book in Microbiology of Foods.* The Garrard Press, Champaign. Ill.

Tarodyné, Okályi E.–Mocsári, L. (1965): Élelmiszer üzemekben használható agar-lenyomatos higiénás vizsgálati módszer (An Agar-Replica Method for Hygienic Tests Performable in Food Plants). *Élelmiszer Ipar,* Budapest, **19,** pp. 372–374.

Taylor, A. J.–Gilbert, R. J. (1975): *Bacillus cereus* Food Poisoning: a Provisional Serotyping Scheme. *J. Med. Microbiol.,* **8,** pp. 543–550.

28*

ten Cate, L. (1964): A Note on a Simple and Rapid Method of Bacteriological Sampling by Means of Agar Sausages. *J. Appl. Bact.*, London–New York, **28**, pp. 221–223.

Thatcher, F. S.–Clark, D. S. (eds) (1968): *Microorganisms in Foods*. Univ. of Toronto Press, Toronto.

Thatcher, F. S.–Clark, D. S. (1974): *Sampling for Microbiological Analysis: Principles and Specific Applications*. University of Toronto Press, Toronto.

The Oxoid of Culture Media, Ingredients and Other Laboratory Services. (3rd Ed.) (1967). Oxoid Limited, London.

Thompson, D. I.–Black, L. A. (1967): The Use of the 0·01 ml Loop in the Technol. *J. Milk Fd. Technol.*, Shelbyville, **30**, pp. 273–276.

Thompson, D. L.–Donelly, O. B.–Black, L. A. (1960): A Plate Loop Method for Determining Viable Counts of Raw Milk. *J. Milk. Food Technol.*, **23**, p. 167.

Tippett, L. H. O. (1950): *Technological Applications of Statistics*. Wiley and Sons, Inc. New York, Williams–Norgate Ltd.

Tolle, A.–Zeidler, H.–Heeschen W. (1968): Die elektronische Mikrokoloniezählung—ein Verfahren zur Beurteilung der bakteriologisch–hygienischen Qualität der Rohmilch. *Milchwissenschaft*, Nürnberg, **23**, pp. 65–70.

Tomkins, R. C. (1951): The Microbiological Problems in the Preservation of Fresh Fruits and Vegetables. *J. Sci. Food Agric.*, **2**, p. 381.

Ubrizsy, G.–Vörös, J. (1968): *Mezőgazdasági mykologia* (Agricultural Mycology). Akadémiai Kiadó, Budapest.

Vajda, Ö. (1959): Új, gyors ipari módszer élőcsíraszám meghatározására (A New Rapid Industrial Method for Assessing Viable Count). *Élelmiszer Ipar*, **13**, pp. 148–152.

Vas, K. (1962): *Az élelmiszeripari mikrobiológia néhány általános problémája* (Some General Problems in Food-Industrial Microbiology. Manuscript). Felsőoktatási Jegyzetellátó Vállalat, Budapest.

Vas, K. (1972): Az élelmiszeripari mikrobiológiai vizsgálati módszerek egységesítésének jelentősége (The Importance of Microbiological Tests in the Food-Industrial Microbiology). *Söripar*, Budapest, **18**, pp. 105–108.

Vedamuthu, E. R.–Reinbold, G. W. (1967): The Use of Candle Oats Jar Incubation for the Enumeration, Characterization and Taxonomic Study of Propionibacteria. *Milchwissenschaft*, Nürnberg, **22**, pp. 428–431.

Vitéz, I. (1971): *Fizikai, kémiai, kombinált sterilizési módszerek és sterilizési vizsgálatok. A Biokémia Modern Módszerei, I* (Physical and Chemical Combined Sterilization. Methods and Investigations. Methods of Modern Biochemistry, I.) Magyar Kémikusok Egyesülete Biokémiai Szakosztálya, Budapest.

Vörös, J.–Léránth, J. (1974a): Review of the mycoflora of Hungary, Part XI. Deuteromycetes: Moniliales. *Acta Phytopath. Acad. Sci. Hung.*, Budapest, **9**, pp. 99–123.

Vörös, J.–Léránth, J. (1974b): Review of the mycoflora of Hungary, Part XII. Deuteromycetes: Moniliales and Myceliales. *Acta Phytopath. Acad. Sci. Hung.*, **9**, Budapest, pp. 333–361.

Vörös, J.–Ubrizsy, G. (1960) *A penészgombák (Mucorales, Hyphomycetes). Magyarország kultúrflórája* (Moulds *[Mucorales, Hyphomycetes]*. The Culture Flora of Hungary). Vol. I, No. 8, Akadémiai Kiadó, Budapest.

Vosti, D. C.–Hernandez, H. H.–Stard, J. B. (1961): Analysis of Headspace Gases in Canned Foods by Gas Chromatography. *Food Technol.*, **15**, p. 29.

Wedum, A. G. (1953): Bacteriological Safcty. *Am. J. Publ. Hlth.*, **43**, p. 1428.

Wetherill, G. B. (1969): *Sampling Inspection and Quality Control*. Methuen Publishing, London.

Wheaton, E.–Pratt, G. R. (1961) Comparative Studies on Media for Counting Anaerobic Bacterial Spores, II. *J. Fd. Sci. Champaign.*, **26**, pp. 261–268.

Wilkins, J. R.–Mills, S. M. (1975): Automated single-slide staining device. *Appl. Microbiol.*, **30**, pp. 485–588.

436

Willis, A. T. (1969): Techniques for the Study of Anaerobic Sporeforming Bacteria. In: Norris, J. R.–Ribbons, D. W. (eds), Vol. 3B, pp. 79–115.

Wilson, G. S. (1935): *The Bacteriological Grading of Milk*. H. M. Stationery Office, London.

Wolf, J.–Barker, A. N. (1968): The Genus *Bacillus:* Aids to the Identification of its Species. In: Gibbs, B. M.–Shapton, D. A., pp. 92–109.

Wright, R. C.–Tramer, J. (1961): The Estimation of Penicillin in Milk. *J. Soc. Dairy Technol.,* **14,** pp. 85–87.

Zsolt, J.–Pazonyi, B.–Novák, E.–Pelc, A. (1961): *Az élesztők. Magyarország kultúrflórája* (Yeasts. The Culture Flora of Hungary). Vol. I, No. 9, Akadémiai Kiadó, Budapest.

Zycha, H.–Siepmann, R.–Linnemann, G. (1969): *Mucorales*. J. Cramer Verl., Lehre.

SUBJECT INDEX

439

442

443

—, methyl-red 209, 211
—, slide 226
—, test-tube 226
—, Voges–Proskauer 201, 211
Redofix 300
reduction
— methylene-blue 298
— —, modified 298
—, nitrate 318
rennin 96
resazurin-reduction test 299, 300, 310–312
resuscitation 199
rhizoids 136, 264
Rhizopus 88, 136, 264, 341, 344
— *nigricans* 137, 264
— *oryzae* 137, 264
— *stolonifer* 137, 264
Rhodotorula 135, 261
— *glutinis* 135, 263
rinse method 283
ropiness in bread 349
roux bottle 25
Royce bag 35

Saccharomyces 82, 134, 255
— *bayanus* (= *S. pastorianus*) 135, 255
— *bisporus* 134, 255
— — var. *mellis* 134, 255
— *cerevisiae* (= *S. ellipsoideus*) 135, 256
— *fragilis* 134, 252
— *lactis* 134, 252
— *mellis* 135
— *rouxii* 134, 256
— *uvarum* (= *S. carlsbergensis*) 135, 257
Saccharomycetaceae 84
Salmonella 101, 201, 203, 249, 312–314, 318–320, 322
— *gallinarium* 322
— *typhi* 180, 203
— *typhimurium* 180, 322
salt tolerant 319
samples 43, 110
—, bulked 309
— design 110
— preparation 286, 309
—, processing of 43
—, selection of 116
—, storage of 43
—, transport of 43

sampling 42, 109, 110, 285, 308
—, attribute 113
— by scraping 309
— by swabbing 310
— by the contact adhesion technique 310
— for microbiological grading 109
— instruments 43
— methods 311
—, multi-stage 117
— of food 42
—, variable 113
Sarcina 73
— *lutea* 80, 321
Schizomycetes 125
Schizosaccharomyces 82
— *pombe* 291
Schizosaccharomycetes 84
sclerotia 268–272
serological test 54
Serratia 106, 322
— *marcescens* 80
serum
— O polyvalent screening 201, 204
— *Salmonella* O polyvalent screening 201
— Vi 203
Shigella
— *dysenteriae* 180
— *sonnei* 180
short biochemical test series 203
size 69
—, determining 69
slime-producing 317
sonication 44
Sordaria 184
Spheriales 166
spices 319
Spirochetales 125
spirillum 74
spirohaetes 74
spoilage of food 184
— — canned 184
— — raw 184
—, cause of 331
—, microscopic investigation of 329
— of canned products 327
—, processed 184
—, sulphide 347
sporangia 88
sporangiola 136, 265

Trichothecium 266
— *roseum* 137, 267
trituration 44
TTC test 293, 302, 310
tube-plate method 172
tube
—, Browne 34
—, Durham 26
—, Einhorn's 26
—, Pankhurst's anaerobic 62
—, test 26, 311
turbidimetric method 149
Tween derivate 102
tyndallization (see sterilization)

undivided surface 165
unheated products 309

vegetative reproduction 82
Velutina 139, 272
ventilation 21
vesicle 139, 269–272
Vi antigen 202, 203
vibrator, ultrasonic 44
Vibrio 74
— *cholerae* 180
— *parahaemolyticus* 180

Voges–Proskauer
— reaction 198, 201, 205, 209, 211
— reagent 243
— test 99
vortex stirrers 46

wash method 283
water
— drinking 274
— grading 274
Weinzirl test 295
Whirlimixer-type mixer 44

zoogloea 75
zoöspores 87
zygogamy 136
Zygomycetes 136
zygophores 88
zygospore 88, 263–265

Xanthomonas 106, 184

yeasts 63, 82, 250, 296
— in liquid culture 85
—, microscopic examination of 86
— morphology 82
—, spore formation in 84
—, sporogenesis of 86
—, thallus of 82